CHANGING CLIMATES, ECOSYSTEMS AND ENVIRONMENTS WITHIN ARID SOUTHERN AFRICA AND ADJOINING REGIONS

Palaeoecology of Africa

International Yearbook of Landscape Evolution
and Palaeoenvironments

ISSN 2372-5907

Volume 33

Changing Climates, Ecosystems and Environments within Arid Southern Africa and Adjoining Regions

Editor

Jürgen Runge

Centre for Interdisciplinary Research on Africa (CIRA/ZIAF), Johann Wolfgang Goethe University, Frankfurt am Main, Germany

In collaboration with

Marion K. Bamford

Evolutionary Studies Institute and School of Geosciences, University of the Witwatersrand, Johannesburg, South Africa

Linda Basson

Zoology and Entomology, University of the Free State, Bloemfontein, South Africa

Joachim Eisenberg

Institute of Physical Geography, Johann Wolfgang Goethe University, Frankfurt am Main, Germany

CRC Press
Taylor & Francis Group
Boca Raton London New York

CRC Press is an imprint of the
Taylor & Francis Group, an **informa** business
A BALKEMA BOOK

Front cover: Deelpan, situated about 50 km west of Bloemfontein, South Africa (Photo: L. Scott, October 2014). For a detailed look on Deelpan see the contribution of Karl W. Butzer and John F. Oswald (Chapter 3).

This edition of Palaeoecology of Africa was generously supported by the "Georg und Franziska Speyersche Hochschulstiftung", Konrad-Adenauer-Straße 15, 60313 Frankfurt am Main, Germany.

Published by: CRC Press/Balkema
P.O. Box 11320, 2301 EH Leiden, The Netherlands
e-mail: Pub.NL@taylorandfrancis.com
www.crcpress.com – www.taylorandfrancis.com

ISBN 13: 978-1-138-02704-6 (hbk)

Typeset by V Publishing Solutions Pvt Ltd., Chennai, India

Library of Congress Cataloging-in-Publication Data

Changing climates, ecosystems, and environments within arid southern Africa and adjoining regions/editors, Jürgen Runge, Centre for Interdisciplinary Research on Africa (CIRA/ZIAF), Johann Wolfgang Goethe University, Frankfurt am Main, Germany; in collaboration with Marion K. Bamford, Evolutionary Studies Institute and School of Geosciences, University of the Witwatersrand, Johannesburg, South Africa, Linda Basson, Zoology and Entomology, University of the Free State, Bloemfontein, South Africa, Joachim Eisenberg, Institute of Physical Geography, Johann Wolfgang Goethe University, Frankfurt am Main, Germany.

pages cm.—(Palaeoecology of Africa, international yearbook of landscape evolution and palaeoenvironments, ISSN 2372-5907 ; volume 33)

Summary: Discusses climate change impact on both the region of southern Africa, and the nation of South Africa.

Includes bibliographical references and index.

ISBN 978-1-138-02704-6 (hardcover : alk. paper)—ISBN 978-1-315-73830-7 (ebook)
1. Climatic changes—Africa, Southern. 2. Global warming—Africa, Southern. 3. Climatic changes—South Africa. 4. Global warming—South Africa. 5. Arid regions—Africa. 6. Biotic communities—Africa, Southern. I. Runge, Jürgen, 1962-editor. II. Bamford, M.K., editor. III. Basson, Linda, 1958- editor. IV. Eisenberg, Joachim, 1972-editor.

QC903.2.A35C47 2015
577.2'20968—dc23

2015025688

Contents

Foreword

This editon (Volume 33) of the series 'Palaeoecology of Africa – International Yearbook of Landscape Evolution and Palaeoenvironments' is presenting the outcome of a 'tribute conference' to the internationally recognized South African researcher and palynologist Professor Louis Scott from the Botany section, Department of Plant Sciences at the University of the Free State, Bloemfontein, South Africa. Louis Scott has recently retired, but he is still continuing his active research career. The 11 conference proceedings and articles published here highlight and celebrate Professor Scott's contribution to palaeoscience and to the natural sciences in general. The conference was organized in July 2014 by the National Museum, Bloemfontein and the University of the Free State and was attended by over 60 delegates from numerous countries. The conference focused on both past and present environments, ecosystems and climates of the semi-arid to arid regions of southern Africa (and also some adjoining regions), areas that serve as major foci of Prof. Scott's research. The contributions in this book are covering numerous aspects, ideas and methods on former climates, vegetation cover in tropical and subtropical ecosystems, and interdisciplinary questions of landscape dynamics linked to climate change. There are also some more applied papers on recent questions tackling Global Change issues such as mitigation measures in land use (agroforestry) and the conservation of rock art as a cultural heritage.

Within the University of Frankfurt Physical Geography working group (IPG) and the Centre for Interdisciplinary Research on Africa (ZIAF, www.ziaf.de), all manuscripts have been reviewed and corrected several times. The editorial process was also assisted by two guest editors, namely Marion K. Bamford from the Evolutionary Studies Institute and School of Geosciences, University of the Witwatersrand, Johannesburg, and by Linda Basson from the Department of Zoology and Entomology, University of the Free State, Bloemfontein, South Africa. Formatting of the papers for PoA layout and style was done reliably by Joachim Eisenberg (IPG). Apart from the cartographic art work done by the principle authors themselves, Joachim Eisenberg revised numerous figures and assisted by carrying out additional cartographic work for the book.

The CRC Press/Taylor & Francis team in Leiden (The Netherlands) and especially Senior Publisher Janjaap Blom and his team helped streamline the editing process. Financial support for this edition of Palaeoecology of Africa was generously given by the "Georg and Franziska Speyersche Hochschulstiftung", Konrad Adenauer Straße 15, 60313 Frankfurt am Main, Germany. To these workers and to other colleagues who indirectly contributed and helped to make this book a reality, I am greatly obliged.

Jürgen Runge
Frankfurt
May 2015

Contributors

Andri C. van Aardt

Department of Plant Sciences, University of the Free State, PO Box 339, Bloemfontein, 9301, South Africa, Email: vanaardtac@ufs.ac.za

Marion K. Bamford

Evolutionary Studies Institute and School of Geosciences, University of the Witwatersrand, P Bag 3, WITS 2050, Johannesburg, South Africa, Email: Marion.bamford@wits.ac.za

Jennifer Botha-Brink

National Museum Bloemfontein, PO Box 266, Bloemfontein 9300, South Africa; Department of Zoology and Entomology, University of the Free State, Bloemfontein 9300, South Africa, Email: jbotha@nasmus.co.za

C. Britt Bousman

Department of Anthropology, Texas State University, San Marcos, Texas 78666, USA; GAES, University of the Witwatersrand, Johannesburg, South Africa, Email: bousman@txstate.edu

James S. Brink

Florisbad Quaternary Research Department, National Museum, PO Box 266, Bloemfontein, South Africa; Centre for Environmental Management, University of the Free State. Bloemfontein, South Africa, Email: jbrink@nasmus.co.za

George A. Brook

Department of Geography, University of Georgia, Athens, GA 30602, USA, Email: gabrook@uga.edu

Karl W. Butzer

Department of Geography and the Environment, University of Texas at Austin, Austin, Texas, USA, Email: karl.butzer@austin.utexas.edu

Michael Chazan

Department of Anthropology, University of Toronto, Canada; Evolutionary Studies Institute, University of the Witwatersrand, Johannesburg, South Africa, Email: mchazan@me.com

Michaela Ecker

Research Laboratory for Archaeology and the History of Art, University of Oxford, Oxford OX1 3QY, UK, Email: michaela.ecker@keble.ox.ac.uk

Rainer Grün

Research School of Earth Sciences, The Australian National University, Canberra ACT 0200, Australia, Email: Rainer.Grun@anu.edu.au

Wondimu Tadiwos Hailesilassie

National Meteorological Agency, Addis Ababa, Ethiopia, Email: wonde721@gmail.com

Peter Holmes

Department of Geography, University of the Free State, PO Box 339, Bloemfontein, 9300, South Africa, Email: holmespj@ufs.ac.za

Liora Kolska Horwitz

Natural History Collections, Faculty of Life Sciences, The Hebrew University, Jerusalem 91904, Israel, Email: lix1000@gmail.com

Makarius Peter Itambu

Department of Archaeology and Heritage, University of Dar es Salaam, PO Box 35050, Dar es Salaam, Email: mcpeter7@gmail.com

Zenobia Jacobs

Centre for Archaeological Science, School of Earth and Environmental Sciences, University of Wollongong, Wollongong 2522, Australia, Email: zenobia@uow.edu.au

Julia A. Lee-Thorp

Research Laboratory for Archaeology and the History of Art, University of Oxford, Oxford OX1 3QY, UK, Email: julia.lee-thorp@rlaha.ox.ac.uk

Alvord Nhundu

Anthropology and Archaeology Department, University of Pretoria, South Africa, Email: nhundual@gmail.com

John F. Oswald

Department of Geography and Geology, Eastern Michigan University, 205 Strong Hall, Ypsilanti Michigan 48197, USA, Email: oswa0018@yahoo.com

André Piuz

Musée d'Histoire Naturelle, CH-1208 Genève, Switzerland, Email: andre.piuz@ville-ge.ch

Pieter J. du Preez

Department of Plant Sciences, University of the Free State, PO Box 339, Bloemfontein, 9301, South Africa, Email: dpreezpj.sci@mail.uovs.ac.za

Lloyd Rossouw

Department of Plant Sciences, University of the Free State, PO Box 339, Bloemfontein, 9301, South Africa; Archaeology Department, Museum, PO Box 266, Bloemfontein, South Africa, Email: lloyd@nasmus.co.za

Jürgen Runge

Centre for Interdisciplinary Research on Africa (ZIAF) and Institute of Physical Geography, Johann Wolfgang Goethe University, Altenhöferallee 1, 60438 Frankfurt am Main, Germany, Email: j.runge@em.uni-frankfurt.de

Louis Scott

Department of Plant Sciences, University of the Free State, PO Box 339, Bloemfontein, 9301, South Africa, Email: scottl@ufs.ac.za

Maitland T. Seaman

Centre for Environmental Management, Faculty of Natural and Agricultural Sciences, University of the Free State, Bloemfontein, South Africa, Email: seamanmt@ufs.ac.za

J. Francis Thackeray

Evolutionary Studies Institute, University of the Witwatersrand, P Bag 3, WITS, Johannesburg 2050, South Africa, Email: francis.thackeray@wits.ac.za

CHAPTER 1

From past to present: Louis Scott tribute conference

Jürgen Runge

Centre for Interdisciplinary Research on Africa (ZIAF) and Institute of Physical Geography, Johann Wolfgang Goethe University, Frankfurt am Main, Germany

Maitland T. Seaman

Centre for Environmental Management, Faculty of Natural and Agricultural Sciences, University of the Free State, Bloemfontein, South Africa

1.1 BACKGROUND

This editon of the series 'Palaeoecology of Africa—International Yearbook of Landscape Evolution and Palaeoenvironments' is presenting the outcome of a 'tribute conference' to the internationally recognized South African researcher and palynologist Professor Louis Scott from the Botany section, Department of Plant Sciences at the University of the Free State, Bloemfontein, South Africa. Louis Scott has recently retired, but he is still continuing his active research career (see 1.5 bibliography). The 11 conference proceedings and articles published here highlight and celebrate Professor Scott's contribution to palaeoscience and to the natural sciences in general. The conference was organized in July 2014 by the National Museum, Bloemfontein and the University of the Free State and was attended by over 60 delegates from numerous countries (Figure 1). The conference focused on both past and present environments, ecosystems and climates of the semi-arid to arid regions of southern Africa (and also some adjoining regions), areas that serve as major foci of Prof. Scott's research.

1.2 LIFE CAREER OF LOUIS SCOTT

Louis Scott (Figure 2) was born on May 12th, 1947 in Edenville, Free State, South Africa. In 1967 he registered for Botany and Zoology at the University of the Free State (UFS) in Bloemfontein. Subsequently, his whole scientific and personal life was bound up with this institution. He specialized in palynology and environmental changes on different temporal and spatial scales in semi-arid and arid landscapes in South Africa. Early in 1971 he obtained a M.Sc. in Botany on the topic of 'Lower Cretaceous pollen and spores from the Algoa Basin (South Africa)'. This work was encouraged and supervised by Professor Dr. Eduard Meinen van Zinderen Bakker (1907–2002), the founder of the series Palaecology of Africa (since 1966 the series has published up to 33 volumes).

Figure 1. Delegates of the Louis Scott tribute conference (7–11 July 2014) on the UFS main campus in Bloemfontein (photo by H. Human).

After a short period (1972–1973) of professional work with the Southern Oil Exploration Corporation (SOEKOR) in Johannesburg, he joined in 1974 the Institute for Environmental Sciences at the University of the Free State (UFS) where he stayed up to 1984.

Scott graduated and obtained his PhD in 1979 for a thesis entitled 'Late Quaternary pollen analytical studies in the Transvaal, South Africa' that was supervised by Dr. Joey A. Coetzee (1921–2007). From 1979–80 he spent a sabbatical at the Paleoenvironmental Laboratory of the University of Arizona (USA). Subsequently he remained at UFS in the Botany section of the Department of Plant Sciences where he has recently (2014) retired as a professor. Following in the footsteps of Profs van Zinderen Bakker and Coetzee, his pioneering precursors and mentors at Bloemfontein, Louis Scott has single-handedly kept South African palynology firmly on the international map. He has also pioneered the reconstruction of past environments in dry areas of Africa based on analysis of pollen from fossil hyrax and hyaena dung. The wide recognition that he enjoys among his peers is evidenced by his B2 rating with the National Research Foundation and his long-standing collaborations with colleagues in numerous countries.

Louis Scott's research interests include stratigraphic palynology, long-term continental environmental change during the Cainozoic, and interpretation of palaeoenvironmental records associated with archaeological sites. His research has contributed insights into the origin of our current environment by identifying long-term patterns of climate change. Results have been applied in numerical models of vegetation change in Africa and globally. The results of these studies are relevant across the fields of botany, geology, climatology, archaeology, anthropology and palaeontology. He has a prodigious publication record. His papers include reconstructions of vegetation and climatic history in various areas of southern Africa, including the Tswaing Crater

Figure 2. Louis Scott in 2006 (photo by M.T. Seaman).

near Pretoria with a record of 200 000 years, and give insights into environmental conditions during the Last Glacial Period and the subsequent development of modern conditions. His observations provide key baseline information, which contributes to understanding past human and environmental contexts and climatic change and the effects of global warming.

Amongst several memberships, Louis Scott was one of the group of palynologists that formed the prestigious African Pollen Database in 1996 in Bierville, France, and was regional co-ordinator for Southern Africa for IGBP/PAGES Pole-Equator-Pole palaeoenvironmental transects (PEPIII). He is a long-serving member of the Southern African Society for Quaternary Research (SASQUA), having been its President in 1991–1993, and has served the International Union for Quaternary Research (INQUA) in various capacities. He has been Guest Editor and Editorial Board Member for many journals including of the Journal of Arid Environments and Palaeoecology of Africa.

1.3 WORDS OF WELCOME AND MESSAGES IN HONOUR OF LOUIS SCOTT (PRESENTED AT THE CONFERNCE DINNER ON JULY 8TH, 2014)

1.3.1 Maitland T. Seaman, Bloemfontein

I have known Louis and Leenta Scott and their children (Fred, Christiaan and Carolien, better known within the family as Bolla, Pallie and Caro, who are now, respectively, a programmer, a paediatrician and an art restorer) for 35 years, having worked with Louis' brother Willem at the CSIR before that, and met brother Fred in the meantime. And when we moved to Bloemfontein in 1980, we (my wife Helga, our daughter Birgit and I) lived in their house for three months while they were in Tucson, Arizona, where Louis was doing his post-doc.

While Louis' career started before that, at the UFS and at SOEKOR (the Southern Oil Exploration Corporation), we might say his career really got going in 1980, when he got back from that post-doc, working with EM van Zinderen Bakker and Joey Coetzee. Once they had retired, Louis came into his own as a palynologist.

So, something about the more private Louis I know, not being a palaeontologist myself. Let me start with the time when we lived in that house of theirs—our nephews and nieces were struck by his paintings, many of which featured weird animals with massive teeth—some of these paintings were above the children's beds—no pampering of children with pictures of bunnies and teddy bears. Some of Louis' paintings now form part of important art collections in South Africa.

Over many years we have enjoyed Louis and Leenta's sociability and open house, many visiting researchers and students and researchers (including some of you) have enjoyed this hospitality, extending in some cases to living for extended periods in their home (thinking of Aharon Horowitz many years ago, and in more recent years also Graciella Gil Romera).

This hospitality would entail enough wine and always a braai on Sunday—Louis, being Louis, he doesn't braai the way most modern South Africans do, he makes a fire on the ground, the traditional way, preserving the basic meaning of a braai, as the pioneers did it. He's laid back, but that doesn't mean he is by any means bland—for instance he is specific in his musical tastes (Bob Dylan being central), he drinks beer from a glass or mug (never a can), he genuinely knows good art from kitsch, and he has strong views on his field of research in particular. He is generally able to distinguish what is important in life from what is not, and he is certainly not out to impress anyone. Conversely he'll make a great effort to help anyone, with no expectation of reward. What he is though is humble, certainly not a self-trumpeter—testimony of his integrity, but not always to his financial benefit.

He comes from a remarkable family. His father Freek Scott, a South African medical student, was caught up in World War II in Holland, where he met Louis' mother, Dora Bossaert, a Flemish physicist also studying in Holland. The couple came out to South Africa with baby Willem. Brothers Louis and Fred were born in South Africa.

Louis's father was a dermatologist and prime mover in the establishment of the Medical Faculty at this University—the medical library is called the Freek Scott Library in his honour. Louis' mother would speak of him, in her ever strong Flemish accent, as "mijn egtgenoot Frederik Petrus Scott".

Dora Scott, who died last year at 99, known to everyone as Tant (or Tante) Dora, was a most remarkable lady, in all senses of the word. What pleasure she gave us all with her intellect, wisdom and humility—I personally loved her dearly. She came from an illustrious family. To illustrate this, I was standing at the Air and Space Museum at the Smithsonian in Washington one day, looking at the Goddard Memorial Trophy for Space Research. The first recipient was understandably Wernher von Braun, and, on a reflex I looked around to tell someone, the second was Karel Bossaert, Louis' uncle, father of the Atlas Rocket, which launched the first Americans into space. When I got home and told Tant Dora, she brought out cuttings about him that she'd kept over the years, and never shown anyone, such was her humility, whence Louis' humility.

Louis' parents were highly influential patrons of the arts in Bloemfontein, whence Louis's interest in painting, his brother Fred's present occupation as art dealer, and daughter Caro's career as art restorer.

Which brings me again to Louis' wife Leenta, a pathologist in her own right. She has the benefits of a most-eclectic youth and a father who was a prominent journalist. She has supported him throughout his career, created their welcoming home with Louis, and given him the freedom to be what he is. We have spent many happy hours with them, in happy open-minded discussion (no winners), always with wine. I'm very glad she's here too tonight to share Louis' spotlight.

The whole family would be, and are, very proud of their Louis. And so are we all. Louis is a good scientist who, though retired, and without a salary, but with the use of his lab at the UFS, continues as before, doing field work, supervising research, collaborating with scholars from all over the world, and of course, publishing valuable articles.

I can't see him doing anything else, it is his life.
Note: this is a TRIBUTE, not a FAREWELL!
Though, maybe, we'll see him doing some more painting and drinking good wine.

Maitland T. Seaman

1.3.2 Mike Meadows, Cape Town

I have known and, indeed, come to love Louis Scott since my arrival in South Africa in 1983. Louis—the years have been kind to you (kinder than they have been to me in terms of retaining a decent head of hair at least!) and the 30 years that have gone by (I can hardly believe that) have seen your flourishing and becoming a real international leader in the field. You were, initially under the guidance of Professors van Zinderen Bakker and Joey Coetzee, responsible for helping me integrate into the landscape—both literally and figuratively—of southern African palaeoecology. I will be forever grateful that the few days I spent at your department in those early months left me with the curiosity and the confidence to explore the possibility of sampling sediments in the Western Cape for pollen.

Louis I think you more than anyone else has demonstrated that simplistic northern hemisphere interpretations of what constitutes sediment suitable for pollen preservation hardly ever work in places like southern Africa! Indeed, you have demonstrated that beautifully stratified highly organic accumulations rich in pollen that is easily identifiable (mainly because that there are so few pollen types there to identify!) are unrealisable dreams in this part of the world. Most palaeoecologists who cut their teeth in temperate latitudes would quite literally turn their noses up at the hyaena coprolites and dassiepis that you have been prepared to work on and provide palaeoenvironmental insights that would otherwise have remained hidden. Your dogged determination to explore the southern African Quaternary using sometimes the most unlikely of sediment sources has been an inspiration to many of us. You have shown that what may look like pretty shitty material (or even shit for that matter) can host a veritable kaleidoscope of palaeoenviromental clues.

But more than anything else, Louis, it is your humility and quiet sense of humour that I celebrate. You are a man with a gentle and totally dignified approach to life in general and to your science in particular. Doubtless you have rare talents, not only in the field of Quaternary Science but also in the world of art. However, you always seem to be quietly self-effacing, seemingly slightly surprised even by your successes and the scores of important papers you have authored.

Louis, enjoy your retirement. I doubt it will result in your backing off or slowing down much as a scientist—thank goodness for that as we will surely still be able to enjoy your company at INQUA and SASQUA conferences (I think it is your round by the way!). But it should at least give you some more time to spend with the family that you love so much and more time too for your creativity. I am so sorry not to be there with you all and raise the toast to your fantastic (ongoing) scientific career but I will do so from afar and wish you a happy, fulfilling and long period of retirement!

Mike Meadows

1.3.3 Graciela Gil Romera, Zaragoza, Spain

You can't imagine how much I would have loved to be there today. But not only me, also Miguel. We are instead on a fieldwork trip to the Pyrenees where I came with my team and my family...somehow distressed by the presence of a four years old Mateo and a four months old Héctor. They are lovely though, you know that.

I am pleased, at least, that I have this little space to express my gratitude to you. This gratitude could be extended to the very many different ways you have helped me both along my research and life careers. I can say many things of Louis Scott; I could say that he is a fantastic researcher as he has the most important features needed to be one: curiosity and enthusiasm. And this is extremely contagious as just by being by his side, one may feel the need to investigate deeper into questions.

I can certainly say that everything I know of African palynology I learnt from him. And if this is so, it is not only because he is one of the most knowledgeable persons I know in that field but also because he's always willing to help, patiently, sympathetically and smiling! So my gratitude in this aspect is total as I would have not chosen this path if it wasn't for his help and encouragement. Despite being retired, Louis is still active, involved in several projects and willing to participate in more.

Then I can say he is a friend, someone who fostered my career from the very beginning, but also someone I can be relaxed with, sharing the joys and sorrows of a field trip to Namibia, watching a movie like the Hitchhikers Guide to the Galaxy and imagining vogons everywhere, discussing art with me, introducing me to biltong degustation and then someone who would treat me and Miguel as part of his family from the very beginning.

Thus, I know I will have the chance to express my gratitude in person at some point, but for now and in front of all your beloved people, I just want to say thank you Louis for letting me be your friend and colleague.

Graciela Gil Romera

1.4 INTRODUCTION TO THE CONFERENCE'S CHAPTERS 2–12

This edition of Palaeoecology of Africa (Vol. 33) presents 11 contributions in honour of Louis Scott. They are covering numerous aspects, ideas and methods on former climates, vegetation cover in tropical and subtropical ecosystems, and interdisciplinary questions of landscape dynamics linked to climate change. There are also some more applied papers included on recent questions tackling Global Change issues such as mitigation measures in land use (agroforestry) and the conservation of rock art as a cultural heritage.

Peter Holmes introduces in chapter 2 the general landscape shape of central South Africa and highlights the role of geomorphology when applying new absolute dating techniques and remote sensing for Quaternary studies. He states that there is a huge potential for future scientific work with regard to palaeo-drainage and pan-lunette suites in the western Free State, South Africa. Chapter 3 by Karl W. Butzer and John F. Oswald introduces the environmental history and discusses possible human impact at Deelpan (see photo on book cover) in the western Free State. They hypothesize that ground-cover deterioration after the youngest pan transgression led to renewed eolian dispersal of Kalahari-type sands, probably before any European settlement in the western Free State. Given the archival and archaeological record for long-term indigenous pastoralism along the middle Riet River, they propose that further research is warranted to examine the dominant view that Boer stock-raisers and British hunters had decimated the ecological resources of the Highveld during the 19th century.

Chapters 4, 5 and 6 are focusing on palaeoanthropological and archaeological questions in South Africa: J. Francis Thackeray (chapter 4) discusses by pairwise comparisons of crania of selected and well-preserved hominin specimens attributed to *Australopithecus africanus* and *Homo habilis* a morphometric approach in the contex of chronospecies and climate change. Michael Chazan (chapter 5) critically discusses archaeological systematics by the example of the Fauresmith industries, obviously developed at the point of transition between the biface technological lineage found in the Acheulean and the prepared core technologies that develop during the Middle Stone Age. Finally, James Brink (chapter 6) does a new reconstruction of the skull of *Megalotragus priscus* (Broom, 1909), based on a find from the Erfkroon fossil site, Free State, offering notes on the chronology and biogeography of this species.

Michaela Ecker and collaborators highlight in chapter 7 how ostrich eggshells can be succesfully used as a source of palaeoenvironmental information for the arid interior of South Africa. Ostrich eggshells from Wonderwerk Cave gave new evidence for the period between 1.96–0.78 Ma years that indicated a generally arid environment from ca. 1.96–1.78 Ma years, and at ca 1 Ma the onset of an even more arid environment.

Chapter 8 by van Aardt et al. (including Louis Scott) gives first and preliminary insights of the Baden-Baden spring mound site located some 70 km northwest of Bloemfontein. The authors underline that the site is rich in archaeological and palaeobotanical material covering the last ~160 ka, and they compare it with other spring, pan and alluvial sites (e.g. Florisbad, Deelpan [see picture on book cover], Erfkroon) in the region.

Marion Bamford (chapter 9) also studied sediments from Wonderwek Cave (Stratum 5, ~15 ka old) using fossil charcoal as a proxy for former environmental conditions. From 134 charcoal pieces she could identify up to six species and two to generic level. The majority of the woods found can tolerate dry conditions and a wide range of temperatures but the presence of *Berchemia discolor* and *Halleria lucida* suggests that conditions between 14,985 and 13,952 years cal. BP may have been slightly wetter than today.

Chapters 10 and 11 are focusing on rock shelters and the cultural heritage of rock art in Iringa (Tanzania) and in the Drakensberg (uKhahlamba-Drakensberg Park [UDP], South Africa). Makarius Peter Itambu (chapter 10) identifies in the case of Iringa rock art that they belong to two different rock art traditions: Hunter-forager and Bantu-speaking art traditions. The former is dominated by naturalistic animal and human figures executed in dark and red pigments, the latter consists of schematic animal and human figures. Alvord Nuhundu (chapter 11) gives an overview on the recent and future effects of ongoing Global Change (temperature and precipitation) on rock art in the UDP World Heritage Site. He also discusses different measures to conserve endangered rock art paintings against the challenge of Global Change. Finally, the contribution of Wondimu Tadiwos Hailesilassie (chapter 12) addresses applied aspects of research on changing climates ("the past is the key to the present, and future") by small farmers' land use, illustrated by a case study (Wolaita zone) in Ethiopia. He can show that agroforestry systems with suitable selection of tree species (e.g. *Grevillea robusta*) can help to mitigate the negative effects of ongoing climate change in the tropics.

BIBLIOGRAPHY OF LOUIS SCOTT

Backwell L.R., McCarthy, T.S., Wadley L., Henderson, Z., Steininger, C.M., de Klerk, B., Barr, M., Lamothe, M., Chase, B.M., Woodborne, S., Susino, G.J., Bamford, M.K., Sievers, C., Brink, J.S., Rossouw, L., Pollarolo, L., Trower, G., Scott, L. and d'Errico, F., 2014, Multiproxy record of late Quaternary climate change and Middle Stone Age human occupation at Wonderkrater, South Africa. *Quaternary Science Reviews,* **99**, pp. 42–59.

Bamford M.K., Neumann, F.H. and Scott, L., accepted, Pollen, charcoal and plant macrofossil evidence of Neogene and Quaternary Environments in Southern Africa. In *Quaternary environmental change in southern Africa: physical and human dimensions,* edited by Knight, J. and Grab, S. (Cambridge: Cambridge University Press).

Bamford, M.K., Neumann, F.H., Pereira, L.M., Scott, L., Dirks, P.H.G.M. and Berger, L.R., 2010, Botanical remains from a coprolite from the Pleistocene hominin site of Malapa, Sterkfontein Valley, South Africa. *Palaeontologia africana,* **45**, pp. 23–28.

Botha, G.A., Scott L., Vogel J.C. and von Brunn, V., 1992, Palaeosols and palaeoenvironments during the 'Late Pleistocene Hypothermal' in Northern Natal. *South African Journal of Science*, **88**, pp. 508–512.

Bousman, C.B. and Scott, L., 1994, Climate or overgrazing? The palynological evidence for vegetation change in the eastern Karoo. *South African Journal of Science,* **90**, pp. 575–578.

Bousman, C.B., Partridge, T.C., Scott, L., Metcalfe, S.E., Vogel, J.C., Seaman, M. and Brink, J.S., 1988, Palaeoenvironmental implications of Late Pleistocene and Holocene valley fills in Blydefontein basin, Noupoort, C.P., South Africa. *Palaeoecology of Africa,* **19**, pp. 43–67.

Brook, G.A., Cowart, J.B., Brandt, S.A. and Scott, L., 1997, Quaternary climatic change in southern and eastern Africa during the last 300 ka; the evidence from caves in Somalia and the Transvaal region of South Africa. *Zeitschrift für Geomorphologie, N.F.*, **108**, pp. 15–48.

Brook, G.A., Scott, L., Railsback, B. and Goddard, E.A., 2010, A 35 ka pollen and isotope record of environmental change along the southern margin of the Kalahari from a stalagmite in Wonderwerk Cave, South Africa. *Journal of Arid Environments,* **74(5)**, pp. 870–884.

Butzer, K.W., Fock, G.J., Scott, L. and Stuckenrath, R., 1979, Dating and contextual analysis of rock engravings in South Africa. *Science,* **203**, pp. 1201–1214.

Carrión, J.S. and Scott, L., 1999, The challenge of pollen analysis in palaeoenvironmental studies of hominid beds: the record from Sterkfontein Caves. *Journal of Human Evolution,* **36**, pp. 401–408.

Carrión, J.S., Brink, J.S., Scott, L. and Binneman, J.N.F., 2000, Palynology and palaeoenvironment of Pleistocene hyena coprolites from an open-air occurrence at Oyster Bay, Eastern Cape coast, South Africa. *South African Journal of Science*, **96**, pp. 449–453.

Carrión, J.S., Scott, L. and Davis, O.K., 1997a, Interés de algunos depósitos biogénicos en la reconstrucción paleoambiental de zonas áridas. El caso de Procavia, Petromus y Neotoma. I. Bases conceptuales y metodológicas. *Cuaternario y Geomorfología,* **11**, pp. 45–50.

Carrión, J.S., Scott, L. and Davis, O.K., 1997b, Interés de algunos depósitos biogénicos en la reconstrucción paleoambiental de zonas áridas. El caso de Procavia, Petromus y Neotoma. II. Datos palinológicos. *Cuaternario y Geomorfología,* **11**, pp. 51–74.

Carrión, J. S., Scott, L. and Marais, E., 2006, Environmental implications of pollen spectra in bat droppings from south-eastern Spain and potential for palaeoenvironmental reconstructions. *Review of Palaeobotany and Palynology,* **140**, pp. 175–186.

Carrión, J.S., Scott, L. and Vogel, J.C., 1999, Twentieth century changes in montane vegetation in the eastern Free State, South Africa, derived from palynology of hyrax middens. *Journal of Quaternary Science,* **14(1)**, pp. 1–16.

Carrión, J.S., Scott, L., Arribas, A., Fuentes, N., Gil-Romera, G. and Montoya, E., 2007, Pleistocene landscapes in Central Iberia inferred from pollen analysis of hyena coprolites. *Journal of Quaternary Science,* **22(2)**, pp. 191–202. doi: 10.1002/jqs.1024.

Carrión, J.S., Scott, L., Huffman, T., and Dryer, C. 2000. Pollen analysis of Iron Age cow dung in Southern Africa. *Vegetation History and Archaeobotany*, 9, pp. 239–249.

Carrión, J.S., Yll, R., Gonzáles-Sampériz, P. and Scott, L., 2004, Advances in the palynology of cave sites from Spain: taphonomical and palaeoecological aspects. In: *Investigaciones en sistemas kársticos españoles,* edited by Andreo, B. and Durán, J.J. (*Hidrogeologicia y aguas subterraneas*), **12**, pp. 351–366.

Chase, B.M., Meadows, M.E., Scott, L., Thomas, D.S.G., Marais, E., Sealy, J. and Reimer, P.J., 2009, A record of rapid Holocene climate change preserved in hyrax middens from SW Africa. *Geology,* **37**, pp. 703–706.

Churchill, S.E., Brink, J.S., Berger, L.R., Hutchinson, R.A., Rossouw, L., Stynder, D., Hancox, P.J., Brandt, D., Woodborne, S., Loock, J.C., Scott, L. and Ungar, P., 2000, Erfkroon: a new Florisian fossil locality from fluvial contexts in the western Free State, South Africa. *South African Journal of Science,* **96**, pp. 161–163.

Chazan, M. Avery, M.D., Bamford M.K., Berna, F., Brink, J., Holt, S., Fernandez-Jalvo, Y., Goldberg, P., Matmon, A., Porat, N., Ron, H., Rossouw, L., Scott, L. and Horwitz, L.K., 2012, The Oldowan horizon in Wonderwerk Cave (South Africa): Archaeological, geological, paleontological and paleoclimatic evidence. *Journal of Human Evolution,* **63(6)**, pp. 859–66.

Chase, B.M., Quick, L., Meadows, M.E., Scott, L., Thomas, D.S.G. and Reimer, P.J., 2011, Late-glacial interhemispheric climate dynamics revealed in South African hyrax middens. *Geology,* **39**, pp. 19–22, doi:10.1130/G31129.1.

Chase, B.M., Scott, L., Meadows, M.E., Gil-Romera, G., Boom, A., Carr A.S., Reimer, P.J., Britton, M.N., Quick, L.J. 2012. Rock hyrax middens: a novel palaeoenvironmental archive in southern African drylands. *Quaternary Science Reviews,* **56**, pp. 107–125.

Cordova, C. and Scott, L., 2010, The potential of Poaceae, Cyperaceae and Restionaceae phytoliths to reflect past environmental conditions in South Africa. *Palaeoecology of Africa,* **30**, pp. 107–133.

Daniau, A.-L., Bartlein, P.J., Harrison, S.P., Prentice, I.C., Brewer, S., Friedlingstein, P., Harrison-Prentice, T.I., Inoue, J., Izumi, K., Marlon, J.R., Mooney, S., Power, M.J., Stevenson, J., Tinner, W., Andrič, M., Atanassova, J., Behling, H., Black, M., Blarquez, O., Brown, K.J., Carcaillet, C., Colhoun, E.A., Colombaroli, D., Davis, B.A.S., D'Costa, D., Dodson, J., Dupont, L., Eshetu, Z., Gavin, D.G., Genries, A., Haberle, S., Hallett, D.J., Hope, G., Horn, S.P., Kassa, T.G., Katamura, F., Kennedy, L.M., Kershaw, P., Krivonogov, S., Long, C., Magri, D., Marinova, E., McKenzie, G.M., Moreno, P.I., Moss, P., Neumann, F.H., Norström, E., Paitre, C., Rius, D., Roberts, N., Robinson, G.S., Sasaki, N., Scott, L., Takahara, H., Terwilliger, V., Thevenon, V., Turner, R., Valsecchi, V.G., Vannière, B., Walsh, M., Williams, N. and Zhang, Y., 2012, Predictability of biomass burning in response to climate changes. *Global Biogeochemical Cycles,* **26**, GB4007, doi: 10.1029/2011GB004249.

Deacon, J., Lancaster, N. and Scott, L., 1984, Summary of the evidence for Late Quaternary climatic change in southern Africa presented at Workshop I, Johannesburg 1983. In *Late Cainozoic Palaeoclimates of the Southern Hemisphere*, edited by Vogel, J.C. (Rotterdam: Balkema), pp. 391–404.

Elenga, H, Peyron, O., Bonnefille, R., Jolly, D., Cheddadi, R., Guiot, J., Andrieu, V., Bottema, S., Buchet, G., Debeaulieu, J.L., Hamilton, A.C., Maley, J., Marchant, R., Perezobiol, R., Reille, M., Riollet, G., Scott, L., Straka, H., Taylor., D., Vancampo, E., Vincens, A., Laarif, F. and Jonson, H., 2000, Pollen-Based Biome Reconstruction for Southern Europe and Africa 18,000 yr BP. *Journal of Biogeography,* **27**, pp. 621–634.

Fernandez-Jalvo, Y., Scott, L. and Andrews, P., 2011, Taphonomy in palaeoecological interpretations. *Quaternary Science Reviews,* **30**, pp. 1296–1302, doi:10.1016/j.quascirev.2010.07.022.

Fernandez-Jalvo, Y., Scott, L. and Denys, C., 1996, Pollen composition in owl pellets and their environmental implications. *Comptes rendus de l'Académie des Sciences, Paris, (Paleontology/Paleobotany) série IIa,* **323**, pp. 259–265.

Fernandez-Jalvo, Y., Scott, L. and Denys, C., 1999, Taphonomy of pollen associated with predation. *Palaeogeography, Palaeoclimatology, Palaeoecology*, **149**, pp. 271–282.

Finné, M., Risberg, J., Norström, E. and Scott, L., 2010, Siliceous microfossils as Late Quaternary paleo-environmental indicators at Braamhoek wetland, South Africa. *The Holocene,* **20(5)**, pp. 747–760, doi:10.1177/0959683610362810.

Fock, G.J., Butzer, K.W., Scott, L. and Stuckenrath, R., 1980, Rock engravings and the Later Stone Age, Northern Cape Province, South Africa: a multi-disciplinary study. In *Proc. 8th Panafrican Congress of Prehistory and Quaternary Studies, Nairobi, 1977*, edited by Leakey, R.E. and Ogot, B.A. (Nairobi: The International Louis Leakey Memorial Institute for African Pre-history), pp. 311–313.

Gil Romera, G. and Scott, L., 2006, Reconstrucción de paleoambientes en zones áridas: investigations en el Desierto del Namib. In *Paleoambientes y cambio climatico. Fundación Séneca, Agencia de ciencia y technología de la Región de Murcia,* edited by Carrión, J.S., Fernandez, S. and Fuentes, N., pp. 191–202. ISBN 84-932456-6-6.

Gil Romera, G., Scott, L., Marais, E. and Brook, G.A., 2006, Middle to late Holocene moisture changes in the desert of northwest Namibia, derived from fossil hyrax dung pollen. *The Holocene,* **16(8)**, pp. 1073–1084.

Gil Romera, G., Scott, L., Marais, E. and Brook, G.A., 2007, Late Holocene environmental change in the northwestern Namib Desert margin: New fossil pollen evidence from hyrax middens. *Palaeogeography, Palaeoclimatology, Palaeoecology,* **249**, pp. 1–17, doi:10.1016/j.palaeo.2007.01.002.

Gil-Romera, G., Neumann, F.H., Scott, L., Sevilla-Callejo, M. and Fernández-Jalvo, Y., 2014, Pollen taphonomy from hyaena scats and coprolites: preservation and quantitative differences. *Journal of Archaeological Science*, **46**, pp. 89–95.

Grab, S., Scott, L., Rossouw, L. and Meyer, S., 2005, Holocene palaeoenvironments inferred from a sedimentary sequence in the Tsoaing River Basin, western Lesotho. *Catena,* **61**, pp. 49–62.

Henderson, Z., Scott, L. Rossouw, L. and Jacobs, Z., 2006, Dating, Palaeoenvironments, and Archaeology: A Progress Report on the Sunnyside 1 Site, Clarens, South Africa. *Archaeological Papers of the American Anthropological Association*, **16**, pp. 139–149, doi:10.1525/ap3a.2006.16.1.139.

Holmgren, K., Lee-Thorp, J.A., Cooper, G.R.J., Lundblad, K., Partridge, T.C., Scott, L., Sithaldeen, R., Talma, A.S. and Tyson, P.D., 2003, Persistent millennial-scale variability over the past 25, 000 years in Southern Africa. *Quaternary Science Reviews,* **22**, pp. 2311–2326.

Horowitz, A., Sampson, C.G., Scott, L. and Vogel, J.C., 1978, Analysis of the Voigtspost site, O.F.S., South Africa. *South African Archaeological Bulletin*, **33**, pp. 152–159.

Jolly, D., Prentice, I.C., Bonnefille, R., Ballouche, A., Bengo, M., Brenac, P., Butchet, G., Burney, D., Cazet, J., Cheddadi, R., Edorh, T., Elenga, H., Elmoutaki, S., Guiot, J., Laarif, F., Lamb, H., Lezine, A.M., Maley, J., Mbenza, M., Peyron, O., Reille, M., Reynaud-Farrera, I., Riollet, G., Ritchie, J., Scott, L., Ssemmanda, I., Straka, H., Umer, M., Van Campo, E., Vilimubalo, S., Vincens, A. and Waller, M., 1998, Biome Reconstruction from pollen and plant macrofossil data for Africa and the Arabian peninsula at 0 and 6 ka. *Journal of Biogeography,* **25**, pp. 1007–1027.

Marais, E., Scott, L., Gil-Romera, G. and Carrión, J.S., 2015, The potential of palynology in fossil bat-dung from Arnhem Cave, Namibia. *Transactions of the Royal Society of South Africa*, http://dx.doi.org/10.1080/0035919X.2014.999734.

Markgraf, V. and Scott, L., 1981, Lower timberline in central Colorado during the past 15 000 years. *Geology,* **9**, pp. 231–234.

Metwally, A.A., Scott, L., Neumann, F.H., Bamford, M.K. and Oberhänsli, H., 2014, Holocene palynology and palaeoenvironments in the Savanna Biome at Tswaing Crater, central South Africa. *Palaeogeography, Palaeoclimatology, Palaeoecology*, **402**, pp. 125–135.

McLean, B. and Scott, L., 1999, Phytoliths in sediments of the Pretoria Saltpan (Tswaing Crater) and their potential as indicators of the environmental history at the site. In *Tswaing—Investigations into the origin, age and palaeoenvironments of the Pretoria Saltpan,* edited by Partridge, T.C. (Pretoria: Council for Geosciences), pp. 167–171.

Mucina, L., Rutherford, M.C., Palmer, A.P., Milton, S.J., Scott, L., Lloyd, W., van der Merwe, B., Hoare, D.B., Bezuidenhout, H., Vlok, J.H.J., Euston-Bron, D.I.W., Powrie, L.W. and Dold, A.P., 2006, Nama-Karoo Biome 7. *The Vegetation of South Africa, Lesotho and Swaziland.* Strelitzia, **19**, edited by Mucina, L. and Rutherford, C.M., 13. 978-1-919976-21-1 (Pretoria: South African National Biodiversity Institute), pp. 324–347.

Mucina, L., Hoare, D.B, Lötter, M.C., du Preez, P.J., Rutherford, M.C, Scott-Shaw, C.R., Bredenkamp, G.J., Powrie, L.W., Scott, L., Cilliers, S.S., Bezuidenhout, H., Mostert, T.H., Camp, K.G.T., Siebert, S.J., Winter, P.J.D., Burrows, J.E., Dobson, L., Ward, R.A., Stalmans, M., Oliver, E.G.H., Siebert, F., Kobisi, K. and Kose, L., 2006, Grassland Biome 8. *The Vegetation of South Africa, Lesotho and Swaziland.* Strelitzia, **19**, edited by Mucina, L. and Rutherford, C.M., 13. 978-1-919976-21-1 (Pretoria: South African National Biodiversity Institute), pp. 348–437.

Neumann, F., Stager, C., Scott, L., Venter, H.J.T. and Weyhenmeyer, C., 2008, Holocene vegetation and climate records from Lake Sibaya, KwaZulu-Natal (South Africa). *Review of of Palaeobotany and Palynology,* **152**, pp. 113–128.

Neumann, F., Scott, L., Bousman, C.B. and van As, L., 2010, A Holocene pollen sequence and vegetation changes at Lake Eteza, KwaZulu-Natal (South Africa). *Review of Palaeobotany and Palynology,* **162(1)**, pp. 39–53, doi:10.1016/j.revpalbo.2010.05.001.

Neumann, F., Scott, L. and Bamford, M., 2011, Climate change and human disturbance of fynbos vegetation during the late Holocene at Princess Vlei, Western Cape, South Africa. *The Holocene,* **21(7)**, pp. 1137–1150.

Neumann, F.H., Botha, G.A. and Scott, L., 2014, 18,000 years of grassland evolution in the summer rainfall region of South Africa—evidence from Mahwaqa Mountain, KwaZulu-Natal. *Vegetation History and Archaeobotany*, doi:10.1007/s00334-014-0445-3.

Norström, E., Scott, L., Partridge, T.C., Risberg, J. and Holmgren, K., 2009, Reconstruction of environmental and climate changes at Braamhoek wetland, eastern escarpment South Africa, during the last 16000 years with emphasis on the Pleistocene-Holocene transition. *Palaeogeography, Palaeoclimatology, Palaeoecology,* **271**, pp. 240–258, doi:10.1016/j.palaeo.2008.10.018.

Nyakale, M. and Scott, L., 2002, Interpretation of Late Holocene pollen in channel fills in the eastern Free State, South Africa. *South African Journal of Botany,* **68**, pp. 464–468.

Olsen, A., Prinsloo, L.C., Scott, L. and Jäger, A.K., 2007, Hyraceum, the fossilized metabolic product of rock hyraxes (*Procavia capensis*), shows GABA-benzodiazepine receptor affinity. *South African Journal of Science,* **103**, pp. 437–438.

Oschadleus, H.D., Vogel, J.C. and Scott, L., 1996, Radiometric date for the Port Durnford peat and development of yellow-wood forest along the South African east coast. *South African Journal of Science,* **92**, pp. 43–45.

Partridge, T.C., Avery, D.M., Botha, G.A., Brink, J.S., Deacon, J., Herbert, R.S., Maud, R.R., Scholtz, A., Scott, L., Talma, A.S. and Vogel, J.C., 1990, Late Pleistocene Climatic Change in Southern Africa. *South African Journal of Science,* **86**, pp. 315–317.

Partridge, T.C., Kerr, S.J., Metcalfe, S.E., Scott, L., Talma, A.S. and Vogel, J.C., 1993, The Pretoria Saltpan: a 200 000 year southern African lacustrine sequence. *Palaeogeography, Palaeoclimatology, Palaeoecology*, **101**, pp. 317–337.

Partridge, T.C., Metcalfe, S.E. and Scott, L., 1999, Conclusions and implications for a model of regional palaeoclimates during the last two glacial cycles. In *Tswaing—Investigations into the origin, age and palaeoenvironments of the Pretoria Saltpan,* edited by Partridge, T.C. (Pretoria: Council for Geosciences), pp. 193–198.

Partridge, T.C., Scott, L. and Hamilton, J.E., 1999, Synthetic reconstructions of Southern African environments during the Last Glacial Maximum (21–18 Kyr) and the Holocene Altithermal (8–6 Kyr). *Quaternary International,* **57/58**, pp. 207–214.

Partridge, T.C. and Scott, L., 2000, Lakes and Pans. In *The Cainozoic of Southern Africa* edited by Partridge, T.C. and Maud, R.R., Oxford Monographs on Geology and Geophysics no. **40** (New York: Oxford University Press), pp. 145–161, ISBN 0-19-512530-4.

Partridge, T.C., Scott, L. and Schneider, R.R., 2004, Between Agulhas and Benguela: responses of Southern African climates of the Late Pleistocene to current fluxes, orbital precession and extent of the Circum-Antarctic vortex. In *Past Climate Variability through Europe and Africa,* edited by Battarbee, R.W., Gasse, F. and Stickley, C.E. (Dordrecht: Springer), pp. 45–68.

Polevova S., Tekleva, M., Neumann, F.H., Scott, L. and Stager, J.C., 2010, Pollen morphology, ultrastructure and taphonomy of the Neuradaceae with special reference to *Neurada procumbens* L. and *Grielum humifusum* E. Mey. ex Harv. et Sond. *Review of Palaeobotany and Palynology,* **160**, pp. 163–171, doi:10.1016/j.revpalbo.2010. 02.010.

Quick, L.J., Chase, B.M., Meadows, M.E., Scott, L., Reimer, P.J. 2011. Palynological evidence from rock hyrax middens. *Palaeogeography, Palaeoclimatology, Palaeoecology,* **309**, pp. 253–270.
Rebelo, A.G., Boucher, C., Helme, N., Mucina, L., Rutherford, M.C., Smit, W.S., Powrie, L.S., Ellis, F., Lambrechts, J.N.J., Scott, L., Radloff, G.T., Johnson, S.D., Richardson, D.M., Ward, R.A., Procheş, S.M., Oliver, E.G.H., Manning, J.C., Jürgens, N., McDonald, D.J., Janssen, A.M., Walton, B.A., le Roux, A., Skowno, A.L. Todd, S.W. and Hoare, D.B., 2006, Fynbos Biome 4. *The Vegetation of South Africa, Lesotho and Swaziland.* Strelitzia, **19**, edited by Mucina, L. and Rutherford, C.M., 13. 978-1-919976-21-1 (Pretoria: South African National Biodiversity Institute), pp. 52–219.
Roberts, D.L., Matthews, T., Herries, A.I.R., Boulter, C., Scott, L., Musekiwa, C., Ntembi, P., Browning, C., Smith, R.H.M., Haarhoff, P. and Bateman, M. 2011. Regional and global context of the Late Cenozoic Langebaanweg (LBW) Palaeontological Site: West Coast of South Africa. *Earth Science Reviews,* **106**, pp. 191–214.
Rossouw, L. and Scott, L., 2011, Phytoliths and pollen, the microscopic plant remains in Pliocene volcanic sediments around Laetoli, Tanzania. In *Palaeontology and Geology of Laetoli, Tanzania: Human Evolution in Context. Volume 1: Geology, Geochronology, Palaeoecology and Palaeoenvironment, Vertebrate Palaeobiology and Palaeoanthropology*, edited by Harrison, T. (Dordrecht: Springer Science+Business Media), pp. 201–215, DOI 10.1007/978-90-481-9956-3_9.
Roberts, D.L., Scicio, L., Herries, A.I.R., Scott, L., Bamford, M.K., Musekiwa, C. and Tsikos, H., 2013, Miocene fluvial systems and palynofloras at the Southwestern tip of Africa: Implications for regional and global fluctuations in climate and ecosystems. *Earth Science Reviews,* **124**, pp. 184–201.
Rutherford, M.C., Mucina, L., Lötter, C., Bredenkamp, G.J., Smit, J.H.L., Scott-Shaw, C.R., Hoare, D.B., Goodman, P.S., Bezuidenhout, H., Scott, L., Ellis, F., Powrie, L.W., Siebert, F., Mostert, T.H., Henning, B.J., Venter, C.E., Camp, K.G.T., Siebert, S.J., Matthews, W.S., Burrows, J.E., Dobson, L., van Rooyen, N., Schmidt, E., Winter, P.J.P., du Preez, P.J., Ward, R.A., Williamson, S.W. and Hurter, P.J.H., 2006, Savanna Biome 9. *The Vegetation of South Africa, Lesotho and Swaziland.* Strelitzia, **19**, edited by Mucina, L. and Rutherford, C.M., 13. (Pretoria: South African National Biodiversity Institute), pp. 438–539.
Sandersen, A., Scott, L., McLachlan I., and Hancox, J., 2011, Cretaceous biozonation based on terrestrial palynomorphs from two wells in the offshore Orange Basin of South Africa. *Palaeontologia africana*, **46**, pp. 21–41.
Scott, L., 1972, Palynology of the Lower Cretaceous deposits (The Uitenhage series) from the Algoa Basin. *Palaeoecology of Africa,* **7**, p. 4.
Scott, L., 1973, Palynology of deposits overlying Lemphane Kimberlite Pipe, Lesotho. In *Lesotho Kimberlites,* edited by Nixon, P.H. (Lesotho National Development Corporation), pp. 168–171.
Scott, L., 1976, Preliminary palynological results from the Alexandersfontein Basin near Kimberley. *Annals of the South Africa Museum*, **71**, pp. 193–199.
Scott, L., 1976, Palynology of Lower Cretaceous deposits from the Algoa Basin (Republic of South Africa). *Pollen et Spores,* **XVIII(4)**, pp. 563–609.
Scott, L. and Vogel, J.C., 1978, Pollen analyses of the thermal spring deposit at Wonderkrater (Transvaal, South Africa). *Palaeoecology of Africa*, **10**, pp. 155–162.
Scott, L. and Klein, R.G., 1981, A hyaena-accumulated bone assemblage from late Holocene deposits at Deelpan, Orange Free State. *Annals of the South Africa Museum,* **86(6)**, pp. 217–227.

Scott, L., 1982a, A 5000-year old pollen record from spring deposits in the bushveld at the north of the Soutpansberg, South Africa. *Palaeoecology of Africa,* **14**, pp. 45–55.
Scott, L., 1982b, Pollen analyses of Late Cainozoic deposits in the Transvaal, South Africa, and their bearing on palaeoclimates. *Palaeoecology of Africa,* **15**, pp. 101–107.
Scott, L., 1982c, Late Quaternary fossil pollen grains from the Transvaal, South Africa. *Review of Palaeobotany and Palynology,* **36**, pp. 241–278.
Scott, L., 1982d, A Late Quaternary pollen record from the Transvaal bushveld, South Africa. *Quaternary Research,* **17**, pp. 339–370.
Scott, L., 1983, Palynological evidence for vegetation patterns in the Transvaal (South Africa) during the Late Pleistocene and Holocene. *Bothalia,* **14(3,4)**, pp. 445–449.
Scott, L. and Hall, K.J., 1983, Palynological evidence for interglacial vegetation cover on Marion Island, sub-Antarctic. *Palaeogeography, Palaeoclimatology, Palaeoecology,* **41**, pp. 35–43.
Scott, L. and Vogel, J.C., 1983, Late Quaternary pollen profile from the Transvaal highveld, South Africa. *South African Journal of Science,* **79**, pp. 266–272.
Scott, L., 1984a, Palynological evidence for Quaternary paleoenvironments in southern Africa. In *Southern African Palaeoenvironments and Pre-history* edited by Klein, R.G. (Rotterdam: Balkema), pp. 65–80.
Scott, L., 1984b, Reconstruction of Late Quaternary palaeoenvironments in the Transvaal region, South Africa, on the basis of palynological evidence. In *Late Cainozoic Palaeoclimates of the Southern Hemisphere*, edited by Vogel, J.C. (Rotterdam: Balkema), pp. 317–327.
Scott, L. and Srivastava, S.K., 1984, Reworked Cretaceous palynomorphs in Quaternary deposits from Central Colorado, USA. *Pollen et Spores,* **XXVI(2)**, pp. 227–240.
Scott, L. and van Zinderen Bakker, E.M., 1985, Exotic pollen and long-distance wind dispersal at a sub-Antarctic island. *Grana,* **24**, pp. 45–54.
Scott, L., 1985, Palynological indications of the Quaternary vegetation history of Marion Island (sub-Antarctic). *Journal of Biogeography,* **12**, pp. 413–431.
Scott, L., 1986a, Pollen analysis and palaeoenvironmental interpretation of Late Quaternary sediments exposures in the eastern Orange Free State, South Africa. *Palaeoecology of Africa,* **17**, pp. 113–122.
Scott, L., 1986b, The Late Tertiary and Quaternary pollen record in the interior of South Africa. *South African Journal of Science,* **82(2)**, p. 73.
Scott, L. and Bonnefille, R., 1986, A search for pollen from the hominid deposits of Kromdraai, Sterkfontein and Swartkrans: some problems and preliminary results. *South African Journal of Science,* **82**, pp. 380–382.
Scott, L. and Thackeray, J.F., 1987, Multivariate analysis of Late Pleistocene and Holocene pollen spectra from Wonderkrater, Transvaal, S. Africa. *South African Journal of Science,* **83(2)**, pp. 93–98.
Scott, L., 1987a, Pollen analysis of hyena coprolites and sediments from Equus Cave, Taung, Southern Kalahari (S. Africa). *Quaternary Research,* **28**, pp. 144–156.
Scott, L., 1987b, Late Quaternary forest history in Venda, Southern Africa. *Review of Palaeobotany and Palynology,* **53**, pp. 1–10.
Scott, L., 1988a, The Pretoria Saltpan: A unique source of Quaternary palaeoenvironmental information. *South African Journal of Science,* **84**, pp. 560–562.
Scott, L., 1988b, Holocene environmental change at western Orange Free State pans, South Africa, inferred from pollen analysis. *Palaeoecology of Africa,* **19**, pp. 109–118.
Scott, L., 1989a, Pollen analysis and palaeoenvironmental significance of Quaternary faecal deposits in Africa. In *Environmental Quality and Ecosystem Stability,* edited by Spanier, E. and Steinberger, Y. (Jerusalem: IV-B ISEEQS Pub.) pp. 65–71.

Scott, L., 1989b, Late Quaternary vegetation history and climatic change in the eastern O.F.S, South Africa. *South African Journal of Botany,* **55(1)**, pp. 107–116.

Scott, L., 1989c, Climatic conditions in Southern Africa since the Last Glacial Maximum, inferred from pollen analysis. *Palaeogeography, Palaeoclimatology, Palaeoecology,* **70**, pp. 345–353.

Scott, L., 1990, Summary of environmental changes reflected by pollen in some Holocene sediments from south Africa and Marion Island, southern ocean. *South African Journal of Science,* **86**, pp. 464–466.

Scott, L. and Cooremans, B., 1990, Late Quaternary pollen from a hot spring in the upper Orange River Basin, South Africa. *South African Journal of Science,* **86**, pp. 154–156.

Scott, L. 1990a, Palynological evidence for late Quaternary environmental change in southern Africa. *Palaeoecology of Africa,* **21**, pp. 259–268.

Scott, L., 1990b, Hyrax (Procaviidae) and dassie rat (Petromuridae) middens in palaeoenvironmental studies in Africa. In *Packrat Middens: the last 40 000 years of biotic change*, edited by Betancourt, J., van Devender, T.R. and Martin, P.S. (Tuscon: University of Arizona Press), pp. 398–407.

Scott, L. and Bousman, C.B., 1990, Palynological analysis of hyrax middens from southern Africa. *Palaeogeography, Palaeoclimatology, Palaeoecology*, **79**, pp. 367–379.

Scott, L., Cooremans, B., de Wet, J.S. and Vogel, J.C., 1991, Holocene environmental changes in Namibia inferred from pollen analysis of swamp and lake deposits. *The Holocene,* **1(1)**, pp. 8–13.

Scott, L. and Cooremans, B., 1992, Pollen in recent Procavia (hyrax), Petromus (dassie rat) and bird dung in South Africa. *Journal of Biogeography,* **19**, pp. 205–215.

Scott, L., 1992, Environmental implications and origin of microscopic *Pseudoschizaea* Thiergart and Frantz ex R. Potonié emend. in sediments. *Journal of Biogeography,* **19**, pp. 349–354.

Scott, L., Cooremans, B. and Maud, R.R., 1992, Preliminary palynological evaluation of the Port Durnford Formation at Port Durnford, Natal coast, South Africa. *South African Journal of Science*, **88**, pp. 470–474.

Scott, L. and Vogel J.C., 1992, Short-term changes of climate and vegetation revealed by pollen analysis of hyrax dung in South Africa. *Review of Palaeobotany and Palynology,* **74**, pp. 283–291.

Scott, L., 1992, Pollen for the people. *South African Journal of Science*, **88**, p. 553.

Scott, L. and Brink, J.S., 1992, Quaternary palaeoenvironments of pans in central South Africa: palynological and palaeontological evidence. *South African Geographer*, **19**, pp. 22–34.

Scott, L., 1993, Palynological evidence for late Quaternary warming episodes in Southern Africa. *Palaeogeography, Palaeoclimatology, Palaeoecology*, **101**, pp. 229–235.

Scott, L., 1994a, Palynology of late Pleistocene hyrax middens, south-western Cape Province, South Africa: a preliminary report. *Historical Biology,* **9**, pp. 71–81.

Scott, L., 1994b, Past vegetation changes in mountainous areas in South Africa as revealed by pollen analysis of hyrax middens. In *Proceedings of the XIII Plenary Meeting of AETFAT, Zomba, Malawi,* edited by Seyani, J.H. and Chikuni, A.C., Volume **2**, 2–11 April 1991, pp 1007–1014.

Scott, L., 1995, Pollen evidence for vegetational and climate change during the Neogene and Quaternary in Southern Africa. In *Palaeoclimate and Evolution with Emphasis on Human Origins*, edited by Vrba, E.S., Denton, G., Partridge, T.C. and Burckle, L.H. (Yale: Yale University Press), pp. 65–76.

Scott, L., Steenkamp, M., and Beaumont, P.B., 1995, Palaeoenvironmental conditions in South Africa at the Pleistocene-Holocene transition. *Quaternary Science Reviews* (Global Younger Dryas Issue), **14(9)**, pp. 937–947.
Scott, L., 1996, Palynology of hyrax middens: 2000 years of palaeoenvironmental history in Namibia. *Quaternary International,* **33**, pp. 73–79.
Scott, L., Fernandez-Jalvo, Y. and Denys, C., 1996, Owl pellets, pollen and the palaeoenvironment. *South African Journal of Science,* **92**, pp. 223–224.
Scott, L., Anderson, H.M. and Anderson, J.M., 1997, Vegetation History. In *The Vegetation of Southern Africa* edited by Cowling, R.M., Richardson, D.M. and Pierce, S.M. (Cambridge: Cambridge University Press), pp. 62–84.
Scott, L., 1997, Quaternary Environment. In *Encyclopedia of Precolonial Africa,* edited by Vogel, J.O. (London: Altamira Press), pp. 42–45.
Scott, L. and Steenkamp, M., 1996, Environmental history and recent human disturbance at coastal Lake Teza, Kwazulu/Natal. *South African Journal of Science*, **92**, pp. 348–350.
Scott, L., 1999a, Palynological analysis of the Pretoria Saltpan (Tswaing Crater) sediments and vegetation history in the bushveld savanna biome, South Africa. In *Tswaing—Investigations into the origin, age and palaeoenvironments of the Pretoria Saltpan,* edited by Partridge, T.C. (Pretoria: Council for Geosciences), pp. 143–166.
Scott, L., 1999b, The vegetation history and climate in the Savanna Biome, South Africa, since 190 000 ka: A comparison of pollen data from the Tswaing Crater (the Pretoria Saltpan) and Wonderkrater. *Quaternary International,* **57/58**, pp. 215–223.
Scott, L., Cadman, A. and Verhoeven, R.L., 1999, Preface (Proceedings of the Third Conference on African Palynology). *Palaeoecology of Africa,* **26**, pp. IX–X.
Scott, L., 2000, Pollen. In *The Cainozoic of Southern Africa* edited by Partridge, T.C. and Maud, R.R., Oxford Monographs on Geology and Geophysics no. **40** (New York: Oxford University Press), pp. 339–350.
Scott, L. and Thackeray, J.F., in press, Palynology of Holocene deposits in Excavation 1 at Wonderwerk Cave, Northern Cape (South Africa). *African Archaeological Review.*
Scott, L., Rossouw, L., Cordova, C. and Risberg, J., in press, Palaeoenvironmental context of coprolites from the last interglacial in Azokh Cave, Caucasus, and their plant microfossils. In *Azokh Cave and the Transcaucasian Corridor. Vertebrate Paleobiology and Paleoanthropology*, edited by Fernández-Jalvo, Y., Andrews, P., King, T. and Yepiskoposyan, L. (Dordrecht: Springer).
Scott, L. and Vogel, J.C., 2000, Evidence for environmental conditions during the last 20 000 years in Southern Africa from ^{13}C in fossil hyrax dung. *Global and Planetary Change,* **26(1–3)**, pp. 207–215.
Scott, L., 2002, Microscopic charcoal in sediments and Late Quaternary fire history of the grassland and savanna regions in Africa. *Journal of Quaternary Science,* **17(1)**, pp. 77–86.
Scott, L., 2002, Grassland development under glacial and interglacial conditions in Southern Africa: review of pollen, phytolith and isotope evidence. *Palaeogeography, Palaeoclimatology, Palaeoecology,* **177(1-2)**, pp. 47–57.
Scott, L. and Nyakale, M., 2002, Pollen indications of Holocene palaeoenvironments at Florisbad in the central Free State. South Africa. *The Holocene,* **12(4)**, pp. 497–503.
Scott, L., Fernandez-Jalvo, Y., Carrión, J.S. and Brink, J.S., 2003, Preservation and interpretation of pollen in hyena corprolites: taphonomical observations from Spain and Southern Africa. *Palaeontologia africana,* **39**, pp. 83–91.
Scott, L., Holmgren, K., Talma, A.S., Woodborne, S. and Vogel, J.C., 2003, Age interpretation of the Wonderkrater spring sediments and vegetation change in the

savanna biome, Limpopo Province, South Africa. *South African Journal of Science*, **99**, pp. 484–488.

Scott, L., 2003, The Holocene of middle latitude arid areas. Chapter 27. In *Global change in the Holocene,* edited by Mackay, A., Batterbee, R., Birks, J. and Oldfield, F. (London: Edward Arnold), pp. 396–405.

Scott, L., 2004, Pollens fossiles dans des environnements du Pléistocène et de l'Holocène en Afrique du Sud. In *Évolution de la Végétation depuis deux millions d'années. Guide de la Préhistoire mondiale* edited by Sémah, A.-M. and Renault-Miskovsky, J. (Paris: Artcom/Errance), pp. 218–225.

Scott, L. and Lee-Thorp, J.A., 2004, Holocene climatic trends and rhythms in Southern Africa. In *Past Climate Variability through Europe and Africa* edited by Battarbee, R.W., Gasse, F. and Stickley, C.E. (Dordrecht: Springer), pp. 69–91.

Scott, L,. Marais, E. and Brook, G.A., 2004, Fossil hyrax dung and evidence of Late Pleistocene and Holocene vegetation types in the Namib Desert. *Journal of Quaternary Science,* **19(8)**, pp. 829–832.

Scott, L., Bousman, C.B. and Nyakale, M., 2005, Holocene pollen from swamp, cave and hyrax dung deposits at Blydefontein (Kikvorsberge), Karoo, South Africa. *Quaternary International,* **129**, pp. 49–59.

Scott, L. and Rossouw L., 2005, Reassessment of botanical evidence for palaeoenvironments at Florisbad, South Africa. *South African Archaeological Bulletin*, **60,** pp. 96–102.

Scott, L., Cadman A. and McMillan, I., 2006, Early history of Cainozoic Asteraceae along the Southern African west coast. *Review of Palaeobotany and Palynology,* **142**, pp. 47–52.

Scott, L. and Woodborne, S., 2007, Pollen analysis and dating of Late Quaternary faecal deposits (hyraceum) in the Cederberg, Western Cape, South Africa. *Review of Palaeobotany and Palynology,* **144(3–4)**, pp. 123–134, doi:10.1016/j.revpalbo.2006.07.004.

Scott, L. and Woodborne, S., 2007, Vegetation history inferred from pollen in Late Quaternary faecal deposits (hyraceum) in the Cape winter-rain region and its bearing on past climates in South Africa. *Quaternary Science Reviews,* **26**, pp. 941–953, doi:10.1016/j.quascirev.2006.12.012.

Scott, L., 2007, Professor Joey Coetzee 1921–2007. *Review of Palaeobotany and Palynology,* **147**, pp. 1–2, doi:10.1016/j.revpalbo.2007.08.001.

Scott, L., Holmgren, K. and Partridge, T.C., 2008, Reconciliation of vegetation and climatic interpretations of pollen profiles and other regional records from the last 60 thousand years in the Savanna Biome of Southern Africa. *Palaeogeography, Palaeoclimatology, Palaeoecology,* **257**, pp. 198–206, doi:10.1016/j.palaeo.2007. 10.018

Scott, L. and Smith, V.R., 2008, Peat and vegetation development on Marion Island, Southern Ocean, during the Late Holocene. *Navorsinge van die Nasionale Museum Bloemfontein,* **24(7)**, pp. 62–70.

Scott, L., Neumann, F.H., Brook, G.A., Bousman, C.B., Norström, E. and Metwally, A., 2012, Terrestrial fossil-pollen evidence of climate change during the last 26 thousand years in Southern Africa. *Quaternary Science Reviews,* **32**, pp. 100–118.

Scott, L., Neumann, F.H., Brook, G.A., Bousman, C.B., Norström, E. and Metwally, A., 2012, Corrigendum to: Terrestrial fossil-pollen evidence of climate change during the last 26 thousand years in Southern Africa [Quaternary Science Reviews 32 (2012) 100-118]. *Quaternary Science Reviews,* **59**, pp. 115–116.

Scott, L., Rossouw, L., Cordova, C. and Risberg, J., accepted, Palaeoenvironmental context of coprolites from the last interglacial in Azokh Cave, Caucasus, and their plant microfossils. In *Azokh Cave and the Transcaucasian Corridor. Vertebrate*

Palaeobiology and Palaeoanthropology, edited by Fernández-Jalvo, Y., Andrews, P., King, T. and Yepiskoposyan, L. (Dordrecht: Springer).

Sciscio, L., Neumann, F.H., Roberts, D.L., Tsikos, H., Scott, L. and Bamford, M., 2013, Fluctuations in Miocene climate and sea levels along the south-western South African coast: Inferences from biogeochemistry, palynology and sedimentology. *Palaeontologia africana*, **48(1)**, pp. 2–18.

Thackeray, J.F. and Scott, L., 2006, The Younger Dryas in the Wonderkrater sequence, South Africa? *Annals of the Transvaal Museum,* **43**, pp. 111–112.

Truc, L., Chevalier, M., Favier, C., Cheddadi, R., Meadows, M.E., Scott, L., Carr, A.S., Smith, G.F. and Chase, B.M., 2013, Quantification of climate change for the last 20,000 years from Wonderkrater, South Africa: Implications for the long term dynamics of the Intertropical Convergence Zone. *Palaeogeography, Palaeoclimatology, Palaeoecology,* **386**, pp. 575 587.

Verschuren, D., Briffa, K., Hoelzmann, P., Barker, P., Scott, L., Barber. K., Snowball, I. and Roberts, N., 2004, Holocene climate variability in Europe and Africa: a PAGES PEP III time-stream 1 synthesis. In *Past Climate Variability through Europe and Africa,* edited by Battarbee R.W., Gasse, F. and Stickley, C.E. (Dordrecht: Springer), pp. 567–582.

Verleye, T. Mertens, K.N., Young, M.D., Dale, B., Esper, O., Holzwarth, U., Lezine A.-M., McMinn, A., Scott, L., Vink, A., Zonneveld, K.A.F. and Louwye, S., 2012, Average process length variation of the marine dinoflagellate cyst *Operculodinium centrocarpum* in the Tropical and Southern Hemisphere Oceans: assessing its potential as a palaeosalinity proxy. *Marine Micropaleontology*, **86–87**, pp. 45–58.

Visser, N.J.N., Beukes, G.J. and Scott, L., 1986, Vivianite in Late Pleistocene swamp deposits near Clarens, S. Africa. *Transactions of the Royal Society of South Africa,* **89**, pp. 395–400.

van Aardt, A., Bousman, C.B., Brink, J.S., Brook, G.A., Jacobs, Z., du Preez, P.J., Rossouw, L. and Scott, L., in press, First chronological, palaeoenvironmental, and archaeological data from the Baden-Baden fossil spring complex in the western Free State, South Africa. *Palaeoecology of Africa,* **33**.

Weinstein-Evron, M., Scott, L. and Horwitz, L., 2010, Special issue on 'Paleoenvironment' in honour of Professor Aharon Horowitz. *Journal of Arid Environments,* **74(5)**, pp. 723–724.

CHAPTER 2

Geomorphology and late Quaternary environmental change: An overview of central southern Africa

Peter Holmes
Department of Geography, University of the Free State, Bloemfontein, South Africa

ABSTRACT: Physiographically, central southern Africa can be defined as the catchment of the Orange River and its tributaries. It comprises a number of terrain-morphological units which have the potential to contribute towards elucidating late Quaternary environmental change within the region. This paper provides an overview of geomorphic evidence and suggests how geomorphology has contributed to a better understanding of central southern African palaeoenvironments. Secondly, it suggests where geomorphology might focus in future, so as to remain relevant. Central southern Africa's topography varies from the mountains of the Great Escarpment (in both South Africa and Lesotho), through the undulating plains of the central Free State province, to plains and pans in the west, and lowlands with hills in the south. The predominant geology comprises Karoo Supergroup sedimentary rocks, with igneous intrusions and basalts on a sub-regional scale. The region's geomorphology has provided evidence for late Quaternary change within three major sub-regions. In addition, the drainage lines of the region have contributed both geomorphological and archaeological information.

Firstly, the eastern highlands have provided a setting for cold-climate weathering studies and periglacial processes, as well as featuring in the controversial debate on late Quaternary glaciation of the Drakensberg escarpment. Secondly, the western Free State panfield has contributed significantly to the geomorphic literature on the late Quaternary of the province. Thirdly, there is the southern and south eastern Kalahari. Finally, there are the drainage lines of the central and western Free State, and the Orange and Vaal rivers. The central Free State continues to offer some potential for weathering and slope process studies, but has not yielded much in terms of geomorphic evidence for Quaternary environmental change.

In conclusion, it is argued that what may prevent geomorphology from becoming moribund, or irrelevant with regard to Quaternary studies, are the prospects offered by advances in absolute dating techniques and remote sensing, particularly with regard to palaeo-drainage and pan-lunette suites in the western Free State. This is where the potential for future cutting edge geomorphological research into, in particular the late Quaternary evolution of this landscape appears, to lie.

2.1 INTRODUCTION

Defining the geomorphology of a landscape in terms of regions can be problematic. The occurrence and distribution of landforms are not constrained by any human demarcations or political boundaries. Nevertheless, central southern Africa, here defined as the catchment of the Orange River and its major tributaries (Figure 1),

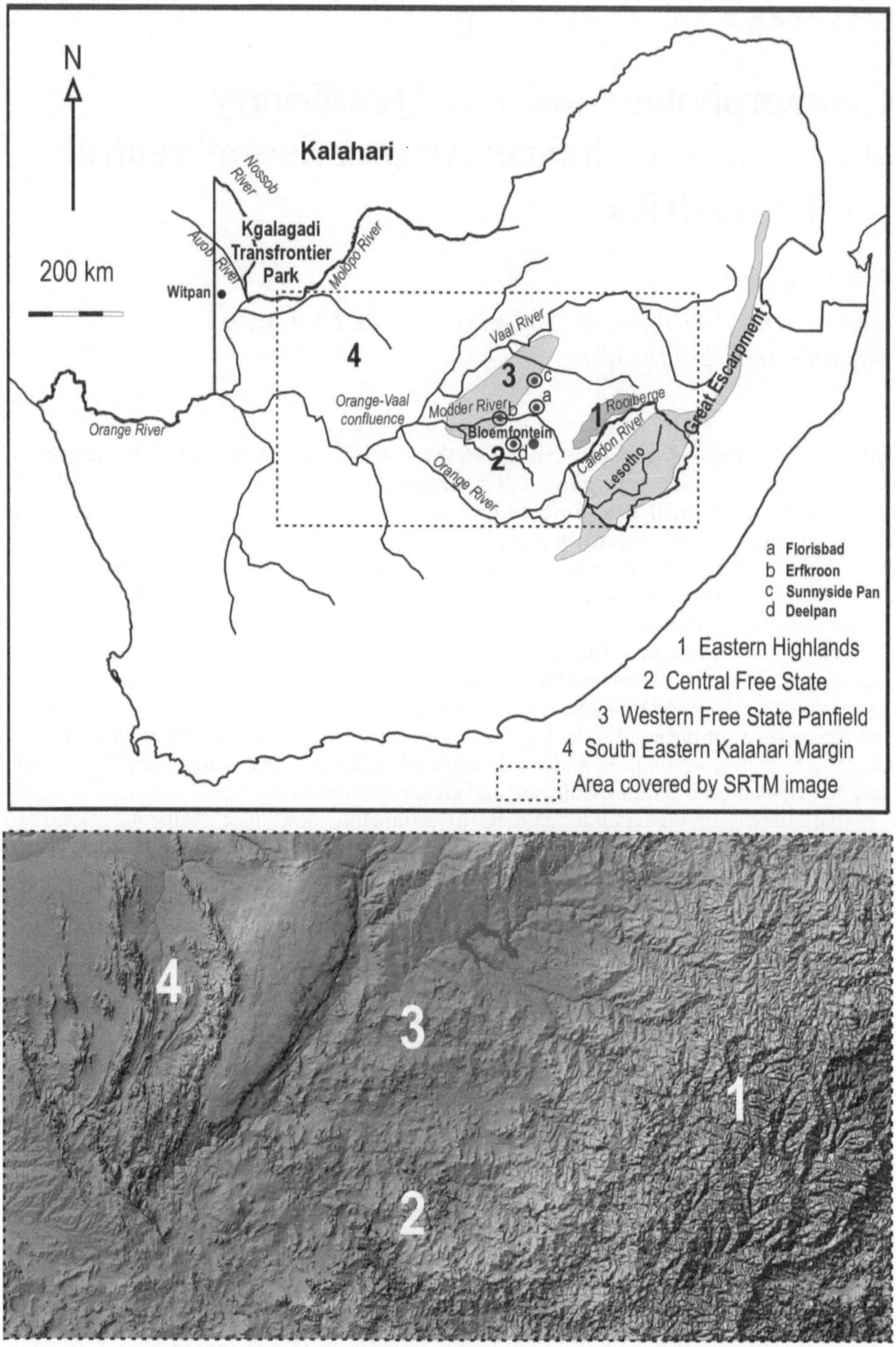

Figure 1. Central southern Africa, showing (a) the main sub-regions, sites and drainage lines discussed in this paper and (b) simple SRTM model of the area under discussion.

provides a reasonably well defined region in terms of its characteristic, but varied suites of landforms and landscapes. From a geomorphological perspective, what evidence has this region yielded which might contribute to our understanding of, in particular, late Quaternary environmental change in the region, and how might it continue to do so in the future?

2.2 BACKGROUND

2.2.1 Climate and vegetation

Central southern Africa falls entirely within the summer rainfall zone. Rainfall varies from ~800 mm in the extreme east, to ~200 mm or less on the south eastern margins of the Kalahari. The 500 mm isohyet runs north–south, through the city of Bloemfontein (Figure 1). Temperatures are extreme, with means of >30°C in large parts of the region in summer, while winter minima of −10°C are common, particularly in the highlands of the extreme east. The vegetation varies from dry Highveld Grassland to Kalahari Hardveld Bushveld vegetation types. As it is a transition zone, the central part of central southern Africa is potentially climatically sensitive to environmental change.

2.2.2 Geology and geomorphology

From both a geological and a geomorphological perspective, the current southern African landscape is largely a product of its ancient Gondwana heritage. Tankard *et al.* (1982) identified five stages in the geological development of the subcontinent. The position of central South Africa, situated as it is close to the geographic centre of the subcontinent, was profoundly influenced by these events. The landscape still reflects aspects of these developments, though the subsequent breakup of Gondwanaland has greatly modified it (Maud, 2012).

The entire region, with the exception of Kalahari Group sands in the west, is dominated by Karoo Supergroup sandstones, siltstones and mudstones (Holmes and Barker, 2006). There are numerous dolerite intrusions, resulting in local metamorphism and lithologically controlled, flat-topped mesa and butte type landforms in the central and eastern parts.

Central southern Africa falls entirely within the catchments of the westward-draining Vaal and Orange Rivers, as well as the non-perennial Molopo River, which drains the arid western part of the region (Figure 1). The greater Orange River catchment covers 953,200 km^2. With the break-up of Gondwana, and the concomitant removal of an important sediment source south of the present coastline, the nature of southern Africa's drainage systems changed, from primarily depositional, to erosional environments (Helgren, 1979). The formation of the Great Escarpment and the establishment of an erosion base level in the proto-Atlantic ocean form the point of departure for most studies dealing with the origins of drainage systems in southern Africa. Central southern Africa is no exception. There is little doubt that the Orange River has undergone major course changes (Du Toit, 1910; Helgren, 1979; Partridge and Maud, 1987; Holmes, 1987; Barker, 2002). Partridge and Maud (1987) suggest that the Orange River only established its present course during the late Pliocene-early Pleistocene.

The significance of lithological controls on the gross geomorphology of central southern Africa should not be underestimated. This is discussed in detail by Holmes and Barker (2006). Kruger's (1983) terrain morphological map of South Africa provides a useful physiographic classification of the region. In essence the macro-scale terrain morphology can be described as follows. Firstly, the eastern highlands have provided a setting for cold-climate weathering studies and periglacial processes, as well as featuring in the controversial debate on late Quaternary glaciation of the Drakensberg escarpment. Secondly, the central Free State has offered some potential for weathering and slope process studies. Thirdly, the western Free State panfield has contributed significantly to the geomorphic literature on the Quaternary of the province. Fourthly, there is the southern and south eastern Kalahari. Finally, there are the drainage lines of the central and western Free State, as well as the Orange and Vaal Rivers.

Eastern highlands

The eastern highlands comprise the landward flanks of the Drakensberg and Maluti Mountains of the Great Escarpment (Figures 1 and 2). The eastern highlands have provided a setting for cold-climate weathering studies and periglacial processes. They have also, along with the Eastern Cape Drakensberg, been the focus of considerable controversy regarding the interpretation of the role of various high altitude landforms in providing geomorphic evidence for Quaternary glaciation in central southern Africa. For the most recent summary of this debate, and reference to its literature, see Telfer *et al.*, 2014). Issues around prerequisite conditions for glaciation have more to do with moisture availability than temperature. Secondly, the issue of equifinality must be considered. Equifinality asserts that different processes can give rise to similar landforms. At high altitudes, cold climate weathering may produce rock debris and depositional features such as ridges which cannot be unequivocally linked to, for example, glacial action. The same may be said about erosional features (loosely described using the neutral term *hollows*). Hedding (2014) devised a model, based on empirical work, which appears to refute the geomorphic evidence for true glaciation. The model is based on a set of preconditions for nivation and resultant geomorphic responses on Marion Island and Antarctica. Hedding (2014) concluded that a number of geomorphic features in the Golden Gate area, hitherto ascribed to glacial action, did not fulfil the necessary criteria for classification as glacial landforms.

Unlike true glaciation, the evidence for periglacial conditions is reflected in numerous landforms ranging from the macro- to micro-scale which have resulted from cryogenic activity in the region. These are summarised and described by Grab *et al.* (2012). Both high altitude erosional as well as depositional features suffer from the disadvantage of not being amenable to absolute dating. Relating such features to specific cold phases, and correlating the evidence with that from other geomorphic environments that are readily dateable remains a challenge.

There is also evidence for aeolian deposition in the Drakensberg foothills. Telfer *et al.* (2014) described sand ramps comprised of fine sands and coarse silts which mantle lower hillslopes in the Rooiberge (Figure 1). They ascribe these to increased aeolian sediment mobility during late Quaternary cold (periglacial) phases, though detailed OSL dating was not undertaken as part of this study.

Critics might argue that the debate around evidence for true glaciation in the Eastern Highlands has diverted attention from issues around periglacial environments. It may be opportune, in light of recent research to refocus on these high altitude environments.

Figure 2. Eastern Highlands. Summits comprise Drakensberg Formation basalt, overlying Clarens Formation sandstone. Note erosional features (indicated by arrows) which are a topic of debate in this landscape.

Central Free State province

The central Free State province (Figure 1) has not yielded significant and unequivocal geomorphic evidence for late Quaternary environmental change in central southern Africa. Physiographically, the region does not lie at sufficient altitude to carry the stamp of past nivational or cryogenic activity. Neither does it have the geological underpinnings, nor wear the mantle of soil and sand which the western Free State possesses and which promotes pan and lunette formation.

Western Free State

The western Free State, here defined as the sub-region to the west of the 500 mm isohyet (see Figure 1: Bloemfontein is on the isohyet, which runs roughly north–south), has proved to be a significant source of data on both late Quaternary, as well as contemporary environmental change within central southern Africa. This area includes the renowned Florisbad archeozoological site (Figure 1), which has contributed to a better understanding of the complex interactions between aeolian and fluvial processes in controlling the geomorphic development of the Soutpan environment.

The gently undulating landscape of the western Free State is home to the largest concentration of pans (playas) in southern Africa. Holmes *et al.* (2008) published what is to date the most comprehensive account on the western Free State panfield, including 68 new Optically Stimulated Luminescence (OSL) ages from pan fringing lunettes at five sites. The western Free State panfield is adjacent to the Kalahari region (Figure 1), where OSL ages from lunettes have previously been reported. It should also be noted that, although the present-day Kalahari is situated mainly to the north-west, during periods of high aridity, the western Free State would have been more proximal in terms of the environmental conditions it experienced (Thomas and Shaw, 2002).

Lunettes typically occur on the southern and south-eastern margins of Free State pans. The sediments comprising these characteristically crescent-moon shaped features have been emplaced by wind. Lunettes, therefore, reflect phases of aeolian activity during which environmental conditions (sediment supply, local moisture budget, vegetation and, possibly, anthropogenic influences) were conducive to deposition. The morphology, sedimentology and age of lunettes at three primary and two secondary sites were reported by Holmes *et al.* (2008). The lunettes form distinct topographic features, with heights of up to 5 m above the pan floor (Figure 3), and all have been dissected by gully erosion. Sediment in the sand size class dominates in the lunettes, often overlying clay-rich basal or pan floor sediments. OSL dating of sediments from the lunettes revealed phases of lunette building from ~0.07 to 18 ka ago. The association of the Free State lunettes with phases of lunette dune building in the Kalahari desert to the west was also examined. Lunettes in the western Free State are currently not in an accretion phase, but are subject to degradation by fluvial erosion, with sediment being recycled into the pans (Figure 3).

Figure 3. Typical, dissected lunette dune flanking a western Free State pan (see Figure 1). Photo taken from the pan floor in foreground. Note the series of holes on the face of the lunette where samples have been systematically taken for OSL dating.

Following the investigations of Lancaster (1989) and the subsequent application of OSL dating to pan-margin lunettes, these features have played an important role in late Quaternary palaeoenvironmental research. Within central southern Africa, the dating of periods of lunette development, from pan systems includes the Karoo (Thomas *et al.*, 2002), Free State (Holmes *et al.*, 2008) and southern Kalahari (Telfer and Thomas, 2007). It is apparent that the palaeoenvironmental interpretation of pan and lunette sediments is best achieved when undertaken in conjunction with other records of environmental change from the same vicinity, and when the full geomorphic considerations leading to landform development can be assessed (Thomas and Shaw, 2012).

The Florisbad site (Figures 4, 5 and 6) warrants special mention, not only because of its rich fossil faunal assemblages and the discovery of a hominid cranium (*Homo helmei*) in 1932 and its subsequent description (Dreyer, 1935), but also because of its unique geomorphic setting. Various hypotheses have been put forward for the development of the unique spring mound at Florisbad (Brink, 1987; Butzer, 1988; Douglas *et al.*, 2010). There is still no consensus with regard to an explanation for this feature though, increasingly, the role of aeolian action in contributing to its formation has been recognised. Rabumbulu and Holmes (2012) were able to establish depth to bedrock of the spring mound via a series of boreholes drilled on the upwind flank of the crest of the spring mound (Figure 5). There is a series of highly degraded lunettes flanking Soutpan (Figure 5). Only the Florisbad spring mound remains intact. This suggests that the copious supply of water from the Florisbad spring must have played a role in at least preserving, if not primarily creating, the spring mound. For a detailed discussion, refer to Douglas at al. (2010).

Southern Kalahari

In the southern Kalahari, Thomas (1988) and Thomas and Wiggs (2012) describe dunes and dune ridges up to 25 m high and 250 m wide. The dunes of the southern Kalahari are not constrained by any political boundaries, extending as they do into Botswana and Namibia. Morphological variability is marked (Bullard, 1994). Dune morphology varies between near-straight, widely spaced dunes in upwind locations (Figure 7) through to sinuous dunes, which are more closely spaced, even merging. Thomas and Wiggs (2012) have proposed the term network dunes. Between the Auob and Nossob River valleys (Figure 1), several small patches of distinct parabolic dunes have been shaped out of the linear ridges (Thomas and Shaw, 1991).

Linear dune formation is not occurring in the southern and south eastern Kalahari today. Though droughts and fire regularly reduce vegetation cover on the dunes of the southern Kalahari (Bullard *et al.*, 1996; Wiggs *et al.*, 1995; Thomas and Leason, 2005), sand transport on dune surfaces is relatively limited. Furthermore, it is primarily restricted to crestal locations (Livingstone and Thomas, 1993; Wiggs *et al.*, 1995). The dunes of the south eastern Kalahari are therefore commonly regarded as relict geomorphic features dating from earlier periods that were more favourable to widespread aeolian action (Thomas and Wiggs, 2012). This of course raises the question as to when these dunes were constructed. In this regard, Lawson and Thomas (2002) published a number of OSL ages for dunes in the southern Kalahari. Similarly, Telfer and Thomas (2007) presented 71 OSL ages from the linear dunes of the southwestern Kalahari at Witpan (see Figure 1). The earliest sand accumulation recorded at Witpan is at 104 ka. Further evidence of dune activity is recorded at 77–76, 57–52 and 35–27 ka. Following the Last Glacial Maximum there is near continuous evidence of dune-building, with a peak of accumulation recorded from 15 to 9 ka at five individual sites. Telfer and Thomas (2007) suggest that these dunes may not be far from their threshold of activation.

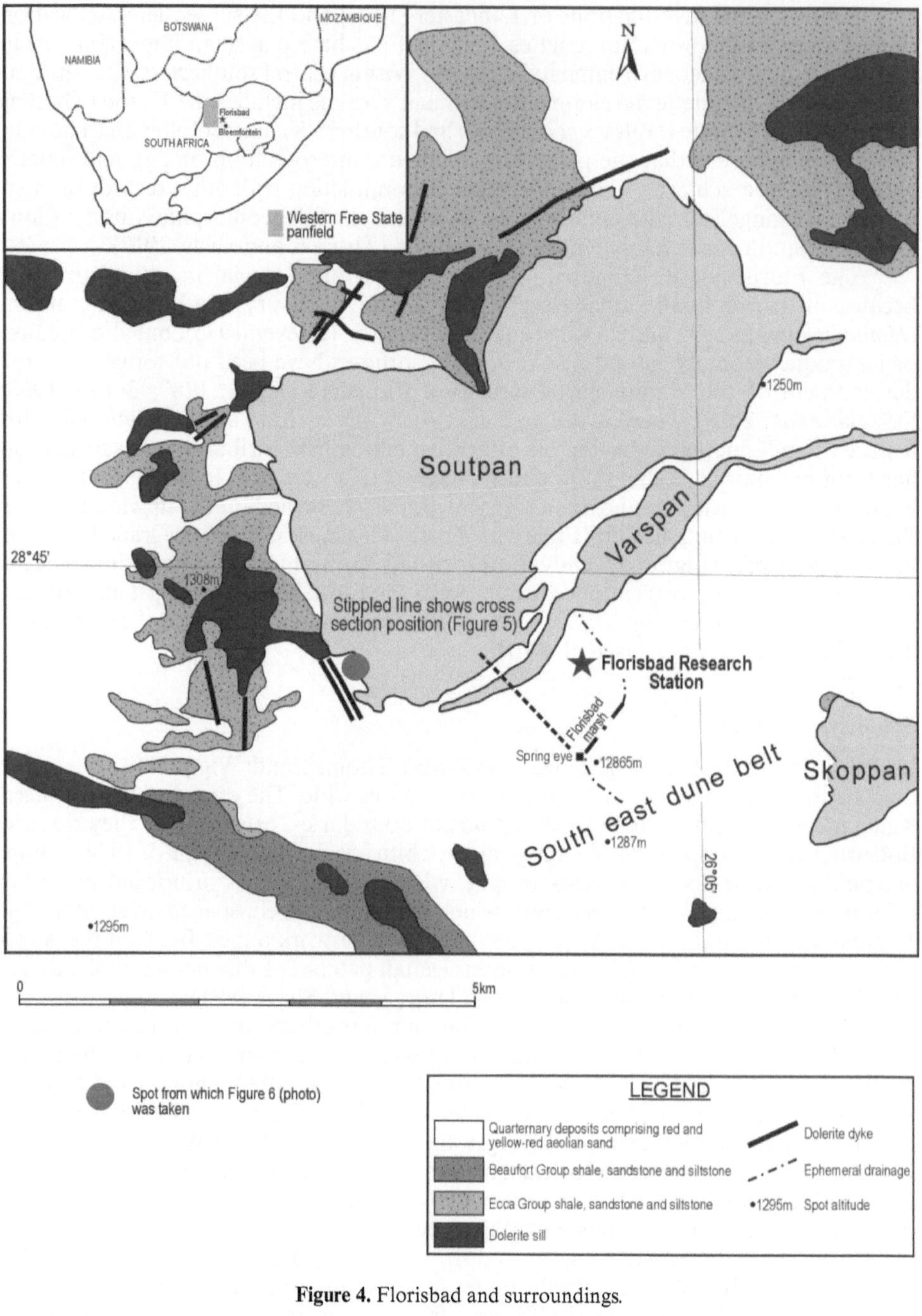

Figure 4. Florisbad and surroundings.

Orange River and tributaries (drainage lines)

The Orange River, and in particular two of its major tributaries, the Vaal and the Caledon Rivers (Figure 1), were amongst the earliest landscape features in the region to receive attention from archaeologists and geomorphologists. From just prior to the Second World War, and on into the 1980s, arguably more geomorphic research was

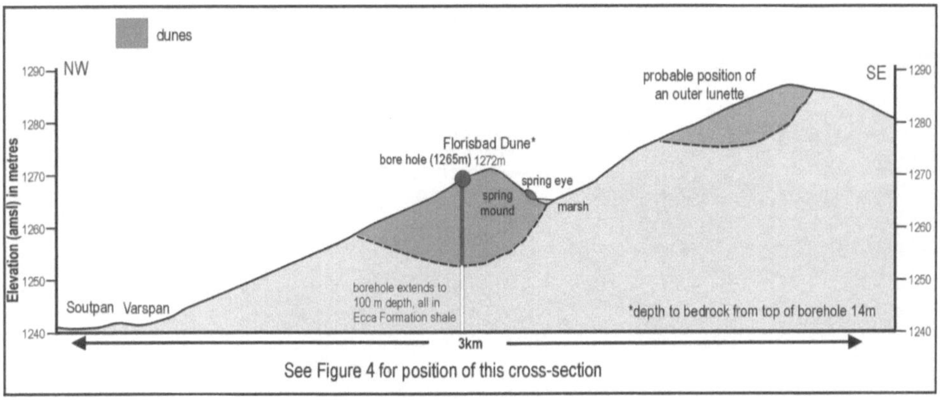

Figure 5. Cross-section from Soutpan, through the Florisbad spring mound to the east. See Figure 4 for position of cross-section.

Figure 6. Soutpan. Photo taken from the base of a flanking lunette, looking north. Florisbad is to the east (see Figures 4 and 5).

undertaken along these drainage lines than in any other region of southern Africa (Söhnge *et al.,* 1937; MacFarlane, 1945; Cooke, 1947; Breuil *et al.,* 1948; Van Riet Lowe, 1952; Visser and Van Riet Lowe, 1955; Partridge and Brink, 1967; Butzer *et al.*, 1973; Helgren, 1979; Holmes, 1987; Holmes and Reynhardt, 1989).

Most of this work reveals the intimate connection between river gravels (Figure 8) and terraces as repositories of archaeological evidence. Palaeofloods, themselves an

Figure 7. Linear dunes of the southern Kalahari (Kgalagadi Transfrontier Park). Note the degree of vegetation on these dunes.

Figure 8. Highly degraded (due to diamond mining operations) alluvial gravels on a terrace of the Vaal River near its confluence with the Orange River (see Figure 1).

indicator of environmental disturbance, if not environmental change, and efforts to apply absolute dating to inorganic river gravels may also be mentioned (Zawada, 1996; Bateman and Holmes, 1999). Latterly, due to advances in OSL dating techniques, it has become possible to date fossil bearing overbank deposits at Erfkroon on the Modder River (Figure 1) (Bousman *et al.*, 2014).

2.3 FUTURE DEVELOPMENTS

Opportunities for research into late Quaternary environmental change from central southern Africa have by no means been exhausted. Increasingly, process studies focusing on aeolian deflation and sediment transfer have a role to play in understanding how lunettes, linear dunes and sand sheets formed in the past. Wiggs and Holmes (2011) published a paper on dust mobility based on empirical work in the area north of Bloemfontein. Similar empirical work on Deelpan, on the eastern edge of the Free State panfield is in preparation.

Secondly, with the availability of cosmogenic nuclide dating, systematic and definitive dating of the sediments of the Florisbad spring mound and other aeolian features should now be possible. Past dating attempts, using radiocarbon, OSL, and even Electron Spin Resonance (ESR) has tended to be ad-hoc, or sample, rather than site specific (see for example Grün *et al.*, 1996).

Thirdly, coring of pan floor and lunette dune sediments and detailed geochemical analyses, coupled with the dating of such sediments in order better to understand the geomorphic evolution of these features warrants investigation. River terraces, gravels and overbank deposits also hold potential geomorphic evidence for Quaternary environmental change. In places, neotectonics prevent a simple correlation of river terraces and gravels based on altitudinal similarities (MacFarlane, 1945; Holmes and Reynhardt, 1989).

Finally, geomorphology in central southern Africa has a triple role to play.

2.3.1 A context for other disciplines

Geomorphology provides a context for many other investigations into Quaternary environmental change in central southern Africa. This is particularly true of archaeological and palynological research in the region. River gravels, cave deposits, alluvial overbank deposits and lunette dunes are frequently the environments for preserved archaeological and palynological evidence. The geomorphic settings of sites such as Florisbad, Erfkroon and Deelpan, to name three examples, must be properly elucidated in order to permit meaningful interpretations of the evidence found therein.

2.3.2 Stand-alone geomorphic work

Stand-alone geomorphic evidence, particularly with regard to fluvial and depositional environments, offers the opportunity for interpretation with regard to Quaternary environmental change based on the sedimentology and geochemistry of such sites. In addition, dating of inorganic sediments relies on techniques such as OSL dating which is, typically within the domain of geomorphology and geomorphic expertise.

2.3.3 Process studies

Process studies aimed at an improved understanding of both past and present fluvial and aeolian processes relies heavily on geomorphic expertise (Figure 9). Environments under investigation include river terraces, gravels and overbank deposits, as well as pan and lunette sequences, and linear dunes. Weathering studies have also proved invaluable in determining how processes operate in cold climate environments, as well as contributing either to rock art deterioration or preservation (Sumner *et al.*, 2012).

Figure 9. Process studies: gathering wind data from the floor of Deelpan to the west of Bloemfontein (Figure 1). Note the dissected lunette dunes in the background.

It is worth noting that interdisciplinary and multidisciplinary research into Quaternary palaeoenvironments of central southern Africa has increased significantly in the last decade. It stands to reason that the fields of archaeology, geology, palaeogeohydrology, geomorphology, palaeobotany, palaeozoology and palynology can benefit from mutual interaction and cooperation. An examination of recent literature into Quaternary palaeoenvironmental research in central southern Africa will inevitably reveal that the highly cited work is multi-authored. In conclusion, within the context of palaeogeomorphology, remote sensing studies, and in particular technologies such as ground penetration radar, have an important role to play.

2.4 CONCLUSION

Because of its physical geography, central southern Africa affords a unique opportunity to investigate Quaternary environmental change from a number of sub-regions within a relatively constrained spatial milieu. The east–west distance from the western flanks of the Great Escarpment to the eastern most linear dunes of the Kalahari (Figure 1) is less than 400 km. From the summit of the Great Escarpment to the valley of the Orange River there is a not insignificant altitudinal variation of ~2200 m. The rainfall gradient is ~600 mm. Conversely, there are limited geological substrates (primarily shales and sandstones in the east, and Kalahari sands in the west), which means that the influence of geology as a variable in process studies is more easily controlled.

Geomorphic studies into Quaternary environmental change in central southern Africa have focused on cold climate environments in the extreme east, through the pan and flanking lunettes of the western Free State panfield, and on into the linear dunes of the Kalahari to the west. In addition, the drainage lines of the Orange, Vaal and Caledon Rivers have yielded geomorphic evidence for Quaternary environmental change from their complex but well understood terrace and gravel sequences. The role of geomorphology in providing a context for the region's rich archaeological heritage should not be underestimated. Similarly, of late, advances in the dating of inorganic sediments, particularly using OSL have facilitated the interpretation of depositional sequences from everything from alluvial overbank deposits, lunette dunes, through to linear dunes. Cosmogenic nuclide dating, with its ability to date sequences extending further back into the Quaternary should, in the future, also play a role. Finally, there is a need for more detailed process studies, linking in particular current aeolian processes to both geomorphic palaeoenvironments as well as present day environmental change. Such studies might usefully include research into palaeodrainage lines and pan connectivity, which are still in an early stage.

Geomorphology, both as a discipline on its own, as well as providing support for those working in other fields of Quaternary environmental change, clearly has a future. By contributing to a better understanding of how central southern Africa's Earth surface systems function in a period of accelerated environmental change, geomorphology also has a full and meaningful role to play.

ACKNOWLEDGEMENTS

This paper has drawn in no small measure from the contributions of a number of colleagues who contributed to *Southern African Geomorphology* (Holmes and Meadows, 2012). The paper is dedicated to my student and friend, Rod Douglas, whose untimely passing in early 2014 has left central southern African geomorphic research much the poorer. Thanks to anonymous referees for suggestions and ideas which have improved this paper.

REFERENCES

Barker, C.H., 2002, 'n Morfometriese ondersoek na landskapontwikkeling in die Sentraal-Vrystaat. PhD Thesis, University of the Free State.

Bateman, M.D. and Holmes, P., 1999, Orange River alluvial terraces: is luminescence dating a useful tool? *South African Journal of Science,* **95**, pp. 57–58.

Bousman, B., Brink, J., Bateman, M.D., Meier, H., Trower, G., Grün, R., Codron, D., Rossouw, L., Bronk Ramsey, C. and Scott, L., 2014, *Middle and Late Pleistocene Terraces and Archaeology in the Modder River Valley, South Africa.* 14th Congress of the Pan-African Archaeological Association of Prehistory, Johannesburg.

Breuil, H., Van Riet Lowe, C. and Du Toit, A.L., 1948, *Early man in the Vaal River Basin. Archaeological Survey,* Archaeological Series, no **6.** (Pretoria: Department of the Interior).

Brink, J.S., 1987, The archaeozoology of Florisbad, Orange Free State. *Memoirs van die Nasionale Museum, Bloemfontein,* **24**, pp. 1–151.

Bullard, J.E., Thomas, D.S.G., Livingstone, I, and Wiggs, G.F.S., 1994, Analysis of linear sand dune morphological variability, southwestern Kalahari Desert. *Geomorphology,* **11**, pp. 189–203.

Bullard, J.E., Thomas, D.S.G., Livingstone, I. and Wiggs, G.F.S., 1996, Wind energy variations in the Southwestern Kalahari desert and implications for linear dunefild activity. *Earth Surface Processes and Landforms,* **21**, pp. 263–278.

Butzer, K.W., Helgren, D.M., Fock G.J. and Stukenrath, R., 1973, Alluvial terraces of the Lower Vaal River, South Africa: a reappraisal and reinvestigation. *Journal of Geology,* **81**, pp. 341–362.

Butzer, K.W., 1988, Sediment interpretation of the Florisbad spring deposits. *Palaeoecology of Africa,* **19**, pp. 181–189.

Cooke, H.B.S., 1947, The development of the Vaal River and its deposits. *Transactions of the Geological Society of South Africa,* **46**, pp. 243–260.

Douglas, R.M., Holmes, P.J. and Tredoux, M., 2010, New perspectives on the fossilization of faunal remains and the formation of the Florisbad archaeozoological site, South Africa. *Quaternary Science Reviews,* **29**, pp. 3275–3285.

Dreyer, T.F., 1935, A human skull from Florisbad, Orange Free State, with a note on the endocranial cast, by C.U. *Ariëns Kappers, Koninklijke Akademie van Wetenschappen te Amsterdam,* **38**, 3e12.

Du Toit, A.L., 1910, The evolution of the river system of Griqualand West. *Transactions of the Royal Society of South Africa,* **1**, pp. 347–362.

Grab, S.W., Mills, S.C. and Carr, S.J., 2012, Periglacial and Glacial Geomorphology. In *Southern African Geomorphology; Recent Trends and New Directions,* edited by Holmes, P.J. and Meadows, M.E. (Bloemfontein: Sun), pp. 229–262.

Grün, R., Brink, J.S., Spooner, N.A., Taylor, L., Stringer, C.B., Franciscus, R.B. and Murray, A., 1996, Direct dating of the Florisbad hominid. *Nature,* **382**, pp. 500–501.

Hedding, D.W., 2014, On the identification, genesis and palaeoenvironmental significance of pronival ramparts. PhD thesis. (Pretoria: University of Pretoria).

Helgren, D.M., 1979, *River of Diamonds.* (Chicago: University of Chicago Press).

Holmes, P.J., 1987, A palaeogeomorphological investigation into the Late Cenozoic terraces and deposits in the vicinity of Aliwal North. MSc dissertation. (Pretoria: University of South Africa).

Holmes, P.J. and Reynhardt, J.H., 1989, Late Cenozoic alluvial gravels in the vicinity of Aliwal North. *South African Journal of Science,* **85**, pp. 65–68.

Holmes, P.J. and Barker, C.H., 2006, Geological and geomorphological controls on the physical landscape of the Free State. *South African Geographical Journal,* **88**, pp. 3–10.

Holmes, P.J., Bateman, M.D., Thomas, D.S.G., Telfer, M.W., Barker, C.H. and Lawson, M.P., 2008, A Holocene late Pleistocene aeolian record from lunette dunes in the western Free State panfield, South Africa. *The Holocene,* **18**, pp. 1193–1205.

Kruger, G.P., 1983, *Terrain morphological map of southern Africa.* (Pretoria: Soil and Irrigation Research Institute Department of Agriculture).

Lancaster, N., 1989. *The Namib Sand Sea.* (Rotterdam: Balkema).

Lawson, M.P. and Thomas, D.S.G., 2002, Late Quaternary lunette dune sedimentation in the southwestern Kalahari desert, South Africa: luminescence based chronologies of aeolian activity. *Quaternary Science Reviews,* **21**, pp. 825–836.

Livingstone, I. and Thomas, D.S.G., 1993, Modes of linear dune activity and their palaeoenvironmental significance: an evaluation with reference to southern African examples. In *The dynamic and context of aeolian sedimentary systems*, edited by Pye, K. (London: Geological Society Special Publication, **720)**, pp. 91–101.

MacFarlane, D.R., 1945, The Orange River high level gravels at Aliwal North in relation to crustal movement and the Stone Age. *South African Journal of Science,* **41**, pp. 415–428.

Maud, R.R., 2012, Macroscale geomorphic evolution. In *Southern African Geomorphology; Recent Trends and New Directions*, edited by Holmes, P.J. and Meadows, M.E. (Bloemfontein: Sun), pp. 5–22.

Partridge, T.C. and Brink, A.B.A., 1967, Gravels and terraces of the Lower Vaal River Basin. *South African Geographical Journal,* **49**, pp. 21–38.

Partridge, T.C. and Maud, R.R., 1987, Geomorphic evolution of southern Africa since the Mesozoic. *South African Journal of Geology,* **90**, pp. 179–208.

Rabumbulu, M. and Holmes, P.J., 2012, Depositional environments of the Florisbad Spring site and surrounds: a revised synthesis. *South African Geographical Journal,* **94**, pp. 191–207.

Söhnge, P.G., Visser, D.J.L. and Van Riet Louw, C., 1937, *The Geology and Archaeology of the Vaal River Basin.* Memoir, **35**, Parts 1 and 2. (Pretoria: Geological Survey).

Sumner, P.D., Hall, K.J., Meiklejohn, K.I. and Nel, W., 2012, Weathering. In *Southern African Geomorphology; Recent Trends and New Directions,* edited by Holmes, P.J. and Meadows, M.E. (Bloemfontein: Sun), pp. 71–91.

Tankard, A.J., Jackson, M.P.A., Eriksson, K.A., Hobday, D.K., Hunter, D.R. and Minter, W.E.L., 1982, *Crustal evolution of southern Africa: 3.8 billion years of earth history.* (New York: Springer-Verlag).

Telfer, M.W. and Thomas, D.S.G., 2007, Spatial and temporal complexity of lunette dune development, Witpan, South Africa: Implications for palaeoclimate and models of pan development in arid regions. *Geology*, **34**, pp. 853–856.

Telfer, M.W. and Thomas, D.S.G., 2007, Late Quaternary linear dune accumulation and chronostratigraphy of the southwestern Kalahari: implications for aeolian palaeoclimatic reconstructions and predictions of future dynamics. *Quaternary Science Reviews,* **26**, pp. 2617–2630.

Telfer, M.W., Mills, S.C. and Mather, A.E., 2014. Extensive Quaternary aeolian deposits in the Drakensberg foothills, Rooiberge, South Africa. *Geomorphology,* **219**, pp. 161–175.

Thomas, D.S.G., 1988, Analysis of linear dune sediment-form relationships in the southern Kalahari Dune Desert. *Earth Surface Processes and Landforms,* **13**, pp. 545–553.

Thomas, D.S.G., Holmes, P.J., Bateman, M.D. and Marker, M.E., 2002, Geomorphic evidence for late Quaternary environmental change from the eastern Great Karoo margin, South Africa. *Quaternary International,* **89**, pp. 151–164.

Thomas, D.S.G. and Leason, H.C., 2005, Dunefield activity response to climate variability in the southwest Kalahari. *Geomorphology,* **64**, pp. 117–132.

Thomas, D.S.G. and Shaw, P.A., 1991, *The Kalahari Environment.* (Cambridge: Cambridge University Press).
Thomas, D.S.G. and Shaw, P.A., 2002, Late Quaternary environmental change in central southern Africa: new data, synthesis, issues and prospects. *Quaternary Science Reviews,* **21**, pp. 783–797.
Thomas, D.S.G. and Shaw, P.A., 2012, Terminal Basins: Lacustrine and Pan Systems. In *Southern African Geomorphology; Recent Trends and New Directions,* edited by Holmes, P.J. and Meadows, M.E. (Bloemfontein: Sun), pp. 166–187
Thomas, D.S.G. and Wiggs, G.F.S., 2012, Aeolian systems. In *Southern African Geomorphology; Recent Trends and New Directions,* edited by Holmes, P.J. and Meadows, M.E. (Bloemfontein: Sun), pp. 139–164.
Van Riet Lowe, C., 1952. The Vaal River chronology, an up to date summary. *South African Archaeological Bulletin,* **7(20)**, pp. 135–149.
Visser, D.J.L. and Van Riet Lowe C., 1955, *The Geology and Archaeology of the Little Caledon River Valley.* Department of Mines, Pretoria.
Wiggs, G.F.S., Livingstone, I., Thomas, D.S.G., and Bullard, J.E., 1995, Airflow characteristics over partially vegetated linear dunes in the southwest Kalahari. *Earth Surface Processes and Landforms,* **21**, pp. 19–24.
Wiggs, G.F.S. and Holmes, P.J., 2011, Dynamic controls on wind erosion and dust generation on west-central Free State agricultural land, South Africa. *Earth Surface Processes and Landforms,* **36**, pp. 827–838.
Zawada, P.K., 1996, *Palaeoflood hydrology of selected South African rivers.* PhD thesis, University of Port Elizabeth, Port Elizabeth.

CHAPTER 3

Dry Lakes or *Pans* of the western Free State, South Africa: Environmental history of Deelpan and possible early human Impacts

Karl W. Butzer

Department of Geography and the Environment, University of Texas, Austin, Texas, USA

John F. Oswald

Department of Geography and Geology, Eastern Michigan University, Ypsilanti, Michigan, USA

ABSTRACT: The dry lakes of the grassy high plains known as the Highveld have long attracted the interest of scientists, much like the non-outlet playas of the West Texas Caprock. This investigation reconnects to a joint-study with Louis Scott on various pans in the Free State and Northern Cape. It expands the original reports with new data, laboratory analyses, fresh attention to micro-stratigraphy and palaeosols, reinforced by diagnostic laboratory analyses. The changing sedimentology and pedogenic imprints were recorded for 5 late Holocene profiles along the pan's windward shores, as well as a core into a 4000 yr old spring seep. Prior to a long hiatus, end-Pleistocene and early Holocene sediments are represented by hardpans of both calcrete and silcrete, related to different geochemical regimes that require more attention. Central to late Holocene environmental history were episodic flood surges across the Deelpan floor with construction of backshore features that included lunettes. Mixed facies indicate two major transgressive phases based on sediment parameters and redox zones that unexpectedly point to extended times with wet soil and lacustrine microenvironments. While some recent pan floods were sporadic, others were bundled in more complex, moister climatic anomalies. Current focus on high-resolution corings should be complemented by geomorphologic and pedogenic study of three-dimensional processes that better illuminate sequential change. Ground-cover deterioration after the youngest transgression led to renewed eolian dispersal of Kalahari-type sands, probably before any European settlement in the western Free State. Given the archival and archaeological record for long-term indigenous pastoralism along the middle Riet River, further research is warranted to examine the dominant view that Boer stock-raisers and British hunters had decimated the ecological resources of the Highveld during the 19th century.

3.1 INTRODUCTION

Dry lakes dominate a number of semiarid world landscapes. There is no generally accepted technical designation but many regional names with different linguistic roots: playa lakes (Mexican, Spanish), alkali or salt flats (American English), sebkha (French, Arabic), pan (Afrikaans), and so forth. Pans in enclosed basins tend to be elliptical and hold water seasonally or after sporadic heavy rains. Typically floored by clays, they are infused to varying degree by salts of several kinds. While the episodic waters are fresh they attract a myriad of insects, birds, ungulates and people in ecological succession, and the waters recharge shallow aquifers that may promote clusters of hygrophytic vegetation such as sedges, reeds, true grasses, cattails and salt brush. As the waters evaporate and turn brackish, or are infected with botulism (Dr. Richard Liversidge, pers. comm.), antelopes or cattle stop visiting, leaving only salt brush and hardy animal forms burrowing in the mud. Sodium salts settle out or effloresce on the faces of contacting sediments.

In southern Africa, pans characterize several regions, one of the most striking of which is the western Free State and adjoining Northern Cape with shales (Dwyka and Ecca) of the Karoo Supergroup. Like the word *koppie* (single or multiple hills), *pan* gained wider attention during the Anglo-Boer War, creating sustained interest among geomorphologists (DuToit, 1907; Geyser, 1950; Van Eeden, 1955; Wellington, 1971; DeBruiyn, 1972; Butzer *et al.*, 1973; Butzer, 1974, 1984; LeRoux, 1978; Holmes *et al.*, 2008; Lyons *et al.*, 2014). But going beyond deductive interpretations proved difficult, in view of few and poor exposures, the enigmatic calcrete hardpans, and few options for AMS dating in an environment that preserves limited organic matter.

Working with a shoestring budget on the geoarchaeology of pans in the Kimberley area during the 1970s and 80s (Butzer, 1974, 1984), I began a long collaboration with Louis Scott, who added a vital palynological component to this research (Scott, 1988; Scott and Brink, 1992; Scott and Nyakale, 2002). After scanning the Free State pans from a rented Piper aircraft in 1977, I opted to search for one with more evidence of water. We visited Deelpan briefly, and Scott decided to begin an exploratory stratigraphy, subsequently collecting a suite of sediment samples and locating an important fossil site (Scott and Klein, 1981). This site gained wide attention (Scott *et al.*, 2003) and was excavated by Brink (2005). Completion of the emerging manuscript became possible when John Oswald agreed to undertake the computer cartography of the analytical graphics indispensable for three-dimensional documentation.

3.2 RESEARCH METHODS: DEELPAN AS A HEURISTIC PUZZLE

Deelpan proved to be a good choice. It is a large pan (~490 ha, 25°45′ S, 29°11′ E), 48 km W of Bloemfontein (Figure 1). With a flat floor at ~1278 m, it is encircled by outcrops of dolerite (USA: diabase) intrusives that rise 20–40 m above the pan floor in the NW and E, and about half that to the W. In effect, Deelpan forms an elliptical basin underlain by Ecca Shales but constrained by igneous rocks.

Long stretches of late Quaternary nearshore and backshore deposits with a face of 2 to 7 m in height are exposed along the eastern and southern margins, reflecting winds with a dominant vector of N 20° W. These beds range from lacustrine to eolian and colluvial, with a great deal of mixture as a result of repeated erosion, transport, and accumulation. The advantage of such wide exposures is that many beds or pedogenic features can be followed semi-continuously from site to site. Pedogenic markers include humic palaeosols; structural soils (mainly prismatic, columnar or angular blocky); redox phenomena (oxidation/reduction horizons of reddish or light greenish gray); saline or alkaline enrichment, and calcic or petrocalcic (hardpan) calcretes. There can also be substantial components of reworked "Kalahari-type" sands (see Van Rooyen, 1972: chap. 5; Helgren, 1979: 48–59). Indeed the Google Earth imagery confirms that the whole landscape is veneered by such sands; the area partly shown in Figure 1 is particularly deeply mantled.

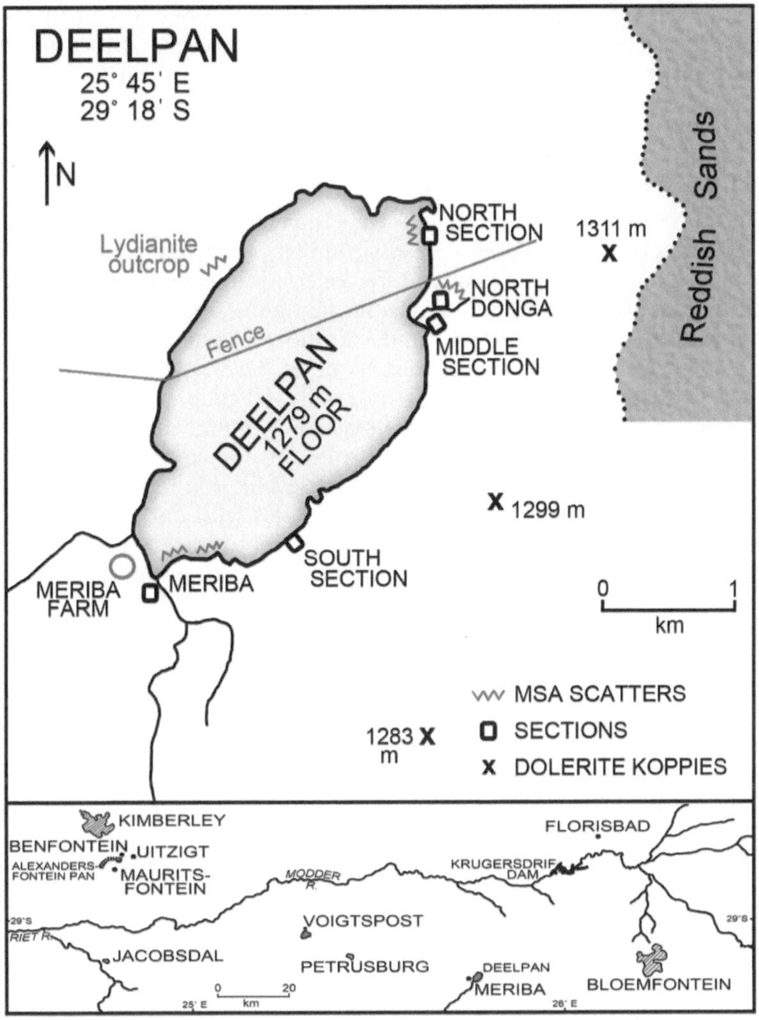

Figure 1. Location of Deelpan with key landmarks. The shoreline represents the +4 m high stand of 1974. The Google Earth image of May 2009 shows a dry pan, without some of the earlier lake embayments, but with disturbance of the floor by putative salt-making operations.

The processes involved in creating this inductive record were interactive, requiring more than drawing stratigraphic boundaries and offering interpretations. They demanded detailed field recording of bedding, lithology, and soil properties, but in conjunction with a range of laboratory analyses. Fundamental here are hydrometer testing combined with wet sieving (7 standard sieves: 37 μ to 6.35 mm, or +4.75 to –2.90 f), followed by computation of Folk sediment parameters (Folk and Ward, 1957) for the sieve fractions under 2.00 μ (–1.00 f). Although laser granulometry is popular at the moment, providing mass data but with a possibility of systematic error (such as faulty calibration or underestimation of sands), I prefer the time consuming hydrometer-and-wet-sieving approach because it allows the observer to follow sample behaviour throughout the processing routine.

Calcium Carbonate Equivalents (CCE) represent a proxy of $CaCO_3$ that can be approached by a combination of Chittick (apparatus complete) readings and bulk loss in HCl, prior to the hydrometer procedure. What are variously called colloidal, amorphous or opaline silica is prominent but commonly neglected in African research of palaeo environments despite the importance of distinguishing *calcretes* from *silcretes.* Biogenic and soluble in sodium hydroxide but not in acid, this fraction was subsequently estimated from the fraction dissolved in NaOH and collected on filter paper. When necessary, the sieve fractions were then reweighed prior to further manipulation. In southern Africa silcretes are environment, time, and geochemistry-related, being mobilized in pHs above 9, under different conditions than petrocalcic horizons.

Folk indices are not routinely shown in the figures, except as necessary for specific documentation. "Kalahari-type sands" do not of course imply derivation from Botswana, but have been recycled numerous times. Overlying shale lithologies, they serve to draw attention to ongoing eolian or colluvial processes. Their presence was estimated microscopically, by attached red plasma or orange discoloration, which can be reversed by immersion in hyperalkaline solutions. Since the quartz sands of the western Free State and Northern Cape are multicyclic, rounding or frosting proved of no import. Organic carbon from selected samples with dark Munsell colors was determined by the Walkley-Black (acid digestion) procedure.

Isotopic dating in mainly sandy deposits was deficient during the time of original fieldwork, and OSL had not yet become readily available. The chronological frame is rudimentary, but our priority was an integrated micro-stratigraphy grounded in multiple sections, laboratory analysis of a comprehensive suite of 80 samples, and a serious treatment of pedogenic features and palaeosols. The study was further contextualized by our multiple sites between Kimberley and Bloemfontein, supported by a total of 350 textural studies. Given the sometimes excessive, current focus on dating methods, we continue to opt for pragmatic interpretation based equally on inductive and deductive inputs.

In the absence of long-term weather or hydrological records, the most recent datum for Deelpan was the flood surge of March 1974. According to the late Mrs. G.R. DeVilliers, of the farm Meriba, it had rained 12 inches (30.5 cm) over 3 weeks, and the water in Deelpan rapidly rose 14 feet (4.2 m). This height also happens to match the highest level of exposed calcretes above the pan floor. Mrs. DeVilliers, whose husband was prominent in Bloemfontein and formerly commuted regularly by light aircraft from the estate's own airfield, was cognizant that this extreme event was regional. We were able to confirm its parallels in Kimberley and Douglas on the middle Vaal, where the 1974 flood event was the strongest perturbation since the late 1890s (A. Lahoud, pers. comm. 1975; see also Nash and Endfield, 2002). Scott swam across Deelpan along the submerged fence in February 1978 and found the water still more than 1.5 m deep near its center. But in 1980 it was essentially dry.

Measured indicators for intermediate water depths at Deelpan include: *no* older organic soil or calcrete below +2.5 m; *no* reduced soils in profiles below +2.0 m; *no* lichens on W shore rocks below +2.7 m, but lime coatings up to +1.2 m. Mrs. DeVilliers also spoke of 18 inch (45 cm) fluctuations during "normal" years, identifying conditions in 1983 as a very extreme drought. This suggests that a dry pan bed is not necessarily a perennial condition.

Visible life forms in 1982 included the gastropod *Succinea* cf. *striata,* which is also common in some Late Quaternary beds; medium-sized barbels (*Barbus* sp.), both skeletal and decomposing; as well as burrowing frogs. Mrs. DeVilliers thought these might have been flushed out of the Meriba farm tank, where they are present all year. We are grateful for both her accurate observations on the local ecology and her recounting of the farm's history. Good informants are vital in accessing a complex geoarchive.

3.3 RESULTS

3.3.1 The North Section: Wind and water

The North Section (Figure 2) profile exceeds 6 m and includes intermediate-level crossbeds with stringers of bedrock-derived grit or fine to medium-grade gravel. There also are several hyena burrows (crotovinas) with ungulate bone and remains of clawless otter (*Aonyx capensis*), an aquatic predator (Scott and Klein, 1981) represented by 2 specimens (Brink, 2005). The beach zone is strewn with Middle Stone Age (MSA) artifacts that at adjacent locations have eroded out of relict calcrete hardpans near the base of the profile. There were no sediments or inclusions amenable to reliable AMS dating, but the sediments and soils are highly informative of interactive change. The increase of the sands from 60 to 90%, and a corresponding decline of CCE from 15 to 3% from base to top, are dramatic. The zone of inflection included large-scale crossbeds, conspicuous oxidation mottling, and a unit rich in pelletal organic loam.

The low-grade calcrete of unit 7 is a prismatic soil, with prominent terminal rhizoliths (carbonate root drip). It may have undergone partial solution prior to wave-truncation, with fresh carbonates subsequently descending from the backshore environment.

Units 6 and 5 are mainly laminated, with zones of prismatic structure, and range from 80 to 280 cm in thickness. The gravel of unit 4 came from the backshore, eroded by small *dongas* (an African term for gullies). In unit 3 the amount and caliber of grit and gravel tapers off, but worn fragments of calcrete appear, hinting at deeper donga incision. This would argue for repetitive high lake levels, and in conjunction with two prismatic soils, standing waters responsible for redox discoloration of units 7 (upper) through 2b, and modest activation of the small dongas. These features suggest a surprisingly protracted time of moister climate.

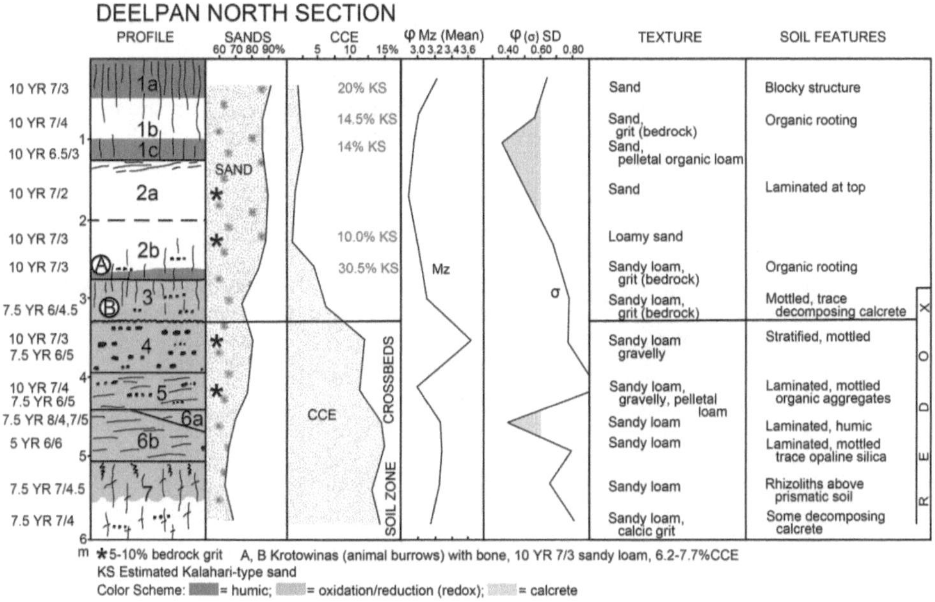

Figure 2. Analytical data for the Deelpan North Section.

The columns recording the Folk indices of mean size (Mz) and sorting (sigma) for the 37 μ to 2 mm fraction provide greater resolution. Units 2b to 1b are coarse sands with good sorting that argue for a primarily eolian origin and deflated from a mainly regressive lake. Unit 3 is finer and only moderately sorted, suggesting a mixed, aeolian and lacustrine facies, while 4 and 5 are moderately coarse with fluvial interbedding, as parts of a different mixture. The crossbed 6a expands laterally to 280 cm in height and is very well sorted, unlike 6b, which is again poorly sorted. In view of the indices and the field evidence, units 6 to 3 can be interpreted as transgressive shore accumulations with lacustrine, fluvial and aeolian components. This bundle, also marked by heavy redox discoloration and frequent shifts of kurtosis (not shown in Figure 2), is assigned to a strong initial transgression followed by oscillating and often high lake levels.

The 2.7 m of sands above unit 2b imply a mainly regressive lake. But a fair amount of variability does suggest further, but modest flood events in the pan. Notable is the appearance of a component of Kalahari-type sands suggesting an incomplete ground cover with a shrubby vegetation and a discontinuous grass sward.

3.3.2 The Hyena burrows

Two burrows (or crotovinas) used by brown hyenas (*Hyaena brunnea*) are found in Figure 2 near the base of unit 2b ("A") and of unit 3 ("B"), below large fissures in the sediments. With spatial dimensions of 30 by 40 cm the bones appear to be refuse that fell down such fissures. The matrices are pale brown (10 YR 6-7/3), sandy loams with weak prismatic structure, and CCE's of 6.2 and 7.7%, slightly greater than that of the adjacent sediments. The bones were gradually breaking free, to tumble down the slope. The list is dominated by a coterie of grassland grazers, including springbok, wildebeest, blesbok, and zebra (Scott and Brink, 1992; Brink, 2005). This rich assemblage and the overwhelming dominance of grass pollen in the 55 hyena coprolites (Scott *et al.*, 2003) argues against a deteriorating ground cover at this particular time. Excavation revealed a complex net of larger, interlocking fissures (Brink, 2005; Louis Scott, pers. comm. 2014)—as at Swartklip, near Cape Town (see Butzer, 2003).

Of particular interest are the specimens of Cape clawless otter, an aquatic predator. This would argue for almost continuous periods of fresh water in the pan, which is confirmed by several centimetres of gray loam that lines the floor of burrow "D," some 30 m south of the North Section. Two bone collagen dates of 120 ± 45 BP (Pta-6348) and 150 ± 20 BP (Pta-6346) are difficult to calibrate (~1812 AD ± 107) because of multiple intercepts and the poor reliability of collagen dating. Further AMS assays on humic soil are currently underway.

Locality "D" is important not only as a faunal site. It may also provide ages for the last major transgression of Deelpan and for the pollen documentation of an intact grass cover, shortly before eolian dispersal of Kalahari-type sands. Uncertain is the potential status of ground cover during the senseless slaughter of game animals by 19th century hunters (Pringle, 1982).

3.3.3 The Middle Section

The North Section was complemented by study of a nearby Middle Section (Figure 1), not illustrated here. The basal unit becomes a lenticular calcrete, with finer to medium sand (64%, with 31% CCE), which includes opaline silica. A mottled, prismatic pedogenic unit follows, but with a convoluted base. Next is a laminated, organic sandy loam that thickens laterally and suggests a lake transgression. Kalahari-type sands appear suddenly in this unit and remain present in overlying finer sands with low-angle, eolian crossbeds (88% sand and 27% of Kalahari type). Although abbreviated, the Middle Section replicates the triad of (a) basal calcrete, (b) mottled transgressive beds with a strong soil (units 4 to 2a), and (c) increasingly coarse sand, with prominent eolian components. The prismatic soil embeds Later Stone Age (LSA) lithics of Smithfield type.

3.3.4 The North Donga: Three Calcrete stages

Light on the genesis of massive calcretes is shed by the North Donga exposures (Figures 1 and 3). The basal unit 4 consists of mottled, decalcified calcrete with a sandy loam residual. With strong, terminal rhizoliths and 12% CCE, it represents an extension of unit 7 of the North Section. Unit 3 is a low-grade calcrete with reduction mottling (5 Y hue); the residual is a sandy loam with dispersed fine gravel. Two thin lenses of partly decalcified loams with calcic concretions follow (units 2a and 2b), and then a hard, prismatic calcrete with a laminar crust and a sandy loam residual (unit 1). MSA artifacts, almost exclusively made of lydianite, are dispersed within this petrocalcic horizon.

The North Donga sequence illustrates several stages of calcrete accretion/dissolution in the region (Figure 3).

1) Unit 1 was formed by post-depositional cementation of a structural soil that preserves humic residues in the form of pelletal mud and a reddish brown, post-acid slurry. 2) Unit 3 is an intermediate-grade calcic mass that has consolidated a body of older sediment, with existing soil structure, through layered horizontal enrichment.

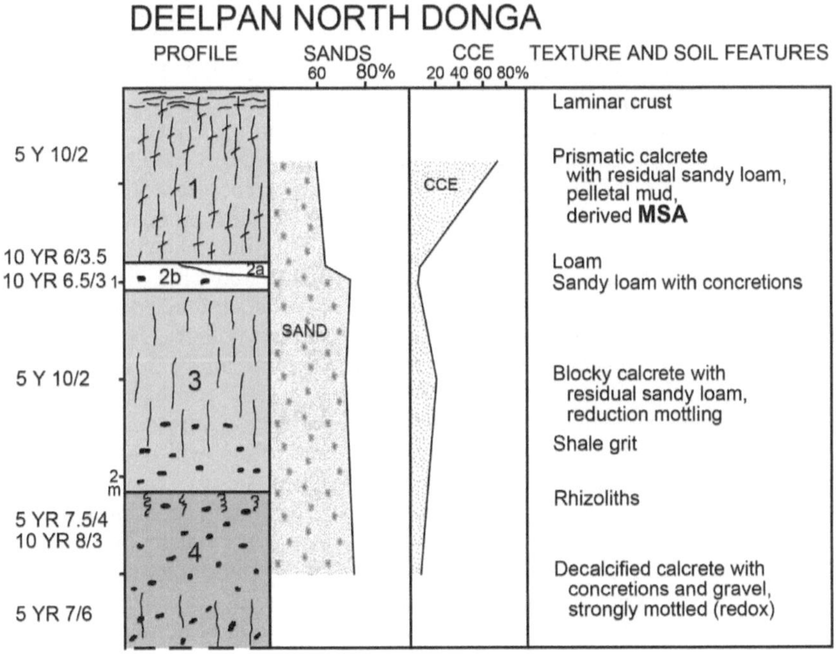

Figure 3. Analytical data for Deelpan North Donga.

3) Unit 4, with prominent oxidation and reduction mottling that suggests partial decalcification by subsurface through-flow. In effect the North Donga records a sequence of subsoil pedogenic evolution.

Periods of carbonate enrichment may have been facilitated by seasonally cooler waters that enhance solubility (highest at +4°C), so during the end-Pleistocene or Early Holocene. Other possibilities include strong, periodic contrasts of subsoil vapor pressure, or organic acids that alternatingly mobilized and precipitated bicarbonates. It is unlikely that such carbonate processes took place in the deep subsoil at the same time as chelation or hydrolysis in the subaerial micro-environment. It is likely that mildly acidic waters attacked the more porous calcretes at depth, during the waning of such a postulated calcrete cycle. The dolerite bedrock exposed upstream probably did not contribute acidic donga discharge or spring seeps, since it is dominated by K-feldspars and calcic pyroxenes. But different communities of vegetation might have contributed acids. A more serious problem of interpretation is that the gleyed reddish units, which involve mobilization of ferrous minerals requiring chelation and a measure of eluviation, demand more time than an implicit model of episodic short-term, high lake stands would allow. This favours iron mobilization as primarily a result of through-flow.

3.3.5 The floor of Deelpan: Erosion *versus* deposition

Perhaps counterintuitively, the pan is not drowning in allogenic sediment. Episodic flood surges that churned into the pan are most directly recorded by low-energy silts, derived from shale watersheds and some airborne dust. One shoal of undated lacustrine mud was a reduced, light (greenish) gray to white silt loam, only moderately calcified, and with an angular blocky structure (Figure 4, unit 1). Another example included an interbedded loam (with 50% sand) and a sandy loam (85% sand), with a coarse tail of dolerite minerals (plagioclase, pyroxene, epidote, microcline) and fragments of shale or meta-sandstone. These record either different flood loads or stages of sedimentation. The floods enter Deelpan from two shallow channels at the southern end that have built up a marshy delta extending 250 m onto the pan floor.

Unit 3, from a 70 cm deep pit, also has two facies, the first a silt (with 9% sand), the second a loamy sand (85% sand), with columnar structure and abundant salts. Sodium salts were concentrated later during a pan cycle, with modest energy switches continuing on the floor, probably in response to the agitating pressure of flood influx.

The underlying Ecca Group shale (unit 4) is in a mixed state of decomposition. The fissile bedding breaks up by salt hydration and the sand-sized fraction ranged from 1 to 74%. Textural class varies from silty clay to sandy loam. High CCE enrichment indicates a combination of weathered plagioclase and allogenic calcium bicarbonate circulating in the phreatic zone well below the pan floor.

Most of these bottom facies are not replicated in the surface profiles studied, although part of the silt and clay deflated from the pan floor will have contributed to formation of the various prismatic subsoil horizons. The paucity of quartz sands among the pan floor sediments is due to the breakdown of Ecca shale into silt and clay, with limited quartz sand. The stop-and-go mechanisms of pan formation involve short episodes of flood ingression, followed by extended periods of puddling or desiccation, during which deep hydrolysis and salt weathering continue to break down bedrock within the phreatic horizon. When the pan floor was dry, wind deflated some of the flood and bedrock residues, but equally so, wind action may also have deposited quartz sand and silt across the pan perimeter. Much of the quartz in the exposures was not mobilized

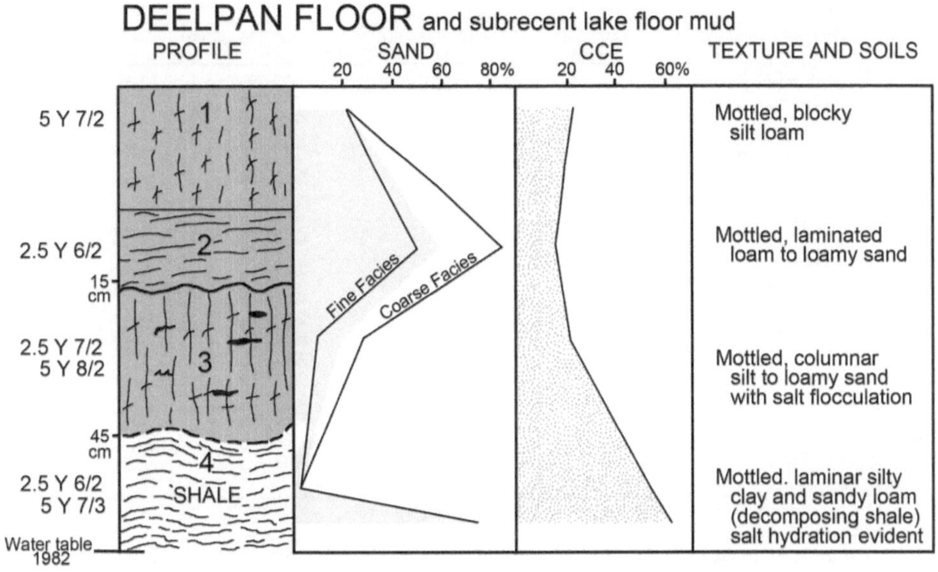

Figure 4. The floor and the excavated pit in the center of Deelpan includes autochthonous shale products and allogenic quartz. "Sand" here is primarily a size class, since most quartz is allogenic, derived from eolian sand and suspended fractions.

inside the pan, but on higher ground, including the lunettes around it, if not as long-distance eolian dust. Aridity served as the context for an ephemeral lake, the floor of which was slowly lowered by hydration, with deflation of selected size classes. The loci of pan formation were controlled by the low gradients of the Highveld landscape, the susceptibility of shale to hydration, and the macro-scale geometry of patterned, intrusive igneous rocks. Such loci were further influenced at smaller scales by surface and groundwater ponding behind dikes and sills of the Karoo volcanics (David Helgren, pers. comm. 2014). With the steeper valley gradients west of Kimberley, this interpretation would need modification, because elongated proto-valleys here developed as stepped basins or *vloers* (see Wellington, 1971: 26, 75, 314; Butzer, 1974; Helgren, 1979: Figures 2–16).

3.3.6 The South Section: Further precisions

Turning to the South Section (Figure 5), the sand curve identifies two major episodes of vigorous flooding, with mobilization of calcic grit from the pan floor. The Folk indices for mean size (Mz) and sorting (sigma) show moderately good sorting but no clear relationship to the sand maxima. Only one sample is better sorted than 0.6, while most of the profile sands studied at Deelpan have better sorting than 0.8. That seems to preclude alluvial or lacustrine sediments, and argues for a common dominance of eroded and reworked aeolian sand.

To define the lacustrine sequence of the South Section, units 3b to 2b have been locally eroded by wave action up to about +3 m, and the internal structure of 3b shows several local incidences of steep (up to 7 degree dips) wave truncations. This lakeshore context is amplified by a bundle of large crossbeds, and strong peaks of coarse sand (dominance of 100–500 μ fraction) in units 4 and 3a.

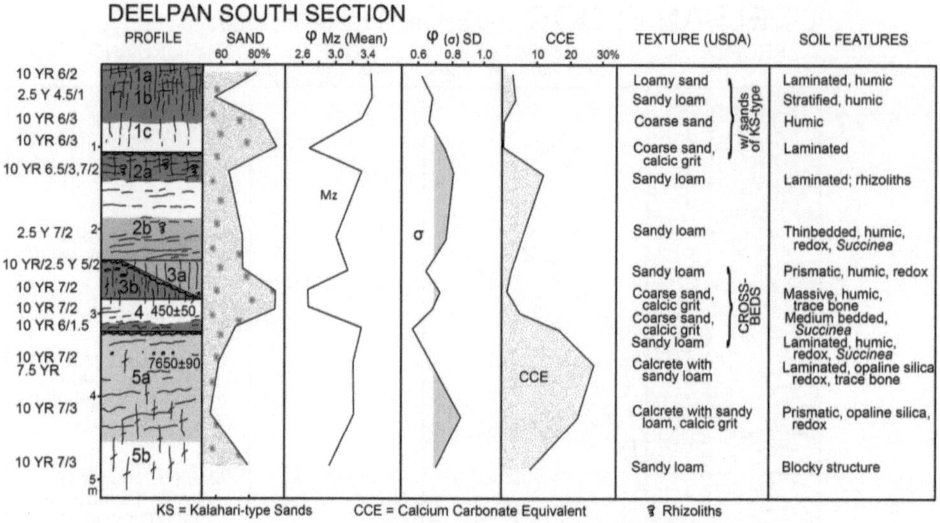

Figure 5. Analytical data for Deelpan South Section. In the Middle Section (not illustrated) LSA artifacts are found in the equivalent of unit 2a as well as scattered nearby.

The AMS assay, based on soil humates, appears unexpectedly young, placing the contact of 3b and 3a at 450 ± 50 BP (cal. 1440 ± 60 AD; Pta-3354), so that this fairly violent lake transgression and the associated redox horizon to ~3 m would be little older than ~1400 AD. If that is the case, the upper half of the South Section may have been deposited in as little as 500 years. OSL dating at other Free State pans has proven quite successful for the last millennium (Holmes *et al.*, 2008), whereas deep alluvial profiles at present provide little information for a Holocene record, other than a case for exceptionally moist climate ~9300 BP (OSL) (see Lyons *et al.*, 2014).

Extending this scenario, the second, final peak of coarse sand in Figure 5 may have happened in the order of two or three centuries ago, coeval with mobilization of Kalahari-type sands, that indicate a declining ground cover—even before the time of general Boer (Afrikaner) settlement in the western Free State (ca. 1848 AD). This opens important problems for interdisciplinary research.

The sharp break between units 4 and 5 could represent up to five millennia since the 7650 ± 70 BP (cal. 6450 ± 80 BC; Pta-3345) on ostrich eggshell from unit 5a. In other words, the calcrete of unit 5 formed during the Early Holocene in an older, prismatic palaeosol, under conditions with abundant carbonate solution (with cool waters) but mobilization of opaline (colloidal) silica (with pH values no less than 9.0), perhaps sequentially. That is consonant with dates on humus-enriched soils directly above calcretes at Meriba (see Figure 6), or with three humus dates of 13,800–14,900 cal. yr at the Benfontein Trench, near Alexandersfontein (Kimberley), for an older group of what are best called silcretes (Butzer, 1984; and unpublished). But a better chronometric framework is essential.

3.3.7 A spring geoarchive: Meriba

The preceding graphics and discussions addressed the nearshore and backshore or lunette environments, and the floor of an episodic lake. Another site, initially studied by Louis Scott, is a small marsh fed by a spring seep and shallow aquifer. This is located east of the farm Meriba, near the seasonal stream running to Deelpan. Scott extracted a 1 m core until he struck calcrete, and I processed 8 samples from this core (Meriba I) (Figure 6). He also tested a second site nearby (Meriba II, not illustrated), but only 4 samples were analyzed for pollen and sediments. The disintegrating calcretes linking the two spring seeps are informative and shown in Figure 7.

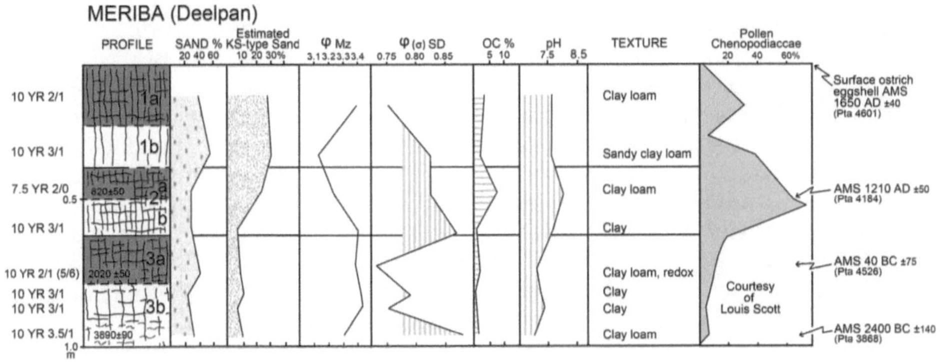

Figure 6. Analytical data for Meriba (Deelpan) Core. Only units that have black Munsell colors are highlighted in brown. After its mid-19th century predecessor was burnt down during the Anglo-Boer War, the first building of the modern farm Meriba was constructed in 1903.

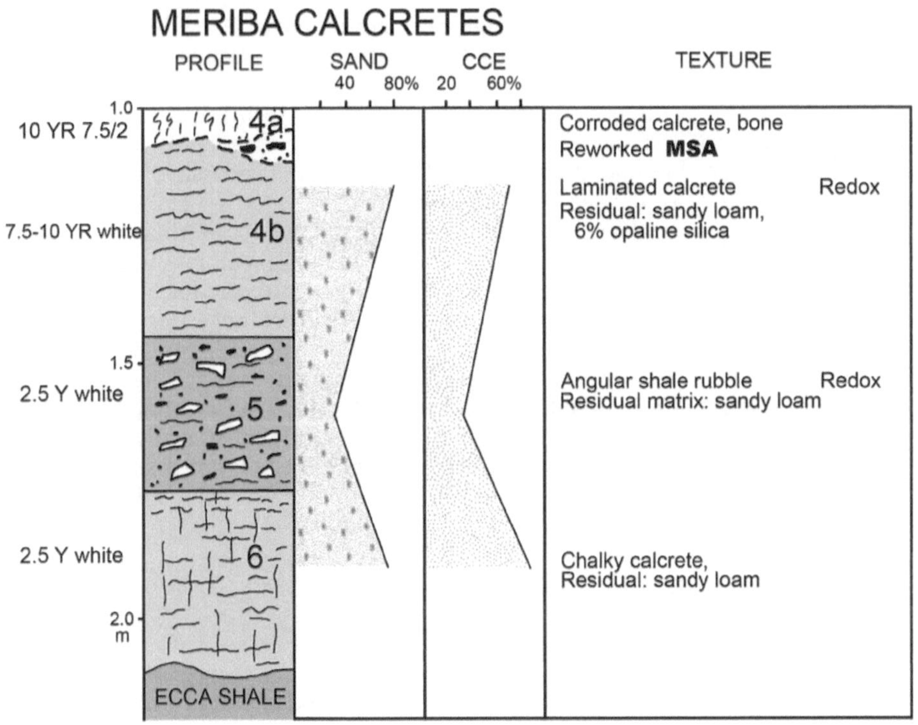

Figure 7. Composite profile for the Calcretes at Meriba, that have broken up into weathered blocks.

The Meriba I core is divided into two or three sections, depending on the criteria preferred.

The basal unit (3b) is clayey (38–56%), with low organic carbon (under 1%), low CCEs (1–4%, not shown in Figure 6), and lower pH values (7.0–7.5) than the remainder of the column. Organic structure is present but most plant debris has been reduced to mull humus, a fact not verified by the OC readings. Greater age, compression and alteration are suggested.

Unit 3a is coarser (40% sands) as CCE jumps to 12%, but 2b returns to a clayey texture as mull humus fades out and pH increases. There was a shorter period of increased energy that has an AMS date of 2020 ± 50 BP (cal.~40 ± 75 BC, Pta-4526).

Unit 2b shows this systemic change continuing after 820 ± 50 BP (cal. ~1210 ± 50 AD, Pta-4184), as the mean pH rises to 7.95 and OC briefly exceeds 6% (the maximum of the method), with densely-packed plant residues but no mull humus. While the sand fraction remained low (28%) and with the spring failing, Kalahari-type sands rapidly acquired prominence, indicating a deteriorating ground cover and accelerated eolian activity. This drastic switch, centered near –50 cm in the core and at the 2b/2a boundary, coincides with a major pollen event, during which the once prevailing grasses (Poaceae) were temporarily replaced by the intrusion of salt brush (Chenopodiaceae) (Scott, 1988; Scott and Brink, 1992; Scott and Nyakale, 2002).

With unit 1b comes a second fundamental change. Spring activity was strongly revived, with a maximum of sand fraction (51%) and Kalahari-type sands, but phasing out of halophytes in favor of sedges (Cyperaceae) (Scott and Brink, 1992; Scott and Nyakale, 2002). That more or less persisted to the top of the core, which the other profiles would suggest is complete. But ostrich eggshell at the surface with a calibrated date of 1650 ± 40 AD creates some ambiguity. Standard methods did pick up the magnitude of a major, unidentified change, but were unable to recognize the biotic, i.e. palynological transformation from grass to salt brush. Thin sections and micromorphology might identify such changes of the dense vegetative mat.

Returning now to discussion of the South Section, it seems that the flood-driven convolutions dated 450 BP in the South Section may have been roughly coeval with the salt brush invasion at Meriba I dated 820 BP. For example, it is possible that the marsh at Meriba was almost eradicated by an unusually violent flood surge. In that case the salt brush and sedges could be part of a recolonizing succession. Meriba II has a single basal date of 2950 ± 80 BP (cal. ~1200 ± 230 BC Pta-4183; with multiple calibration intercepts) at –95 cm. Different gaps in the two spring records are likely.

The calcretes shown at Meriba in Figure 7 require little discussion here. Since the youngest hardpans in the North Gully, South, and Meriba profiles are of similar age, there appears to be a pattern of fundamental environmental change, namely distinct geochemical systems before about 6800 BP and after 4400 BP cal. Similar gaps are apparent in the detailed record of the Benfontein Trench (Kimberley district) (Butzer, 1984), and the challenge is to apply the micro-stratigraphic and processual insights of Deelpan to testing alternative explanations for the absence of such records.

3.4 CONCLUDING DISCUSSION

3.4.1 Synthesis

The complex stratigraphic record assembled here is fundamental to understand the environmental history encapsulated in a single Free State pan. It applies a range of analytical techniques, in combination with exacting attention to field relationships, to allow deductive and inductive perspectives on environmental response to a non-equilibrium climate (Butzer and Helgren, 2005). The goal is not to peg defined periods as "wet" or "dry" but to conceptualize the interaction of multiple variables in the course of short-term intra-annual cycles and long-term trends or occurrences of extreme events (Butzer, 2011: Figure 1).

These play out as equilibrium shifts; episodes of erosion or deposition; expansion, contraction, and filling or thinning of vegetation assemblages; and the nature or rates of pedogenic processes. The role of fire (Scott, 2002) and the impact of human land-use, settlement and mining (Showers, 2010) may be ancillary or fundamental. Add to that the role of well-meaning environmentalists who insist on giving their own political or cultural interpretations of systemic change and response, sometimes with a loss of clarity about cause and effect.

Deelpan at its current state of resolution suggests a tripartite evolution of sediments and soils, interrupted by a major hiatus of tangible processes or deposits. The salient features are: (A) Terminal beds of dominantly eolian sands, with incipient humic soils; (B) An intermediate body of redox units, with mixed facies and loamier textures; and (C) Basal calcrete horizons that were post-depositionally calcified, but sometimes also partly dissolved.

3.4.2 The upper units

(A1) The topmost units (South 1a,1b; Middle 1; North 1a; Meriba 1a) consist of uncompacted sands and loamy sands, with an appreciable component of Kalahari-type sands but mixed textures that lack clear eolian signals. I argue for deflation from an incomplete ground cover as confirmed by the modern karoid pollen rain (Scott, 1982: Figure 3). Flooding of the pan appears to have been rare, with excessive rains and lacustrine conditions during the mid-1970s and the late 1890s.

(A2) Significant here was a transgression (South 1c) with wave action to +3 m, recorded by peaks of both the bulk sand fraction and its mean coarseness (Mz). Prior to this event the Deelpan lake had fluctuated near intermediate levels, with cutting of small gullies and light gray eolian sands. Interbeds include thin (1–3 cm), hard, laminated loamy sand with strong rhizoliths and *Succinea* (North 2b, 2a; South 2b, 2a). These beds suggest standing water amid aquatic vegetation in the backshore zone, as a key part of the structure of a lunette.

(A3) The hyena burrows in North 2b and 3 were infilled with sediment from North 2b. With a suite of grassland ungulates, the 55 hyena coprolites overwhelmingly record grassland pollen (Scott *et al.*, 2003). Given the 2 clawless otters, North and South 2b argue for a fairly persistent but oscillating lake, no later than ~1800 or earlier than 1650 AD, with the coincident activation of Kalahari-type sands.

3.4.3 Redox sands

(B1) The redox unites (North 7 to 3; Middle 4–2a; South 4–7b) begin with alteration of North 7 top, and then continue upwards with large crossbeds of laminated sands and sandy loams. South 4 is highlighted by a peak bulk sand fraction and a mirror image of very coarse sand (Mz), indicating a maximal transgression. These 75 cm of initially light gray coarse sand are mixed with calcic and bedrock grit, implying a high energy mixed facies that repeatedly changes in detail. The upper half is interdigitated with a large crossbed structure of grayish brown sandy loam (South 3a), but the main profile continues with a zone of organic, prismatic structure, that includes abundant snails and several internal beach truncations with 7 degree dips, at elevations of ~2.3 to 2.6 m. The three pollen peaks of salt brush (50, 64, and 89% Chenopodeaceae) compare with that of the modern pollen rain from the adjacent shore vegetation (Scott, pers. comm., 2014; 1982: Figure 3). The AMS date on humic soil at the contact of South 3a and 3b is 450 ± 50 BP (cal. ~1450 AD, Pta-3354), or judging by the salt brush maximum at Meriba, ~1210 AD. Pending further AMS dates this suggests an age of very roughly 1200–1400 AD. The termination of the redox overprint coincides with a gravelly sandy loam (North 4, 130 cm) and a thin-bedded sandy loam with *Succinea* (South 2b, 40 cm). As a provisional working hypothesis, Deelpan was partially flooded for several centuries ~1400 AD. Following this moist anomaly, blowing Kalahari-type sand became prominent. For analogous emphasis on mixed facies and systemic transitions see Helgren (1979: 317–26)

(B2) Incorporating Meriba I into this summation is difficult because sedimentary counterparts for North and South are absent. But this spring was active in the approximate time range 2400 to 50 BC, perhaps implying that older Deelpan records were minimal or destroyed by lacustrine erosion. A general hiatus between 6000 and 2400 BC preceded summary phase B2.

3.4.4 Calcretes and Silcretes

(C) The basal strata around Deelpan consist of calcretes in the order of 50 cm to several meters thick. Post-acid residuals mainly are sandy loams, with consistent Folk indices and some dispersed grit of calcite, pelletal mud, or bedrock minerals. Differences of facies such as laminar flow, or prismatic to blocky and chalky structure reflect subsoil processes, and calcite induration ranges from 30 to 90%. A number of indurated calcretes have a strong silcrete component (opaline silica). Other calcretes are in process of dissolution or have been decalcified except for calcite platelets. In other words the Deelpan calcretes are secondary and represent mixed eolian or colluvial facies that have been calcified by lateral movement of subsoil water.

Silcretes in South Africa are best developed in the Pleistocene Vaal River gravels, and tend to occur as calcrete-silcrete duricrusts. Their evolution is not fully clarified, but case studies in Botswana and Australia (Nash and Hopkinson, 2004; Nash and McLaren, 2007; Watts, 1978) suggest that throughflow and reducing conditions in a pre-existing calcrete allowed depletion of Al, Fe and Ca, but concentration of opaline silica in response to an increasingly arid soil environment and higher pH.

At Meriba the stratigraphic sequence includes eroded residues of a dissolved calcrete over two intact such levels, separated by an indurated rubble of shale. A more complex sequence that includes silcretes (17 to 42% NaOH soluble) was studied in the Benfontein Well Trench (Figure 1), near Kimberley (Butzer, 1984: Table 12; and unpublished): BT 9, white, banded calcrete, 7035 and 7905 BP (uncal. SI-3569, SI-3050, on carbonates);

BT 8, grayish brown, vertically crystallized tufa, 11,025 and 11,250 BP (SI-1116, SI-3048, on carbonates mixed with 23% opaline silica); BT 5, pink Kalahari-type sands; BT 4, grayish brown silty sand, 13,800 and 13,950 BP (SI-2583B, SI-2583, on humus) with fresh Albany (early LSA) artifacts; BT 2B, silty silcrete, 14,900 BP (SI-2584, on humus); and at the base, BT 1 A, a silty silcrete in contact with weathered dolerite and shale contacts. Both the carbonate and humic dates are in correct vertical position. This sequence would appear to begin midway during deep-sea MIS 2, and the facies continue to change after 14,000 BP, shifting through a puzzling number of distinct geomorphologic and geochemical phases that call for specialized study. It is not implied that the calcrete/silcrete sequences at Benfontein are identical to incipient forms at Deelpan, but that time and climatic change are important variables. Much patient work remains to be done before the complex evolutionary history of the Free State pans is fully resolved.

3.4.5 Conclusions on environmental history

In sum, several major conclusions can be singled out:

1. There are tangible records of lake transgressions embedded in near and backshore sediments, that prove to represent more than eolian lunettes. These are best developed on open downwind sections, but fade in the lee of fringing koppies.
2. Although the most recent Deelpan floods were isolated events, the sediment parameters, redox, and soil criteria also identify pan floods more tightly bundled within oscillating, moist climatic anomalies. This points to periods of severe recurrent rainstorms during part of the late Holocene. One extreme weather event may be conspicuous in a geoarchive, but information on recurrent anomalies may be much more difficult to generate (Butzer *et al.*, 2013).
3. The Deelpan record requires rethinking of calcrete genesis and evolution as a relict landscape element encoding a complex palaeoclimatic history that calls for specialist study. This pan, in conjunction with Benfontein, indicates that some apparent calcretes are in fact silcretes—with distinct environmental implications.

3.4.6 The role of human impact

For several decades political ecologists have stressed that the South African interior has been intensely degraded by Euro-African land use practices. While there is abundant evidence for ecological damage, the picture is not at all clear. For example, in the eastern Free State most streams have not been incised or gullied, and loop with high sinuosity across marshy, vertic floodplains. But in the western Free State, streams have mainly been incised into such deposits during a long fluvial evolution (Butzer *et al.*, 1979; Butzer, 1984: 18–28). Against this qualified picture of localized but not general degradation, some political ecologists have projected sweeping generalizations of landscape destruction resulting from European colonization, so for example the mainly insightful writings of Showers (2010). The present investigation illustrates how science-grounded environmental history applies geo- and bioscience techniques and systemic analysis in practicable regional contexts (Butzer, 2008, 2011). Political ecologists need to meet their matching partners halfway and make a serious effort to understand each other's methods, reservations, and applications. Divided by different epistemologies, these two approaches to landscape ecology continue to selectively ignore each other's observations or ideas to the detriment of what are or should be shared goals.

This reflective study of Deelpan should redirect rather than contradict a traditional narrative. Remarkable is that we can actually identify sequences of sediments, structural soils, and other evidence of humification and alteration that provide direct information. There is a record of two "catastrophic" flood surges during the last millennium or so. According to this hypothesis, the first may have taken place ca. 1200 AD, with prominent redox apparent in response to a wet soil and pan environment over a few centuries. The second surge might have happened in the 17th or 18th century, when there was no redox formation; but accelerated eolian activity began to attack soils exposed by an incomplete ground cover.

The implications of this scenario are of such potential interest for environmental history of the Highveld that they deserve closer attention (see, for example Helgren, 1978; Scott, 2002). Did Euro-African settlement inaugurate Veld deterioration, or did indigenous pastoralism already do so a century or two earlier? Khoi settlement is archivally documented along the middle Riet River (see Figure 1 and Helgren, 1979: 272–76) during the early 1800s, but was preceded by enigmatic Type-R settlements with pastoralism since as much as 350 years earlier (Maggs, 1971; Humphreys, 2009). Although LSA material at Deelpan is scarce and does not suggest an attractive environment for prehistoric peoples, the question raised here whether veld deterioration predated Boer settlement is relevant and amenable to further research in the western Free State. First asked by Brink (2005) this may be one of the most important implications raised by this study.

The apparent absence of dated mid-Holocene sediments in the Kimberley region (Butzer *et al.*, 1979; and more generally: Deacon, 1984) also includes the archaeological record. The same hiatus is evident at Deelpan, as that record has shown. One clue is that calcrete induration of sediments spanned the Early Holocene and late Pleistocene, highlighting geochemical questions in regard to cooler waters. Why did a spring seep record like that of Meriba only begin around 3000 or 4000 years ago? That issue can best be addressed in other pans of the western Highveld.

This study suggests that micro-stratigraphy and a discriminating application of sedimentology and soil examination can open new vistas in pan interpretation. They can also provide a more dynamic context for the application of palynology. There is a need for further OSL dating programs and a more specialized application of geochemistry, hydrobiology, and soil micromorphology. Finally, resolution of South African historical range ecology can and should be refined by explicitly including human intervention and impacts, as an explicit multidisciplinary paradigm.

ACKNOWLEDGEMENTS

The fieldwork and evaluations reflected in this article were carried out in collaboration with Louis Scott and the results are dedicated to him with friendship and thanks. David Helgren, with whom I shared trips in the Vaal River diggings and later in Australia, provided many suggestions and constructive criticism. Greg A. Botha offered helpful comments on the manuscript, and the late Robert Stuckenrath was a dependable field companion and discussant. Charles Frederick, who accompanied me on extended field studies in Mexico, contributed to the laboratory phase with meticulous attention. Last but not least, Carlos Cordova, who turned his enthusiasm from the New to the Old World, read and also explicated this paper at Bloemfontein in my absence.

REFERENCES

Brink, J.S., 2005, The Evolution of the Black Wildebeest and modern large mammal faunas of central South Africa. DPhil dissertation, University of Stellenbosch.

Butzer, K.W., 1974, Geo-archaeological interpretation of Acheulean calc-pan sites at Doornlaagte and Rooidam (Kimberley, South Africa). *Journal of Archaeological Science,* **1**, pp. 1–25.

Butzer, K.W., 1984, Archeogeology and Quaternary environment in the interior of Southern Africa. In *Southern African Prehistory and Palaeoenvironments,* edited by Klein, G.J. (Rotterdam: A.A. Balkema), pp. 1–64.

Butzer, K.W., 2004, Coastal eolian sands, paleosols and Pleistocene geoarchaeology of The Southwestern Cape, South Africa. *Journal of Archaeological Science*, **31**, pp. 1743–1781.

Butzer, K.W., 2008, Challenges for a cross-disciplinary geoarchaeology: The intersection between environmental history and geomorphology. *Geomorphology,* **101**, pp. 402–411.

Butzer, K.W., 2011, Geoarchaeology, climate change, sustainability: A Mediterranean perspective. *Geological Society of America, Special Paper,* **476**, pp. 1–14.

Butzer, K.W., Butzer, E.K. and Love, S., 2013, Urban geoarchaeology and environmental history at the Lost City of the Pyramids, Giza: Synthesis and review. *Journal of Archaeological Science,* **40**, pp. 3340–3356.

Butzer, K.W., Fock, G.J., Stuckenrath, R., and Zilch, A., 1973, Palaeohydrology of late Pleistocene Lake Alexandersfontein, Kimberley, South Africa. *Nature,* **243**, pp. 328–330.

Butzer, K.W., Fock, G.J., Scott, L. and Stuckenrath, R., 1979, Dating and context of rock engravings in southern Africa. *Science,* **203**, pp. 1201–1214.

Deacon, J., 1984, Later Stone Age people and their descendants in southern Africa. In *Southern African Prehistory and Paleoenvironments,* edited by Klein, R.G. (Rotterdam: A.A. Balkema), pp. 221–328.

DeBruiyn, H., 1972, Pans in the western Orange Free State, *Annals, Geological Survey of South Africa,* **9**, pp. 121–124.

DuToit, A.L., 1907, Geological survey of the eastern portion of Griqualand West. *11th Annual Report, Geological Commission, Cape of Good Hope*, pp. 87–176.

Folk, R.L. and Ward, W.C., 1957, Brazos River bar: a study in the significance of grain-size parameters. *Journal of Sedimentary Research,* **27**, pp. 9–26.

Geyser, G.W.P., 1950, Panne–hul ontstaan en die Faktoren wat daartoe aanleiding gee. *South African Geographical Journal,* **32**, pp. 15–31.

Holmes, P.J., Bateman, M., Thomas, D.S.G., Telfer, M, Barker, C.H. and Lawson, M.P., 2008, A Holocene-late Pleistocene aeolian record from lunette dunes of the western Free State panfield, South Africa. *The Holocene,* **18(8)**, pp. 1193–1205.

Helgren, D.M., 1978, Acheulian settlement along the lower Vaal River, South Africa. *Journal of Archaeological Science,* **5**, pp. 39–60.

Helgren, D.M., 1979, *River of Diamonds: An alluvial History of the Lower Vaal Basin, South Africa*, (Chicago: University of Chicago Press).

Humphreys, A.J.B., 2009, A Riet River retrospective. *South African Humanities,* **21**, pp. 157–175.

LeRoux, J.S., 1978, The origin and distribution of pans in the Orange Free State. *South African Geographer,* **6**, pp. 167–176.

Lyons, R., Tooth, S. and Duller, G.A.T., 2014, Late Quaternary climatic changes revealed by terrace paleosols: A new form of geoproxy for the South African interior. *Quaternary Science Reviews,* **95**, pp. 43–59.

Maggs, T., 1971, Pastoral settlements on the Riet River. *South African Archaeological Bulletin,* **26**, pp. 37–63.
Nash, D.J. and Endfield, G.H., 2002, A 19th century climate chronology of the Kalahari region of central southern Africa derived from missionary correspondence. *International Journal of Climatology,* **22**, pp. 821–841.
Nash, D.J. and Hopkinson, L., 2004, A reconnaissance laser Raman and Fourier transform infrared survey of silcretes from the Kalahari Desert, Botswana. *Earth Surface Processes and Landforms,* **29,** pp. 1541–1558.
Nash, D.J., and McLaren, S.J., eds, 2007, *Geochemical Sediments and Landscapes* (Oxford: Blackwell).
Pringle, J.A., 1982, *The Conservationists and the Killers* (Cape Town: T.V. Bulpin).
Scott, L., 1988, Holocene environmental change at western Orange Free State pans, South Africa. *Palaeoecology of Africa,* **17**, pp. 113–22.
Scott, L., 2002, Charcoal in sediments and Late Quaternary fire history of grassland and savanna regions in Africa. *Journal of Quaternary Science,* **17**, pp. 77–86.
Scott, L. and Brink, J.S., 1992, Quaternary palaeoenvironments of pans in central South Africa: palynological and paleontological evidence. *South African Geographer,* **19**, pp. 22–34.
Scott, L., Fernández-Jalvó, Y, Carrión, J. and Brink, J.S., 2003, Preservation and interpretation of pollen in hyena coprolites: Taphonomic observations from Spain and southern Africa. *Palaeontologia Africana,* **39**, pp. 83–912.
Scott, L. and Klein, R.G., 1981, A hyena-accumulated bone assemblage from Late Holocene deposits, Orange Free State. *South African Museum,* **86(6)**, pp. 217–22.
Scott, L. and Nyakale, M., 2002, Pollen indications of Holocene palaeoenvironments at Florisbad spring in the central Free State, South Africa. *The Holocene,* **12(4),** pp. 535–541.
Showers, K.B., 2010, A history of African soil: Perceptions, use and abuse. In *Soils and Societies: Perspectives from Environmental History*, edited by McNeill, J.B. and Winiwarter, V. (London: White Horse Press), pp. 118–176.
Van Eeden, O.R., 1955, Die ontstaan van panne in Suid-Afrika. *Tegnikon* (July, 1955), pp. 206–215.
Van Rooyen, T.H., 1972, *Soils of the Central Orange River Basin.* D.Sc. Thesis, University of the Orange Free State, Bloemfontein.
Watts, S.H., 1978, A petrographic study of silcrete from inland Australia. *Journal of Sedimentary Petrology,* **48**, pp. 987–994.
Wellington, J.H., 1971, *Southern Africa: A Geographical Study (I. Physical Geography),* (London: Cambridge University Press).

CHAPTER 4

Homo habilis and *Australopithecus africanus*, in the context of a chronospecies and climatic change

J. Francis Thackeray
Evolutionary Studies Institute and School of Geosciences, University of the Witwatersrand, Johannesburg, South Africa

ABSTRACT: Pairwise comparisons of crania of selected well-preserved hominin specimens attributed to *A. africanus* and *H. habilis* are made using a morphometric approach to assess the probability that they are conspecific in the context of the hypothesised species constant (T = –1.61) and a statistical (probabilistic) definition of a species (Thackeray, 2007). In response to criticisms of Thackeray's method (Gordon and Wood, 2013), comparisons are made in this study where specimen A is on the x axis and specimen B on the y axis, and *vice versa*. The method is applied not only to Plio-Pleistocene hominins but also to two species of chimpanzees (*Pan troglodytes* and *Pan paniscus*). The results obtained for specimens attributed to *A. africanus* and *H. habilis,* are almost identical to the mean value of a morphometric statistic obtained by Thackeray (2007) for modern conspecific specimens and associated with an approximation of a biological species constant (T = –1.61). The morphometric data from hominins can be assessed in the context of analyses of cranial measurements of chimpanzees (*P. troglodytes* and *P. paniscus*) whose distributions in the past are likely to have been affected by changes in climate, notably in cold dry episodes, affecting habitats, in turn influencing gene pools and "incomplete lineage sorting", such that boundaries between taxa are not necessarily clear. The question arises as to whether *A. africanus* and *H. habilis* relate to a chronospecies, across the transition from *Australopithecus* to *Homo*.

4.1 INTRODUCTION

Fifty years ago, the hominin species *Homo habilis* was described by Leakey, Tobias and Napier (1964) on the basis of cranial and postcranial remains from Bed I at Olduvai Gorge in Tanzania, dated at about 1.8 million years ago (mya). The reaction was highly critical. Sceptics included W.E. Le Gros Clark, K. Oakley, B.G. Campbell, D. Pilbeam, E.L. Simons, F.C. Howell, J. Robinson and L. Brace (Tobias, 1992). Some noted that the new material was similar to South African Plio-Pleistocene fossils attributed to *Australopithecus africanus* from sites such as Taung (described initially by Dart, 1925) and Sterkfontein (described by Broom and Robinson since 1936, see Broom *et al.*, 1950). Critics claimed that a new species (*habilis*) in the genus *Homo* was unwarranted. Robinson (1965) argued that the specimens attributed to *H. habilis* should instead be considered to be australopithecines. It was not until some 25 years later that Tobias (1991) published two detailed volumes on the hominin fossils attributed to *H. habilis* from Olduvai Gorge. By that time these specimens were generally considered as a valid species. Tobias (1992) himself triumphantly wrote a retrospective assessment of the controversy, concluding with the words "Today it is widely accepted as a good taxon and one that represents a critical stage in the evolution of modern man". However, there is reason to re-assess the status of certain specimens attributed to *Homo habilis* and *A. africanus*.

Thackeray (1997) wrote a letter to *Nature* entitled "Probabilities of conspecificity", appealing for a statistical approach whereby hominin taxonomy could be assessed in terms of measurements of pairs of specimens, using least squares linear regression to quantify the degree of scatter around a regression line associated with the general equation $y = mx + c$, where m is the slope and c is the intercept. Pairs of crania of modern conspecific mammals, birds and reptiles were analysed in this way. In a pioneering study involving fourteen researchers at the Transvaal Museum (currently called the Ditsong National Museum of Natural History, in Pretoria), Thackeray *et al.* (1997) reported central tendency of the log-transformed standard error of the m-coefficient, referred to as "log se_m", based on pairwise comparisons of conspecific specimens in museum collections of extant taxa (vertebrates and invertebrates). In a more extensive study, Thackeray (2007) showed that this central tendency was reflected by a mean log se_m value of −1.61 which was claimed to be an approximation for a biological species constant (T), expressed through evolutionary time and geographical space, associated with a statistical (probabilistic) definition of a species that could be applied to fossils, including specimens attributed to *Australopithecus* and early *Homo*.

4.2 OBJECTIVE

On the occasion of the 50th anniversary of the announcement of *H. habilis*, it is pertinent to apply Thackeray's morphometric approach to certain specimens attributed to this species, and to other fossils attributed to *A. africanus*. The objective of this study is to undertake pairwise comparisons of crania of selected well-preserved specimens attributed to *A. africanus* and *H. habilis,* and to assess the probability that they are conspecific in the context of the hypothesised species constant (T = −1.61) and a statistical (probabilistic) definition of a species based on log se_m values (Thackeray, 2007).

In response to criticisms of Thackeray's method (Gordon and Wood, 2013), comparisons are made in this study where specimen A is on the x axis and specimen B on the y axis, and *vice versa.* The log se_m values are calculated for both pairwise comparisons, and the results are assessed in terms of the difference between these log se_m values

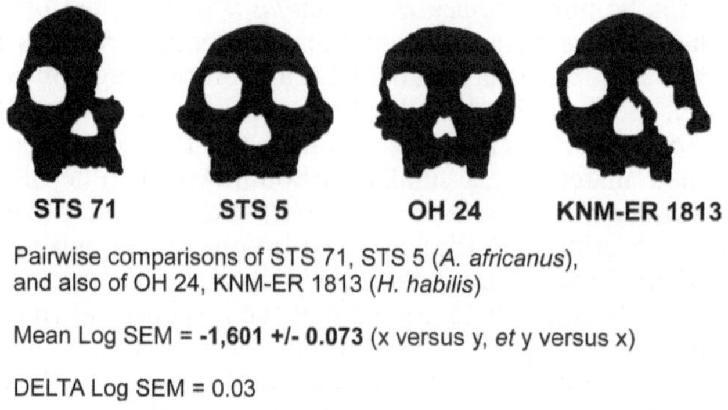

Figure 1. Four Plio-Pleistocene hominin specimens which have been examined in this study. It is suggested that there is not a clear boundary between specimens attributed to *A. africanus* (two specimens on the left) and specimens attributed to *H. habilis* (two crania on the right. It is also suggested that the sequence from left to right relates to a chronospecies from *A. africanus* to *H. habilis* within the period 2.6 and 1.6 million years ago.

(here referred to as delta log se_m for the pair of specimens being compared), as well as the log se_m values themselves. It is recognised that log se_m values and delta log se_m when assessed together, have potential for assessing probabilities of conspecificity (Dykes, 2014).

The morphometric data from hominins can be assessed in the context of analyses of cranial measurements of chimpanzees (*Pan troglodytes* and *Pan paniscus*) whose distributions in the past are likely to have been affected by changes in climate and habitats, in turn influencing gene pools and "incomplete lineage sorting", such that boundaries between taxa are not necessarily clear.

4.3 MATERIALS

The hominin specimens selected for this study are listed below.

- Sts 5. *A. africanus* Sterkfontein, South Africa. *Circa* 2.1 mya
- Sts 71. *A. africanus*. Sterkfontein, South Africa. *Circa* 2.6 mya
- OH 24. *H. habilis*. Olduvai Gorge, Tanzania. *Circa* 1.8 mya
- KNM-ER 1813. *H. habilis*. East Turkana, Kenya. *Circa* 1.6 mya

There has been some debate regarding alternative classifications for these specimens. Clarke (2013) claims that Sts 71 represents not *A. africanus* (as in the case of Sts 5, known as "Mrs Ples") but instead a separate australopithecine species (*A. prometheus*). Wood and Collard (1999) suggested that OH 24 and KNM-ER 1813 should be placed in the genus *Australopithecus* rather than in *Homo*.

The specimens listed above range in age from circa 2.6 mya (Sts 71) to 1.6 mya (KNM-ER 1813), a period of about one million years.

4.4 RESULTS

Results of pairwise comparisons of specimens are given in Table 1, including comparisons where specimen A is on the x axis and specimen B on the y axis, and *vice versa*.

Table 1. Log se_m values based on pairwise regression analyses of cranial measurements of Plio-Pleistocene hominin fossils from South Africa (Sts 5 and Sts 71 attributed to *Australopithecus africanus*) and from East Africa (OH 24 from Olduvai Gorge in Tanzania, as well as KNM-ER 1813 from East Turkana, Kenya, attributed to *Homo habilis*). Log se_m data from Thackeray and Odes (2013).

x axis	y axis	Log se_m
STS 5	STS 71	−1.643
ER 1813	STS 71	−1.507
OH 24	STS 71	−1.566
STS 71	STS 5	−1.691
ER 1813	STS 5	−1.627
OH 24	STS 5	−1.509
STS 71	ER 1813	−1.546
STS 5	ER 1813	−1.603
OH 24	ER 1813	−1.685
STS 71	OH 24	−1.630
STS 5	OH 24	−1.506
ER 1813	OH 24	−1.701

4.5 DISCUSSION

The mean log se_m value for 12 pairwise comparisons of the specimens listed in Table 1 is −1.60 ±0.07. Remarkably, this is almost identical to the mean log se_m values and associated standard deviations obtained from pairwise comparisons of conspecific chimpanzees studied by Gordon and Wood (2013). They calculated the following log se_m values for pair-wise comparisons of conspecific chimpanzees:

- −1.61 (female-female comparisons of *Pan paniscus*)
- −1.62 (male-male comparisons of *Pan paniscus*)
- −1.61 (female-male comparisons of *Pan paniscus*)
- −1.62 (female-female comparisons of *Pan troglodyes*)
- −1.60 (male-male comparisons of *Pan troglodytes*)
- −1.60 (female-male comparisons of *Pan troglodytes*)

The mean log se_m value of −1.60 obtained for specimens attributed to *A. africanus* and *H. habilis* is also almost identical to the mean log se_m value of −1.61 obtained by Thackeray (2007) for modern conspecific specimens in the context of a statistical (probabilistic) definition of a species and a biological species constant (T = −1.61).

Results of pairwise comparisons of specimens in Table 1 include comparisons where specimen A is on the x axis and specimen B on the y axis, and *vice versa*. The mean range of log se_m values for these pairwise comparisons is small (0.03), consistent with conspecific pairwise comparisons of extant species (Dykes, pers. communication). An implication of these results, together with the mean log se_m value of −1.60 ±0.07 for 12 pairwise comparisons of the specimens listed in Table 1, is that these hominin crania attributed to either *A. africanus* or *H. habilis*, have a high probability of conspecificity. Despite variability in size, Sts 71, Sts 5, OH 24 and KNM-ER 1813, when taken together, appear to display a degree of variability that is comparable to what is displayed morphometrically in modern vertebrate species, including chimpanzees such as *Pan troglodytes*, or in *Pan paniscus*.

Wood and Gordon (2013) note that when comparisons are made been certain crania attributed to *P. paniscus* and *P. troglodytes*, a mean log se_m value of −1.6 can be obtained. This observation needs to be understood in the context of the following scenario: in their natural habitats the two extant chimpanzee species, and their ancestral populations, may have come in close proximity in areas when certain rivers (including the Congo) dropped in level during cold and dry seasons, during globally cold "glacial" episodes, corresponding to the establishment of episodic forest refugia in tropical Africa, in areas occupied simultaneously by the two extant chimpanzee species or their ancestral populations within a short time of their recent divergence (circa 2 million years ago). In captivity extant *Pan troglodytes* and *P. paniscus* will interbreed. The latter is sexually promiscuous. From genetic analyses Prüfer *et al.* (2012) demonstrate what they call "incomplete lineage sorting" in the case of chimpanzees. This situation may be due directly to episodic changes in the distribution of forest habitats suitable for chimpanzees of the kind that happened within the last 20,000 years (Hamilton, 1976), and which was repeated at intervals within the last two million years during the Pleistocene.

Within the last million years in tropical Africa, at least 10 episodes of global cooling were associated with aridification and refugiation. Such episodes would have affected the distribution of many primates, and their speciation, in west and central Africa (Colyn, 1991; Colyn *et al.*, 1991; Pilbrow and Groves, 2013). As in the case of closely related Plio-Pleistocene hominin species, boundaries between living primates may not be clearly distinct as a consequence of episodic palaeo-environmental contractions and expansions of habitats, which in turn influence gene pools and ultimately anatomy. These observations are relevant to the results of morphometric analyses of chimpanzee crania reported by Gordon and Wood (2013). Indeed, the "incomplete lineage sorting" identified genetically in *P. troglodytes* and *P. paniscus* (Prüfer *et al.*, 2012) is consistent with the observation that log se_m values for some of the pairwise comparisons of the two chimpanzees do not result in distinct separation.

At intervals of global cooling within the last two million years in Africa, grasslands of the kind associated with wildebeest (*Connochaetes taurinus*) would have expanded at the expense of woodland and forest. The modern distributions of *C. taurinus* in southern and eastern Africa are currently distinct, because miombo woodland acts as a barrier to the movement of wildebeest. However, such barriers would have become fragmented in the cooler or colder episodes of the Plio-Pleistocene, facilitating migrations of wildebeest and other grassland ungulates at intervals of 100,000 years or less. This scenario is relevant to the episodic expansion and contraction of habitats suitable for Plio-Pleistocene hominins such as *A. africanus* and *H. habilis* in eastern and southern Africa, and subsequent episodic gene flow.

4.6 CONCLUSIONS

The results obtained in this study support the view that certain specimens attributed to *H. habilis* and some specimens attributed to *A. africanus* are conspecific. This does not mean that all specimens referred to *H. habilis* should be transferred to *A. africanus*, a nomen which has priority over the former (*A. africanus* was described by Dart in 1925, whereas *H. habilis* was described by Leakey *et al.* in 1964). However, this study does support the view that there is not necessarily a clear boundary between *Australopithecus* and *Homo*.

An implication of the data presented here is that the range of variation in cranial capacity of *A. africanus* is probably much greater than previously thought. Further, the question arises as to whether *A. africanus* and *H. habilis* relate to a chronospecies, a concept which potentially can be assessed through the use of log se_m statistics and "palaeospectroscopy" (Thackeray and Odes, 2013).

A biogeographic and climatic scenario by Bromage and Schrenk (1995) has also suggested an origin of *H. habilis* from *A. africanus*, but an independent origin of *H. rudolfensis* in eastern Africa.

Nomenclature of hominin fossils is fraught with controversy, but a possible solution is through the use of morphometric approaches as one way to define a species in statistical (probabilistic) terms, using not only log se_m but also the range of log se_m values in pairwise comparisons (delta log se_m).

ACKNOWLEDGMENTS

This study is supported by the National Research Foundation and the Andrew W. Mellon Foundation. I am grateful to A. van Arsdale, F. Schrenk, V. Pilbrow, J. Braga, S. Dykes and L. Aiello for useful comments; to G. Viglietti for assistance with Figure 1; and to both J. Runge and J. Eisenberg for facilitating the publication of this volume in honour of Louis Scott who has made substantial and important contributions to an understanding of African palaeo-ecology.

REFERENCES

Bromage, T.G. and Schrenk, F., 1995, Biogeographic and climatic basis for a narrative of early hominid evolution. *Journal of Human Evolution,* **28**, pp.109–114.

Broom, R., Robinson, J.T. and Schepers, G.W.H., 1950, *Sterkfontein Ape-Man, Plesianthropus* (Pretoria: Transvaal Museum).

Clarke, R.J., 2013, *Australopithecus* from Sterkfontein Caves South Africa. *Paleobiology of Australopithecus*. In *The Paleobiology of Australopithecus*, edited by Reed, K., Fleagle, J. and Leakey, R. (Dorderecht: Springer), pp. 105–123.

Colyn, M., 1991, L'importance zoogeographique du bassin du fleuve Zaire pour la speciation: Le cas des primates simiens. *Annales Sciences Zoologique* **264**, Musée Royal de L'Afrique Centrale, Tervuren, Belgique.

Colyn, M., Gautier-Hion, A. and Verheyen, W., 1991, A re-appraisal of paleo-environmental history in central Africa: evidence for a major fluvial refuge in the Zaire basin. *Journal of Biogeography,* **18**, pp. 403–407.

Dart, R.A., 1925, *Australopithecus africanus*: The Man-Ape of South Africa, *Nature,* **115**, pp. 195–199, doi:10.1038/115195a0.

Dykes, S.J., 2014, *A morphometric analysis of hominin teeth attributed to different species of* Australopithecus, Paranthropus *and* Homo. MSc dissertation, University of the Witwatersrand.

Gordon, A.D. and Wood, B.A., 2013, Evaluating the use of pairwise dissimilarity metrics in paleoanthropology. *Journal of Human Evolution,* **65**, pp. 465–477.

Hamilton, A.C., 1976, The significance of patterns of distribution shown by forest plants and animals in tropical Africa for the reconstruction of Upper Pleistocene palaeoenvironments: a review. *Palaeoecology of Africa,* **9**, pp. 63–97.

Leakey, L.S.B., Tobias, P.V. and Napier, J.R., 1964, A new species of the genus Homo from Olduvai Gorge. *Nature,* **202**, pp. 7–9.

Pilbrow, V. and Groves, C., 2013, Evidence for divergence of bonobos (*Pan paniscus*) in the Lomami-Lualaba and Kasai-Sankuru regions based on preliminary analysis of craniodental variation. *International Journal of Primatology,* **34(6)**, pp. 1244–1260.

Prüfer, K., Munch, K., Hellmann, I., Akagi, K., Miller, J.R., Walenz, B., Koren, S., Sutton, G., Kodira, C., Winer, R., Knight, J.R., Mullikin, J.C., Meader, S.J., Ponting, C.P., Lunter, G., Higashino, S., Hobolth, A., Dutheil, J., Karakoç, E., Alkan, c., Sajjadian, S., Catacchio, C.R., Ventura, M., Marques-Bonet, T., Eichler, E.E., André, C., Atencia, R., Mugisha, L., Junhold, J., Patterson, N., Siebauer, M., Good, J.M., Fischer, A., Ptak, S.E., Lachmann, M., Symer, D.E., Mailund, T., Schierup, M.H., Andrés, A.M., Kelso, J. and Pääbo, S., 2012, The bonobo genome compared with the chimpanzee and human genomes. *Nature,* **486**, pp. 527–531.

Robinson, J.T., 1965, *Homo habilis* and the australopithecines. *Nature* **205**, pp. 121–124.

Thackeray, J.F., 1997, Probabilities of conspecificity. *Nature,* **390**, pp. 30–31.

Thackeray, J.F., 2007, Approximation of a biological species constant? *South African Journal of Science,* **103**, p. 489.

Thackeray, J.F. and Odes, E., 2013, Morphometric analysis of early Pleistocene African hominin crania in the context of a statistical (probabilistic) definition of a species. *Antiquity*, **87**, pp. 1–2. http://antiquity.ac.uk/projgall/thackeray335/

Thackeray, J.F., Bellamy, C.L., Bellars, D., Bronner, G., Bronner, L., Chimimba, C., Fourie, H., Kemp, A., Krüger, M., Plug, I., Prinsloo, S., Toms, R., Van Zyl, A.J. and Whiting, M.J., 1997, Probabilities of conspecificity: application of a morphometric technique to modern taxa and fossil specimens attributed to *Australopithecus* and *Homo*. *South African Journal of Science,* **93**, pp. 195–196.

Tobias, P.V., 1991, *The skulls, endocasts and teeth of* Homo habilis, Volume 4 of Olduvai, G., (Cambridge: Cambridge University Press).

Tobias, P.V., 1992, The species *Homo habilis*: example of a premature discovery. *Annales Zoologici Fennici,* **28**, pp. 371–380.

Vervaecke, H., Stevens, J. and Van Elsacker, L., 2004, Pan Continuity: Bonobo-Chimpanzee Hybrids. *Folia Primatologica,* **75**, pp. 42–60. DOI: 10.1159/000073431. Abstracts from the Spring Meeting of the Primate Society of Great Britain, held at University of St. Andrews, Scotland, April 10–11, 2003.

Wood, B.A. and Collard, M.C., 1999, The human genus. *Science,* **284**, pp. 65–71.

CHAPTER 5

The Fauresmith and Archaeological Systematics

Michael Chazan

Department of Anthropology, University of Toronto, Ontario, Canada
Evolutionary Studies Institute, University of the Witwatersrand, Johannesburg, South Africa

ABSTRACT: The Fauresmith Stone Age industry of southern Africa provides a good example of the complexity embedded in archaeological systematics. The goal of this article is to set out a conceptual basis for the Fauresmith as an archaeological construct and the technological attributes that characterize this industry. The Fauresmith is presented here as representing a stage in the development of stone tool technology in the interior of southern Africa that lies at the transition from the Earlier to Middle Stone Age. The Fauresmith is situated as an industry that developed at the point of transition between the biface technological lineage found in the Acheulean and the prepared core technologies that develop during the Middle Stone Age. A high degree of both intrasite and intersite variability is characteristic of this industry—a trait that makes clear definition difficult but that might be explained as an outcome of the dynamics of the evolution of stone tool technology.

5.1 INTRODUCTION

> "The vast accumulations of paleoliths amassed in private collections and public museums during seventy-five years need revision and re-classification. And in the process the very basis of classification must be reviewed and refined."
>
> (Childe, 1944:18–19)

Biological systematics play an essential role in all aspects of palaeontological and palaeoclimatic research (Ridley, 1986). A pollen diagram is unthinkable without an established taxonomy of plant species and communities; even isotopic analyses of zoological specimens requires an initial taxonomic identification of the specimens to genus and if possible to species. When working with extinct taxa there is a degree of ambiguity in the definition of species, a problem that is particularly pronounced in the context of the small sample sizes available for fossil hominins, yet there is clarity at least in the challenges involved. It is not surprising that scientists who work to incorporate the archaeological record into their studies of past environments or hominin evolution assume the same degree of clarity in the systematics of Palaeolithic stone tool industries. In some cases palaeoanthropologists have gone as far as to apply the methods of cladistics to archaeological constructs (Foley and Lahr, 1997). However, there is an essential distinction between palaeontology and archaeology that although obvious is often overlooked. Archaeological systematics while in some ways analogous to biological systematics are applied not to biological organisms—whether plants or animals—but to the products of human activity.

Archaeological systematics work on several levels (see discussion in Clarke, 1968). Typologies work within assemblages to describe individual artifacts and technological designations (i.e., core, flake, retouched piece etc.) can also be applied to classifying objects within an assemblage. At a higher level archaeologists use a range of scales of taxonomic designations to group assemblages into entities that are referred to as technocomplexes, cultures, industries, periods or phases. The scale of such an entity can vary both in spatial extent and temporal duration.

It has been previously argued in looking at the origins of Palaeolithic chronology that these groupings of assemblages are essential heuristic devices that enable us to think about the vast extent and duration of the archaeological record (Chazan, 1995). It can be assumed that many biologists would say the same about species, or at least fossil species. However, the classification of artifact assemblages is not based on a well-defined uniform accepted basis for the construction of archaeological entities. It seems that these entities are in fact heterogeneous and that the lack of a single unified basis for creating groupings is an appropriate response to the nature of the data set. The complexity of archaeological classification is somewhat obscured by the tendency for archaeological taxonomy to be highly conservative. Taxa once introduced are usually maintained, but the basis for this grouping can often shift significantly over time. It is very common to have the same name being used to refer to roughly the same group of assemblages while the meaning attached to this grouping has changed radically. A clear example of this aspect of archaeological systematics is the controversy known as the *Battaille Aurignacien* that erupted in the early 20th Century over the ordering of European Upper Palaeolithic chronology. The result of this controversy was a complete change in the significance and chronological position of the Aurignacian while the taxonomic category *Aurignacian* was maintained (Dubois and Bon, 2006).

The Fauresmith Stone Age industry of southern Africa provides a good example of the kinds of complexity embedded in the labels attached to groupings of archaeological assemblages. The Fauresmith designation has been employed for close to a century and as detailed by Underhill (2011) the defining features of this industry have shifted significantly over time and there is currently a lack of clarity in how this industry is to be defined. As discussed by Underhill (2011) the earliest explicit discussion of the Fauresmith is found in Goodwin and Van Riet Lowe's monograph (1929) on the Stone Age of South Africa. In this work the diagnostic elements of the Fauresmith are a bit unclear but the position of the Fauresmith within the progress of Palaeolithic industries linked to the movement of populations or races is evident.

> "It is difficult as yet to say whether it [the Fauresmith] is an evolved or specialized branch of the Stellenbosch, due partly or entirely to the presence of a useful material as Lowe is inclined to believe, or whether we are here presented for the first time with an infiltration of that racial or cultural impetus which, coming into South Africa, was to give us the industries grouped together as the Middle Stone Age"
>
> (Goodwin and Van Riet Lowe, 1929: 71).

Fieldwork carried out on sites in the Northern Cape Province over the last ten years it was found the designation of Fauresmith to be essential to make sense of the archaeological sites we are investigating (Figure 1). Here the objective is to set out a conceptual basis for the Fauresmith as an archaeological construct and the technological attributes that characterize this industry. The Fauresmith is presented here much in the way it was conceptualized by Goodwin and Van Riet Lowe (1929)—as representing a stage in the development of stone tool technology in the interior of southern Africa that lies at the transition from the Earlier to Middle Stone Age although without the linkage to biological populations of hominins. A high degree of both intrasite and intersite variability is characteristic of this industry—a trait that makes clear definition difficult but it will be suggested that it can provisionally be explained as an outcome of the dynamics of the evolution of stone tool technology.

Although chronological resolution remains limited there is strong reason to place the Fauresmith in the Middle Pleistocene with ages greater than 300 kyr and as early as 500 kyr. Refining the chronology of these sites is a very high priority. This discussion is based primarily on the sites of Wonderwerk Cave (Beaumont and Vogel, 2006, Chazan and Horwitz, 2015), Bestwood 1 (Chazan *et al.*, 2012), and Kathu Pan 1 (Porat *et al.*, 2010), and to a lesser degree Canteen Kopje (McNabb and Beaumont, 2011).

5.2 DIAGNOSTIC CRITERIA FOR THE FAURESMITH

One common element of archaeological practice is to identify 'diagnostic' criteria for identifying a particular industry. Diagnostic criteria are usually technologically distinctive, meaning that their presence implies a common underlying technological system of production. The durability of many archaeological taxa is largely derived from the ability of prehistorians to intuitively identify meaningful diagnostic criteria. In the case of the Fauresmith the ability of early prehistorians to intuitively grasp meaningful similarities vastly exceeded their abilities to put these intuitions into writing. As a result, as pointed out by Underwood (2011), early definitions of the Fauresmith are a mishmash of vague description. However, if one reexamines the type collections made by prehistorians there is a clear sense that they based their categories, and in this case specifically the Fauresmith, on real similarities that they grasped intuitively based on their knowledge of stone tool knapping and of Palaeolithic technology. The development of diagnostic criteria for an archaeological industry is completely inductive, there is no logically necessary set of criteria rather the goal is to draw from experience with archaeological collections those aspects that seem to be diagnostic. These are in a sense a best effort to put into words an intuitive sense based on existing familiarity with collections. New data and new insight are expected to lead to revision and improvement of these criteria.

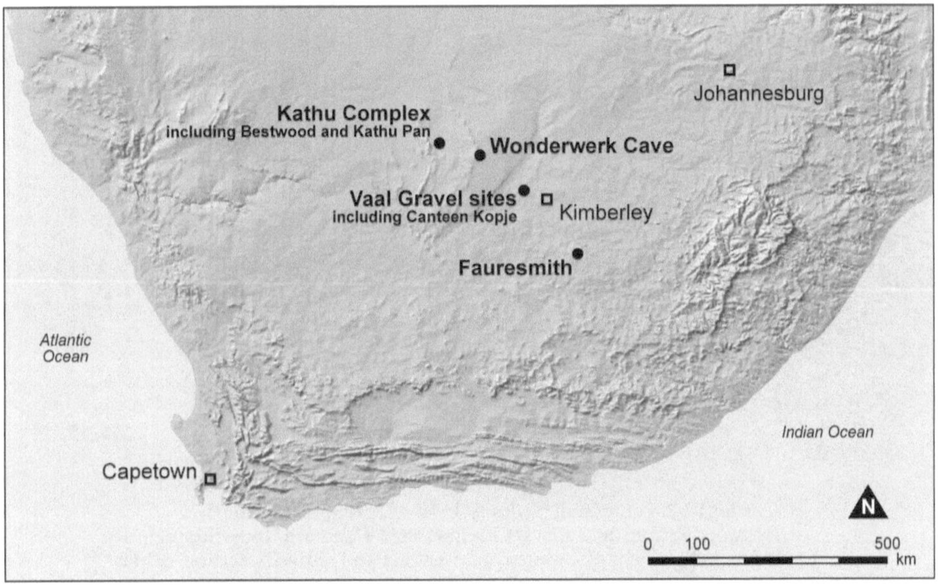

Figure 1. Map showing the location of Fauresmith sites discussed in the text.

Figure 2. Fauresmith blades: a–b: Wonderwerk Excavation 6;
c–d: Bestwood 1, from 2013 Block 1 Excavation; e–f Canteen Kopje Stratum 2a,
Beaumont and McNabb Excavation; g–h: Kathu Pan 1, 2014 Excavation, St. 4a.

Figure 3. Fauresmith cores. a: Wonderwerk Cave, Excavation 6. Globular core, almost polyhedral. Note the series of small blade removals in the right hand view. b: Flat core, cortical on one face with a series of sub-parallel unidirectional removals off the other face.

Figure 4. Fauresmith cores. a: Bestwood 1, from 2013 Block 1 Excavation. Flat core with a series of sub-parallel unidirectional removals of one face, steeply keeled at distal end. Lateral edges acute to flakeing surface. b: Bestwood 1, from 2013 Block 1 Excavation. Complex core worked on three faces. Note blade removal in the left view at the final stage of exploitation off an surface that is extraordinarily steeply convex both laterally and distally. c: Kathu Pan 1, 2014 Excavation, St. 4a. Flat point core with partial cortex on one face. Note absence of preparation on flaking surface. For further images of Kathu Pan 1 cores see Wilkins and Chazan 2012.

The following list of diagnostic criteria for identifying an archaeological industry as belonging to the Fauresmith is proposed:

> Perhaps the most striking aspect of the Fauresmith is the presence of large blades (flakes twice as long as they are wide) roughly rectangular in shape with sub-parallel or convergent flake scars. These blades are occasionally extremely long (>10 cm), however shorter blades are also present (Figure 2).
> Blades are part of a continuum of flake production. The only published metric data is from Kathu Pan 1 where blades are at the end of a normal distribution of L/W ratios (Porat *et al.*, 2010).
> Blades and flakes are produced on a variety of core types that exploit extreme lateral convexity. In some sites such as Wonderwerk Excavation 6 and Bestwood 1 these are globular cores that can resemble polyhedrons while at Kathu Pan 1 only a single surface is exploited resulting in cores that resemble Levallois cores (Figures 3, 4).
> Exploitation is of Type D as defined by Boëda (2013) where flake/blade removals also serve to prepare the surface for the next removal. Kathu Pan 1 does come close to a Levallois system in which there is a distinct stage of preparation but the extreme lateral convexity of the cores results in more continuous production than is found in the Levallois (see the discussion in Wilkins and Chazan, 2012). There is thus lacking from Fauresmith assemblage a high degree of normalization in flake/blade products.
> Tool types include scrapers with well retouched edges. In some assemblages there are also retouched points. Burins, microliths, and endscrapers are absent. There is a core tool component that is highly variable both in form and in frequency. Bifacial tools tend to be poorly made and asymmetrical, but there are exceptions. Bifaces include very small examples but mean biface size is not a useful diagnostic criterion. Choppers and polyhedrons can also be a component of these assemblages.

5.3 BEYOND DIAGNOSTIC: TOWARDS A THEORETICAL GROUNDED DEFINITION

As illustrated in the quote from Goodwin and Van Riet Lowe (1929), a fairly consistent thread in discussions of the Fauresmith is the view that this grouping of assemblages represents a stage in the progress of Palaeolithic stone tool industries. The idea of progress is central to the development of Palaeolithic archaeology as a discipline drawing on a neo-Lamarckian framework that viewed evolution as a whole, and human evolution in particular, as essentially progressive (Bowler, 1986). A weakness of this approach is that developments in stone tool industries are often viewed as a direct extension of the evolution of hominin lineage, which itself is seen as progressive. This tendency is evident in Goodwin and Van Riet Lowe (1929) and, stripped of the emphasis on race, remains current in palaeoanthropology (i.e., Foley and Lahr, 1997). Beginning in the 1960's a sharp reaction to the view of the Palaeolithic as a record of progress and the assimilation of stone tool industries into the biological systematics of the hominin lineage—such that similar stone tools were used as evidence of a single population—led to an emphasis on other potential sources of variability in stone tool industries such as mobility, site function, and raw material (i.e., Kleindienst, 1961). In the case of the Fauresmith, Humphreys (1970) advanced the argument that this 'industry' was in fact simply a variant of the Acheulean that resulted from the characteristics of available raw material.

Throughout the debates concerning the validity of archaeological taxonomy the dynamics inherent in the development of technological systems were often ignored. Changes in lithic industries have been explained largely on the basis of external factors such as change in the cognitive capacity of hominins or shifts in climate and/or adaptation. Recently Boëda (2013) and his students have challenged this reliance on explaining change in stone tool industries on the basis of external factors and argued in favor of integrating the internal logic of the development of technological systems into our study of the Palaeolithic. Boëda (2013) draws on the concepts developed by the sociologist Gilbert Simondon (1958). For Simondon (1958) there is an inherent progressive trend in technological systems towards integration, or in his terms concretization. As Boëda (2013) writes:

> "By abstract we mean technical objects that can be disassembled into several components that are functionally independent from one another. This could involve the simple juxtaposition of elementary functions. This is different from the concrete technical object that is characterized by a series of components that must interrelate in order to be operational. In other words, a concrete object is an object in which none of the parts can be separated one from the other without losing its sense. The concrete object is the result of an evolution that, through a sort of internal convergence of adaptation to itself, culminates in the creation of a synergy between the various components. This synergy results in an increase in structural and functional complexity."
>
> (Boëda, 2013: 40, translation by Chazan).

Boëda applies Simondon's concept—which was developed in the context of the study of modern technology—and uses it to develop a division of the Palaeolithic into technological lineages and it is within the lineages that one sees progressive development towards concretization. It is important to emphasize that Boëda (2013) does not see this process taking place in a vacuum and that he does recognize that external factors will shape the specific manifestation of the lineage. But the value of his ideas is that they challenge us to consider the internal technological dynamics underlying the variability we are able to observe (there are overlaps between this approach and Andre Leroi Gourhan's concept of *tendance*, see Stiegler, 1998 for a discussion).

Using the idea of lineages Chevrier (2012) has studied the bifacial industries of East Africa and the Middle East and found that there is indeed a trend towards increased integration. If the idea of technological lineages as a useful framework for studying Palaeolithic stone tool industry is accepted an interesting question that emerges is what takes place at the transition from one lineage to another. Specifically, by the end of the Acheulean the concretization of the handaxe is complete. The Fauresmith appears just at this moment of transition and can be conceptualized as representing this process of transition between lineages. One question left open in Boëda's theoretical work is the role of new tool functions in driving the process of the development of methods of stone tool production. One critical question is whether new tool functions—such as the use of stone tipped weapons or scrapers for hide working—precedes the development of methods of manufacture that are well adapted for these functions.

With these ideas as a background it is proposed to situate the Fauresmith as an industry that develops at the point of transition between the biface technological lineage found in the Acheulean and the prepared core technologies that develop during the Middle Stone Age. The critical point is that the Fauresmith is defined within the context of the technological development of stone tool industries in southern Africa. The Fauresmith is characterized by a high degree of technological diversity as solutions for tool production were developed to the needs of new tool functions including hide scraping and stone tipped weapons. As such it is expected that a strict list of diagnostic criteria will not apply uniformly across sites. A common element of these new functions is the reliance on hafting. There is a shift from the biface lineage where the prehensile element (the part of the tool that is held in the hand) and the transformative element (the part of the tool that comes in contact with the worked material) are integrated into a single piece. The development of hafts separates the prehensile and transformative elements into two distinct technical components. The creation of composite tools allows for increased flexibility in the component of the tool that transmits energy, allowing for greater force to be exerted at transformative element through enlarged grips or the addition of a spear shaft. During this phase of transition there might be a degree of disequilibrium between the methods of production and the demands of tool function.

This definition of the Fauresmith proposed here is meant not as a definitive statement of fact but rather, as discussed above, as a heuristic devise to help guide research. As such this definition leads to a number of testable corollaries.

1. The Fauresmith is expected to demonstrate a high degree of technological variability that cannot be explained on the basis of raw material.
2. The Fauresmith is expected to demonstrate evidence of use of hafted tools including scrapers and stone tipped weapons. Scrapers with carefully regularized edges are included prominently in Goodwin and Van Riet Lowes (1929: 77, 79) description of the Fauresmith. Recently Wilkins *et al.* (2012) have developed multiple lines of evidence suggesting that pointed flakes at Kathu Pan 1 were used as spear tips.
3. There is expected to be variation in the correlation between tool function and tool form/method of manufacture. Given the instability of this period of technological transition we do not expect a uniform correlation between tool form and function across all Fauresmith assemblages.
4. The Fauresmith should be chronologically intermediary between the Acheulean and the Middle Stone Age, although overlap on both ends are likely. It can be set at this point based on available data set an approximate time range for the Fauresmith between 500–300 kyr (for a discussion of the limitations of currently available age determination for the Fauresmith see Herries, 2011). Specific temporal data are the ESR/U-Series and OSL ages for Kathu Pan 1 at ca. 500 kyr; a minimum OSL age for the Fauresmith at Canteen Kopje at ca. 300 kyr; a minimum U/Th age for Wonderwerk Excavation 6 of 187 kyr, an OSL age for the MSA of Kathu Pan 1 at ca. 300 kyr; and U/Th age for the MSA at Wonderwerk Cave, Excavation that range from 97–220 kyr. (Porat *et al.*, 2010, Chazan *et al.*, 2013, Beaumont and Vogel, 2006). All of these ages are currently the subject of ongoing research but they are adequate for setting the framework of 500–300 kyr as the basis for future testing. The Fauresmith is found stratigraphically overlying the Achelean at Canteen Kopje (and based on early excavations at other Vaal Gravel localities) and at Kathu Pan 1 (Beaumont and Vogel, 2006, McNabb and Beaumont, 2011, Porat *et al.*, 2010). At Wonderwerk Cave the Fauresmith is limited to the back of the cave (Excavation 6) and absent from the Acheulean sequence in Excavation 1. Thus, there is no stratigraphic relationship between these industries at Wonderwerk, however, the most likely scenario is that Excavation 6 postdates the Acheulean sequence of Excavation 1. A small number of possible Fauresmith elements are present in the upper strata of the Acheulean sequence of Excavation 1 (Chazan, in press, specifically a small number of well retouched scrapers and very small bifaces).

5.4 DISCUSSION

The diagnostic criteria and theoretically based definition of the Fauresmith presented here conform fairly well to recent efforts to create a consensus chronology for the Palaeolithic of South Africa (Lombard *et al.*, 2012). In their review article Lombard *et al.* (2012) advocate the use of a hierarchical structure for archaeological nomenclature based on technocomplexes as the highest level of taxonomy and industries, phases, and assemblages as subsequent levels of classification. Lombard *et al.* (2012) include the Fauresmith as an informal designation for the ESA-MSA transition. This scheme is useful but implies a uniformity among entities at the same hierarchical level in the taxonomy. Not surprisingly the review of Stone Age systematics published by Lombard *et al.* (2012) was rapidly followed by an effort to correlate these groupings of assemblages with hominin species (Dusseldorp *et al.*, 2013). While such review and efforts at correlation are useful, and even essential, they should be based on an awareness that technocomplexes, industries etc. are not defined on the basis of a single criterion based on a theoretical concept analogous to the foundation of the species concept on genetics and evolution. Specifically in the case under discussion there is no reason to think that the Fauresmith is an entity of the same order as the Howiesons Poort or the Oakhurst, each of which is categorized as a technocomplex. While a clear structure of nomenclature is extremely useful it should not cloud the complexity of archaeological systematics for the Palaeolithic. These are heuristic devices that help to think about the past and should not be confused with real entities that are the archaeological parallel of species.

In the systematics developed by Lombard *et al.* (2012) there is an allowance for temporally ordered variation that can be recognized as phases. Beaumont and Vogel (2006), building on the work of earlier prehistorians working in the Vaal Gravels, argue for a series of phases within the Fauresmith. The definition of the Fauresmith presented above stresses the significance of diversity as a fundamental aspect of this phase in the evolution of stone tool technology. However, Beaumont and Vogel (2006) offer a compelling counter position according to which the high degree of apparent variability within the Fauresmith actually tracks the change in stone tool through time. Resolving this issue rests on increasing the degree of chronological control over assemblages within the Fauresmith. This work promises to shed light on the dynamics of stone tool industries during periods of major transition. Ultimately understanding the structure of the evolution of hominin technology rests not only on theoretical clarity but also on sound empirical data.

REFERENCES

Beaumont, P.B. and Vogel, J.C., 2006, On a timescale for the past million years of human history in central South Africa. *South African Journal of Science,* **102(5, 6)**, pp. 217.

Boëda, E., 2013, *Technologique & Technologie. Une Paléo-histoire des objets lithiques trenchant*. Archéo-Edtions.

Bowler, P.J., 1986, *Theories of Human Evolution. A Century of Debate, 1844–1944.* Baltimore: Johns Hopkins University Press.

Chazan, M., 1995, Conceptions of time and the development of Paleolithic chronology. *American Anthropologist,* **97(3)**, pp. 457–467.

Chazan, M. and Horwitz, L.K., 2015, *An Overview of Recent Research at Wonderwerk Cave, South Africa*. Proceedings of the 2010 joint meeting of the Panafrican Archaeological Congress and the Society of Africanist Archaeologists, Dakar, Senegal. November 2–4, 2010.

Chazan, M., Wilkins, J., Morris, D. and Berna, F., 2012, Bestwood 1: a newly discovered Earlier Stone Age living surface near Kathu, Northern Cape Province, South Africa. *Antiquity,* **86(331)**: Antiquity Gallery.

Chazan, M., Porat, N., Sumner, T.A. and Horwitz, L.K., 2013, The use of OSL dating in unstructured sands: The archaeology and chronology of the Hutton Sands at Canteen Kopje (Northern Cape Province, South Africa). *Archaeological and Anthropological Sciences,* **5(4)**, pp. 351–363.

Chazan, M., in press, Technological Trends in the Acheulean of Wonderwerk Cave, South Africa. *African Archaeological Review.*

Childe, V.G., 1944, 7. The Future of Archaeology. Man, pp. 18–19.

Chevrier, B., 2012, Les assemblages a pieces bifaciales au Pleistocène inferieur et moyen ancien en Afrique de l'Est et au Proche-Orient. These de Doctorat, University Paris Nanterre.

Clarke, D.L., 1968, *Analytical archaeology.* London: Methuen.

Dubois, S. and Bon, F., 2006, Henri Breuil et les origines de la "bataille aurignacienne". Sur des chemins de la Préhistoire. L'abbé Breuil du Périgord à l'Afrique du Sud, pp. 135–147.

Dusseldorp, G., Lombard, M. and Wurz, S., 2013, Pleistocene *Homo* and the updated Stone Age sequence of South Africa. *South African Journal of Science,* **109(5–6)**, pp. 1–7.

Foley, R. and Lahr, M.M., 1997, Mode 3 technologies and the evolution of modern humans. *Cambridge Archaeological Journal,* **7(1)**, pp. 3–36.

Goodwin, A.J.H. and Lowe, C. van Riet, 1929, The Stone Age cultures of South Africa. *Annals of the South African Museum,* **27**, pp. 1–289.

Herries, A.I.R., 2011, A chronological perspective on the Acheulian and its transition to the Middle Stone Age in southern Africa: the question of the Fauresmith. *International Journal of Evolutionary Biology,* **2011**, ID 961401, p. 25.

Humphreys, A.J.B., 1970, The role of raw material and the concept of the Fauresmith. *The South African Archaeological Bulletin,* **25(99/100)**, pp. 139–144.

Kleindienst, M.R., 1961, Variability within the Late Acheulian assemblage in eastern Africa. *The South African Archaeological Bulletin,* **16(62)**, pp. 35–52.

Lombard, M., Wadley, L., Deacon, J., Wurz, S., Parsons, I., Mohapi, M., Swart, J. and Mitchell, P., 2012, South African and Lesotho Stone Age sequence updated. *The South African Archaeological Bulletin,* **67(195)**, pp. 123–144.

McNabb, J. and Beaumont, P., 2011, A Report on the Archaeological Assemblages from Excavations by Peter Beaumont at Canteen Koppie, Northern Cape, South Africa. Archaeopress.

Porat, N., Chazan, M., Grün, R., Aubert, M., Eisenmann, V. and Horwitz, L.K., 2010, New radiometric ages for the Fauresmith industry from Kathu Pan, southern Africa: Implications for the Earlier to Middle Stone Age transition. *The Journal of Archaeological Science,* **37(2)**, pp. 269–283.

Ridley, M., 1986, *Evolution and classification: the reformation of cladism.* (Longman).

Simondon, G., 1958, Du Mode d'Existence des Objets Techniques (Vol. 1). Aubier-Montaigne.

Stiegler, B., 1998, *Technics and Time: the Fault of Epimetheus (Vol. 1).* (Stanford University Press).

Underhill, D., 2011, The study of the Fauresmith: a review. *The South African Archaeological Bulletin,* **66(193)**, pp. 15–26.

Wilkins, J. and Chazan, M., 2012, Blade production ~500 thousand years ago at Kathu Pan 1, South Africa: support for a multiple origins hypothesis for early Middle Pleistocene blade technologies. *Journal of Archaeological Science,* **39(6)**, pp. 1883–1900.

Wilkins, J., Schoville, B.J., Brown, K.S. and Chazan, M., 2012, Evidence for early hafted hunting technology. *Science,* **338(6109)**, pp. 942–946.

CHAPTER 6

A reconstruction of the skull of *Megalotragus priscus* (Broom, 1909), based on a find from Erfkroon, Modder River, South Africa, with notes on the chronology and biogeography of the Species

James S. Brink
Florisbad Quaternary Research Department, National Museum, Bloemfontein, South Africa
Centre for Environmental Management, University of the Free State. Bloemfontein, South Africa

C. Britt Bousman
Department of Anthropology, Texas State University, San Marcos, Texas, USA
GAES, University of the Witwatersrand, Johannesburg, South Africa

Rainer Grün
Research School of Earth Sciences, The Australian National University, Canberra, Australia

ABSTRACT: A reconstruction of the skull of the giant alcelaphine bovid, *Megalotragus priscus*, is provided based on a brain case and horn cores discovered and excavated at the late Florisian locality of Erfkroon on the Modder River, central Free State Province, South Africa. The sedimentary context of the *M. priscus* specimen can be correlated with fluvial deposits dated previously by luminescence to the Last Interglacial. Electron Spin Resonance (ESR) analyses of dental specimens from various localities at Erfkroon indicate a terminal Middle Pleistocene and Late Pleistocene age for these deposits. The skull reconstruction of *M. priscus* is aided by an upper jaw and mandible from the Late Pleistocene locality of Mahemspan. The *M. priscus* materials from Erfkroon, Mahemspan and other localities allow a re-evaluation of the morphological affinities of the species and it appears to be closer to wildebeest-like alcelaphines (genus *Connochaetes*) than to hartebeest-like alcelaphines (genera *Alcelaphus* and *Damaliscus*). Variability in the fossil horn cores suggests sexual dimorphism and some degree of territorial behaviour. It also suggests geographic variability in the populations of *M. priscus* in central southern Africa during the later part of the Middle Pleistocene and Late Pleistocene, before its extinction at the end of the Late Pleistocene and early Holocene.

6.1 INTRODUCTION

The terraces of the Vaal River are well-known for producing abundant vertebrate fossils that span the Plio-Pleistocene and younger geological ages (Helgren, 1977). The tributaries of the Vaal River, such as the Modder and Riet Rivers, are equally rich in Quaternary fossil vertebrates and archaeological materials. The Modder and the Riet Rivers have lower-gradient longitudinal profiles than the Vaal River (Tooth *et al.*, 2004), flowing over less-resistant bedrock and forming floodplain deposits controlled by igneous barriers, mostly dolerite. These floodplain deposits become eroded as the rivers incise, cutting through dolerite barriers, and in the process the fossil contents of the floodplain deposits are exposed in erosional areas, locally known as dongas. Such erosional areas can be extremely rich sources of vertebrate fossils and archaeological materials. On two adjacent farms on the Modder River, "Erfkroon" and Orangia" (Figure 1; Churchill *et al.*, 2000; Lyons *et al.*, 2014), extensive fossil exposures were discovered in June 1996 during routine reconnaissance by the Florisbad Quaternary Research Department of the National Museum, Bloemfontein. One of the first major discoveries on the farm "Erfkroon" was a skull and horn cores of a giant alcelaphine antelope, *Megalotragus priscus* (Broom, 1909). This find is unusually complete with a well-preserved braincase attached to the horn cores, which provides the opportunity of improved understanding of the skull morphology and affinities of the species. The aim of this contribution is to provide a description of this find, its sedimentary context and age estimates based on Electron Spin Resonance analyses (ESR). A reconstruction of the skull of *M. priscus* is proposed, which is based on the Erfkroon specimen and assisted by a complete mandible and a maxillary fragment from a pan site in the central Free State Province, known as "Mahemspan" (Figure 1). The fossil materials from Mahemspan were discovered and excavated in the late 1930's and 1940's by staff of the National Museum (Hoffman, 1953), and in this contribution we also provide a summary taxonomic list and ESR age estimates for this fossil occurrence.

Although the general awareness of the importance of the tributaries of the Vaal River as sources of Quaternary fossils became somewhat diminished in the second half of the 20th century (Cooke, 1964; Klein, 1984), the early development of the field of Quaternary Palaeontology in South Africa was closely linked to the Modder River. During the 19th and early 20th centuries the Modder River deposits produced the first recorded discoveries of Pleistocene fossils in South Africa. In 1839 a giant long-horned buffalo (*Syncerus antiquus*) was found in the banks of the Modder River and reported to the Geological Society of London (Seeley, 1891). A few years later Broom described the second major find of an extinct Pleistocene mammal from the Modder River sequence—a giant alcelaphine, which he named *Bubalis priscus* (Broom, 1909), now referred to the genus *Megalotragus* Van Hoepen (Gentry and Gentry 1978). The stretch of the Modder River that produced these fossils is situated between the present-day Kruger's Drift Dam and the confluence of the Modder and the Riet Rivers (Figure 1), which is in the same general area where the Erfkroon fossil sites are situated. Thus, it is noteworthy that both the Erfkroon find reported here and the type specimen of *M. priscus* are from essentially the same stretch of fossil deposits of the Modder River and they are probably of comparable geological age. Broom's type specimen from the Modder River lacks much of the distal parts of the horn cores and the braincase is not complete. The Erfkroon specimen complements the type specimen and extends our understanding of the species.

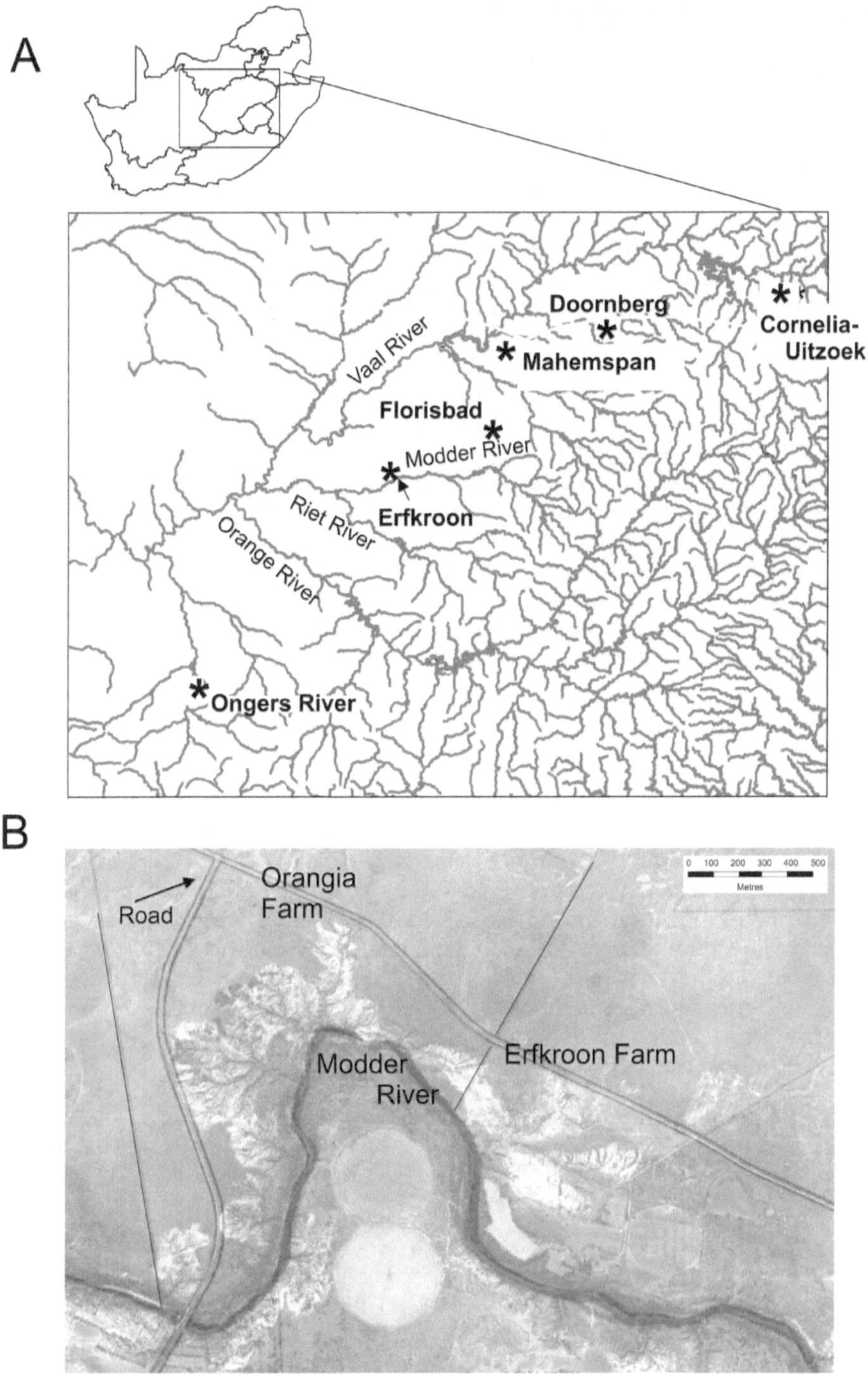

Figure 1. A map illustrating the geographic position of the localities with *Megalotragus*, as referred to in the text (A) and a plan view the fossil-bearing exposures on the farms Orangia and Erfkroon (B).

6.2 MATERIALS AND METHODS

6.2.1 Mahemspan

In the late 1930's and early 1940's E.C. Van Hoepen and A.C. Hoffman from the National Museum, Bloemfontein, excavated a large faunal sample from the lunette of Mahemspan (27°45′50″ S; 26°08′50″ E; Figure 1), situated between Hoopstad and Wesselsbron, Free State Province (Unpublished reports in library of the National Museum). In 1994 one of us (JSB) visited the site and the approximate position of the excavation was relocated through the help of Mrs. De Villiers, the owner of the farm. She was present when Van Hoepen and Hoffman conducted their excavations. Fossil specimens in the same state of preservation as in the collections at Florisbad Quaternary Research Station were found on the surface. Unfortunately a trial excavation did not reveal any *in situ* materials. The site is now covered by aeolian sand, as was the case before the material was exposed by wind action during the intense droughts of the 1930's. The site is now part of a ploughed land.

The *in situ* fossil materials collected by Van Hoepen and Hoffman were found at the base of the lunette and the original death assemblage appears to have been deposited in a marsh-like area in the presence of calcium carbonate-rich groundwater. This is evident from the extensive presence of calcrete deposits in and on the fossils. The fossil matrix, which is still attached to many specimens in the old collection due to cementation by calcium carbonate, is a pale-brown, partly calcretised sand. This was used to establish the background radiation for the ESR measurements.

6.2.2 Erfkroon

After the discovery of fossil-bearing deposits on the farms Erfkroon and Orangia (Figure 1), collectively referred to as "Erfkroon", luminescence dating and sediment analysis (Churchill *et al.*, 2000; Tooth *et al.*, 2013; Lyons *et al.*, 2014; Table 1) suggest that the bulk of the deposits on the farm Orangia represents a Late Pleistocene fluvial terrace. We have named it the "Orangia" Terrace (Figures 2 and 3), and it has both channel and overbank facies. The channel facies sit unconformably on Ecca Bedrock and make up the lower two horizons of the terrace, the Lower Gravel and the Green Sand. The Lower Gravel is composed of small-to-large subrounded to rounded shale and calcium carbonate gravels supported by a dark olive-brown clayey-sand matrix.

Table 1. A comparative list of OSL and IRSL dates (ka) from the Orangia Terrace, as given in previous studies.

	OSL (Lyons *et al.*, 2014)	OSL (Tooth *et al.*, 2013)	IRSL (Tooth *et al.*, 2013)	IRSL (Churchill *et al.*, 2000)
Brown	0.83 ± 0.09 – 6.32 ± 0.67			
Upper Grey	11.3 ± 0.98 – 19.5 ± 1.04			25 ± 1.2
Red	20.0 ± 1.19 – 32.2 ± 1.74	32 ± 2	29 ± 2	
Lower Grey	40.1 ± 2.30 – 43.5 ± 3.18	42 ± 2	42 ± 3	
Green Sand			118 ± 35	113 ± 6
Lower Gravel				163 ± 7

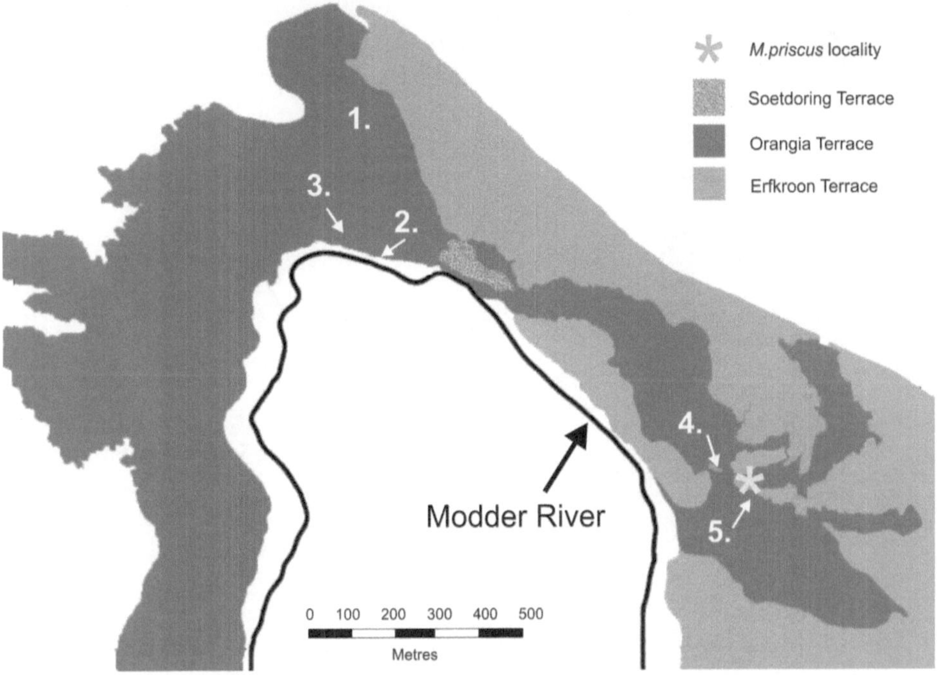

Figure 2. A plan view of the Erfkroon sedimentary deposits, showing the approximate lateral extent of the Erfkroon and Orangia Terraces, the ESR sampling localities and the find locality of the *Megalotragus priscus* specimen. The numbers of the ESR localities correspond to the numbers in Table 4 and in Figure 3.

It is truncated by erosion which forms an unconformity on its upper surface. The Lower Gravel is classified as a C soil horizon. Above this is the Green Sand, which is composed of small Ecca shale pebbles supported in a silty sand to very fine sand matrix. Iron staining is common and highlights tilted lamella and thin beds. Both of these layers reflect fluvial channel deposition. The top of the Green Sand also is truncated by erosion to form another unconformity. This is also a C soil horizon.

Above this unconformity are the overbank facies consisting of the Lower Grey, Red, Upper Grey, and Brown palaeosols. The Lower Grey sediments are yellowish brown to strong brown alternating sand and clay laminae grading up to a clay loam. Calcium carbonate nodules increase upwards in the profile in frequency and size. The Lower Grey grades into the lower part of the Red, which shifts to a yellowish red very firm sandy loam with coarse moderate subangular blocky structure and with no calcium carbonate nodules. It is bound by unconformities on the top and bottom.

The upper part of the Red is a yellowish red sandy loam with coarse, weak subangular blocky structure. It grades into the Upper Grey from a strong brown to light yellowish brown firm loams to sandy loams. Calcium carbonate nodules are absent in the bottom of the Upper Grey, slowly increase in density and size upwards in the profile and then decline at the top of the horizon. The Brown palaeosol overlies the Upper Grey and grades from a strong brown to dark brown loam with declining amounts of calcium carbonate nodules upwards in the profile.

The Orangia Terrace is covered by a thin (~25 cm) aeolian deposit, which corresponds with the "Sandy Cap" of Lyons *et al.* (2014) and we have named it the "Soetdoring Terrace". The Soetdoring Terrace has not been described in detail, but preliminary

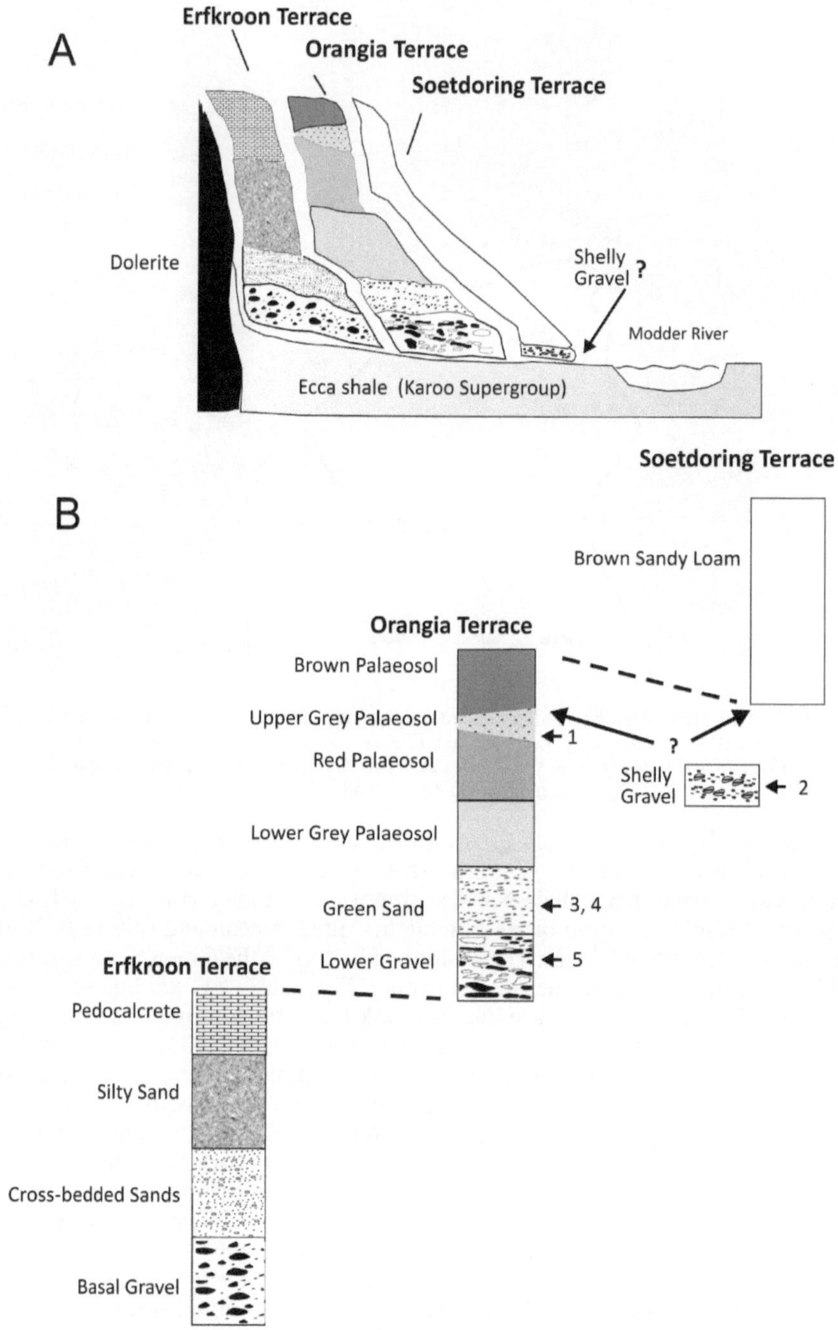

Figure 3. A diagram illustrating our current interpretation of the terrace deposits of the Modder River, as seen on the farms "Erfkroon" and Orangia" (A). The Erfkroon, Orangia and Soetdoring terraces reflect sequential and possibly, marginally overlapping periods of deposition (B). We are at present uncertain of the correct stratigraphic position of the Shelly Gravel as either a lateral facies of the Upper Grey of the Orangia Terrace or the basal horizon of the Soetdoring Terrace. The numbers and arrows indicate ESR sampling points, as illustrated in Figure 2.

observations indicate that it is a light brown sandy loam with at least two buried soils that consist of dark brown to strong brown loams. This deposit becomes deeper towards the present-day channel of the Modder River, where it appears to be underlain by a shelly, fossil-rich gravel deposit, referred to as "Sandy gravel with bivalves" (Lyons *et al.*, 2014). The bivalves in this deposit occur articulated and, therefore, they are untransported and not redeposited and must post-date the depositional event that produced the gravel horizon. For the present purpose we refer to this horizon informally as the "Shelly Gravel" (Figure 3). We are uncertain whether it forms the basal horizon of the Soetdoring Terrace or whether it is a lateral facies of the Upper Grey of the Orangia Terrace.

Upstream from Orangia, on the farm of Erfkroon, the Late Pleistocene overbank component of the sedimentary package is less well developed, but numerous fluvial gravel outcrops proved to be highly fossiliferous. Some of these outcrops are thought to be older than the Orangia Terrace deposits and these are collectively referred to as the "Erfkroon Terrace". These can easily be distinguished by the lack of carbonate clasts so common in the Lower Gravel of the Orangia Terrace. The Erfkroon Terrace deposits include four lithostratigraphic units. From bottom to top these are the Basal Gravel, Cross-Bedded Sand, Silty Sand and Pedocalcrete (Figure 3). These deposits have not been described in detail, but it is important to note that the Basal Gravel completely lacks carbonate clasts, which are a significant component of the Orangia Terrace Lower Gravel. We believe that the overlying Pedocalcrete horizon of the Erfkroon Terrace is the source of the carbonate clasts in the Lower Gravel of the Orangia Terrace.

The *M. priscus* specimen was found embedded in the Lower Gravel at the contact with the overlying Green Sand (Figure 4). Its original orientation of deposition was

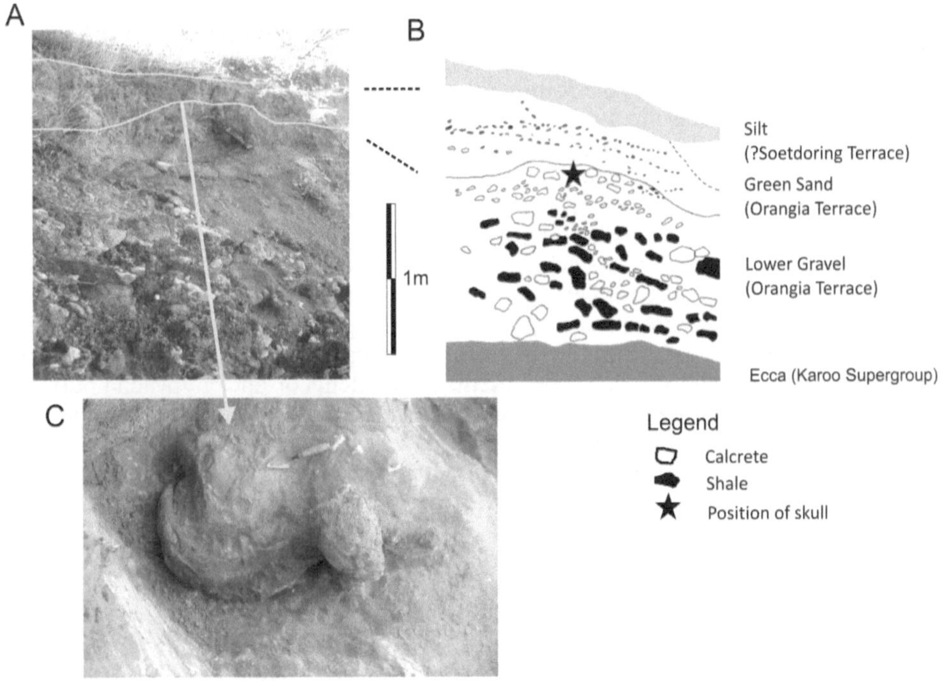

Figure 4. The exposed sedimentary profile of the outcrop, showing the *in situ* position of the skull of *M. priscus* (A), a diagram of the profile, illustrating the *in situ* position of the skull at the contact between the Lower Gravel and the Green Sand (B) and the skull being exposed during excavation (C).

with the horns down and the snout facing upwards. However, the upper dentitions and premaxillae were not preserved and their position would have coincided with the contact between the Lower Gravel and the overlying Green Sand, suggesting that these parts of the skull were eroded away some time after deposition or during the deposition of the Green Sands. The left horn core was broken off and was recovered first. The rest of the specimen was *in situ* and was excavated by first cutting away some of the deposit to create a step. The step was used as a platform from which to work and to allow excavation in plan view. In this way the whole specimen was exposed and consolidated. Wooden struts were used to maintain the connection between the horn cores and the braincase. Later, in the fossil laboratory at the Florisbad Quaternary Research Station the preparation was completed and the left horn core was glued back.

In 2002 and in 2010 we took ESR samples and measurements at various localities in the Orangia Terrace (Figures 2 and 3). In order to establish the gamma dose rate for the various localities, readings were taken with a portable gamma detector. Excavated dental samples of known provenance were used for ESR testing, with the exception of the dental specimens from the Upper Grey palaeosol, which were collected from the surface. From recent test excavations it is clear that these specimens originally derive from the Upper Grey palaeosol.

6.3 RESULTS

6.3.1 Mahemspan

Comparative taxonomic lists are given in Table 2 for the Last Interglacial fossil assemblages from Erfkroon, the fossil assemblage of Mahemspan and for the Florisbad Spring assemblage. The Mahemspan fossil materials are unusually complete. The giant alcelaphine, *M. priscus*, is the predominant element in the collection and is represented by virtually all skeletal elements. The Mahemspan collection of *M. priscus* represents probably the largest and most complete sample of this species in southern Africa. The completeness of the material and the predominance of large-bodied taxa in the assemblage raise the question of selective recovery of the material. However, from the care that was taken in the preparation of the fossils before excavation and the meticulous accessioning of the material it is evident that the excavators took considerable trouble to recover the material as completely as possible and it appears unlikely that recovery was selective. The presence of sun cracks and porcupine gnawing on specimens, the absence of hyaena coprolites and the paucity of carnivores in the fossil assemblage argue against the likelihood that it represents the contents of ancient hyaena burrows. This and the fact the bones were originally deposited in marshy conditions, may point to the Mahemspan assemblage representing natural deaths or carnivore kills on the edge of the pan. This is analogous to the taphonomic reconstruction of the Florisbad Spring assemblage, which is considered to represent the remains of carnivore hunting and scavenging around the ancient spring pools (Brink, 1987, 1988; Grün *et al.*, 1996).

Various authors made reference to the Mahemspan material and formed opinions on its geological age (Cooke, 1974; Gentry and Gentry, 1978). Van Hoepen (1947) and Hoffman (1953) used cranial elements from this assemblage for taxonomic descriptions. Cooke (1974) considered Mahemspan material to be Florisian in age, but somewhat older than the Florisbad spring material. However, based on the shape of the black wildebeest horn cores Gentry and Gentry (1978) suggested an age younger than the Florisbad spring material. The ESR results for an early uranium uptake model (EU) suggest an age of around 12,000 years BP and for a linear uranium uptake model

Table 2. Taxonomic list of Florisian faunas: a comparison between Erfkroon Last Interglacial levels (L/I), Mahemspan and Florisbad Spring*.

	Florisbad spring	Erfkroon L/I	Mahemspan
Primates			
Homo helmei	1		–
Lagomorpha			
Lepus sp.	6	4	–
Rodentia			
Hystrix africae–australis	1	–	–
Pedetes sp. cf. *P. capensis*	8	–	–
Carnivora			
Aonyx capensis	3	–	–
Cynictis penicillata	–	2	–
Galerella sanguinea	1	–	–
Atilax paludinosus	3	–	–
Canis mesomelas	6	2	–
Vulpes chama	–	2	–
Lycaon pictus	3	1	–
Crocuta crocuta	7	–	7
Panthera leo	1	1	1
Perissodactyla			
Equus capensis††	73	6	80
Equus lylei††	61	1	cf.
Equus quagga subsp.	97	2	48
Ceratotherium simum	3	–	–
Artiodactyla			
Hippopotamus amphibious	333	2	4
Phacochoerus africanus/aethiopicus	33	2	17
Taurotragus oryx	24	1	21
Syncerus antiquus††	25	1	28
Kobus leche†	60	2	21
Kobus sp.†	4	–	–
Hippotragus sp.†	16	–	2
Damaliscus niro††	111	11	11
Damaliscus pygargus	9	12	19
Alcelaphus buselaphus	8	–	10
Connochaetes gnou	284	10	24
Connochaetes taurinus	–	–	9
Megalotragus priscus††	30	2	241
Antidorcas bondi††	889	34	1
A. marsupialis	107	2	6
Raphicerus campestris	4	–	–

††Extinct.
†Regionally extinct.
*Faunal lists modified and adapted from Brink (1987, 1994; in press).

Table 3. Electron Spin Resonance results on dental specimens from Mahemspan.

Sample	Dose	De error	U (EN)	U (DE)	TT	U (SED)	Th (SED)	K (SED)	EU Age (ka)	EU-error	LU Age (ka)	LU error
1411						5.5	1.75	0.45				
1412 AM	24.9	0.9	1.86	88.2	1230				**13.6**	0.8	**17.6**	1
1412 AS1	24.3	0.4							**13.3**	0.7	**17.2**	0.8
1412 BM	23.7	0.4	2.04	58	1170				**14**	0.7	**7.6**	0.9
1412 BS1	22.2	0.2							**13.1**	0.6	**16.5**	0.8
1413 AS1	16.7	0.7	0.89	24.5	700				**11.6**	0.7	**13.3**	0.8
1413 BS1	19	0.2	1.08	32.4	730				**12.5**	0.6	**14.7**	0.7
1413 CS1	19.5	0.2	1.36	54.8	830				**11.4**	0.5	**14.2**	0.7
1414 AS1	18.7	0.2	0.98	26.9	730				**12.8**	0.6	**14.8**	0.7
1414 BS1	19.2	0.3	1.33	40.1	700				**11.7**	0.6	**14.2**	0.7
1415 AM	16.7	0.3	0.7	20.94	930				**12.8**	0.6	**14.3**	0.7
1415 AS1	16.8	0.3		21					**12.9**	0.6	**14.4**	0.7
1415 BS1	15.5	0.3	0.6	25.8	830				**11.4**	0.6	**12.9**	0.7
1415 CS1	15.6	0.2	0.58	28.4	1130				**12.1**	0.6	**13.6**	0.7

(LU) of around 13–17,000 years BP (Table 3). These age estimates support the suggestion of Gentry and Gentry (1978) that the Mahemspan assemblage postdates that of the Florisbad Spring (Grün *et al.*, 1996).

6.3.2 Erfkroon

In Table 2 the taxonomic list for Erfkroon represents the fossil assemblages from the Last Interglacial levels. These are the Lower Gravel and Green Sand from the Orangia Terrace, as recovered on the farms Erfkroon and Orangia (see Table 1). The ESR age estimates for the various horizons of the Orangia Terrace are given in Table 4 and are based on the testing of the dental specimens recovered from the farms Erfkroon and Orangia. The numbers of the sample batches in Table 4 correspond to the sampling localities given in Figures 2 and 3. The EU age estimates from the Upper Grey (Orangia Terrace) seem to correspond fairly well with the luminescence estimates, as referred to above, but there is some disagreement between the luminescence and the ESR estimates from the Green Sand of the Orangia Terrace. The ESR estimates for the Shelly Gravel suggest a terminal Late Pleistocene age. We are at present uncertain whether the Shelly Gravel is a lateral facies of the Upper Gray of the Orangia Terrace or the basal horizon of the Soetdoring Terrace. The ESR estimates accord well with either stratigraphic interpretation (Figure 3)

The *M. priscus* skull and horn cores were found in an outcrop which we correlate with the Lower Gravels and Green Sands of the Orangia Terrace (Figure 4). Based on the luminescence dating of the Orangia Terrace this would imply a Last Interglacial age for the *M. priscus* outcrop. We were not able to recover dental specimens from the *M. priscus* outcrop for ESR testing, but the presence of abundant rolled calcrete blocks together with clasts of shale gives confidence to the correlation of this deposit with the Lower Gravel of the Orangia Terrace.

6.3.3 Horns and braincase of *M. priscus*

6.3.3.1 The Erfkroon specimen

The Erfkroon specimen consists of a braincase and both horn cores (Figure 5; Table 5). The left horn core tip is not preserved. The horns are dorso-ventrally compressed near the bases and in mid-course, but become rounded towards the last third of their course. Near the base of the horn on the anterior surface there is a slight swelling, also present in the Florisbad specimen FLO 2274. Horn pedicels are fused and overhang the occipital surface. Horns are bent down sharply and diverge with a mutual angle of around 150°. In mid-course they are sub-horizontal before they curve up and forward and there is clockwise torsion on the right. The horns have faint transverse ridges near the bases, an indication of nodes on the horn sheath. The frontals' suture appears less extremely fused than in *C. gnou*. There is no postcornual fossa. The braincase appears antero-posteriorly shortened in lateral view. This is due mainly to the reduction in the parietals, which are visible only in lateral view so that the occipital makes contact with the frontal. The combined effect of the reduction in the braincase and the posterior projection of the fused pedicels is that the braincase appears partly hidden beneath the horn bases. This configuration of the braincase appears to have caused the nuchal crest to have become inverted to form a concave structure in order to allow sufficient

Table 4. Electron Spin Resonance results on dental specimens from Erfkroon.

Sample	Dose	De error	U (EN)	U (DE)	TT	S1	S2	U (SED)	Th (SED)	K (SED)	gamma	EU Age (ka)	EU-error	LU Age (ka)	LU error
1. Upper Grey, Orangia Terrace															
1920 AS1	56.7	2.1			1470	50	450					42	4	50	6
1920 BS1	33	1.1			1440	40	470	3.8	4.5	0.81	744.5	25.2	2.6	29.5	3.5
1923 AS1	45.5	1.1			1130	70	340	1.45	6.5	1.29	695.9	24.9	2.1	32	3
1923 BS1	49	1.3			1110	90	270					27.2	2.3	35	3
1926 AS1	49.7	0.7			1000	90	330	3	8.5	1	873.8	22.4	1.8	31	2
1926 BS1	44.2	1.4			1130	90	160					22	1.8	29.4	2.8
2. Duke 8, Shelly Gravel, Incertae sedis															
1929 AS1	43.4	1.3	3.7	99	960	50	50				696.7	18	1.2	25.2	1.6
1929 BS1	48.5	1.3	3.8	115	1090	50	50				843.5	19.7	1.2	27.8	1.7
1930A	54.4	0.8	0.88	112	800	40	70				770.1	23.6	1.5	32	2
1930B	41.1	0.6	0.51	112	970	70	160					20.3	1.3	27.1	1.7
1931A	40.6	0.5		181	710	50	70					11.5	0.8	17.6	1.1
1931B	43.9	0.6	5.5	168	820	90	60					13.6	0.9	20.4	1.2
1932A	54.7	0.6	10	144	1040	70	300					14.7	1.1	23.2	1.5
1932B	63.3	1	4.5	147	1010	30	190					20.8	1.4	31	1
1932C	41.6	0.5	1.5	169	1000	160	190					17.1	1.1	24.2	1.5
3. Hippo Site, Green Sand, Orangia Terrace															
1939A	733	17	8	102	1180	90	90				713	158	13	261	19
1939B	825	20	9.1	109	1090	70	70				698	153	12	253	19
1940A	661	13	9.3	91	690	20	20					127	10	209	15
1940B	0		11	90	530	20	20								
1941A	0		2	82	2510	290	290	2.6	6.6	1.07	724.8				
1941B	0		2.3	82	2620	225	225								
1941C	0		2.8	82	2550	290	290								
4. W Site, Green Sand, Orangia Terrace															
1948A	580	13	3.7	65	760	50	50					135	12	201	18
1948B	796	16	13	71	730	20	20					124	11	199	17
1949A	568	10	8	23	900	90	90					135	12	201	18
1949B	522	9	4.5	21	890	110	70					161	14	223	22
1950A	587	11	5	26	730	50	50					164	15	232	22
1950B	407	6	2.3	26	860	100	90					157	14	204	21
5. Duke 13, Lower Gravel, Orangia Terrace															
1953A	1582	65	36	90	650	35	35					123	13	218	23
1953B	1697	30	27	86	580	40	40					154	15	270	25
1954A	601	10	3.5	72	1020	60	150					183	13	277	18
1954B	613	7	4.8	77	1230	100	180					173	12	268	18

area for the attachment of the neck muscles. Although the specimen is somewhat damaged in this region, it appears that the nuchal furrow extends into a very pronounced supramastoid crest anteriorly, while posteriorly it links with the petrosal part of the temporal to form a very strong structure for neck muscle attachment. The occipital condyles are very wide. The basioccipital is short, wide, approaches being rectangular, has no median ridge and has large anterior tuberosities (tubercula muscularia). The basioccipital and the sphenoid are not on the same plane, but have an angle of around 140°.

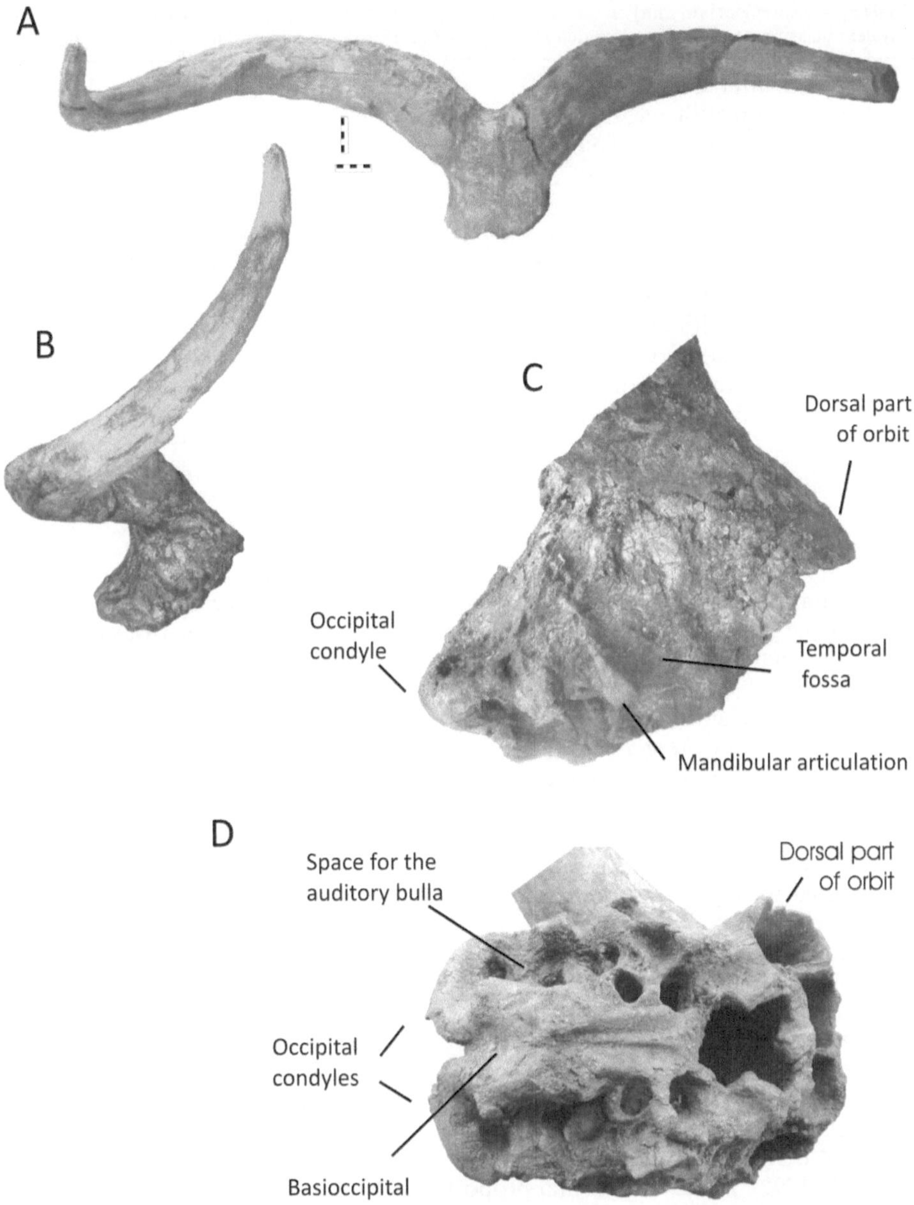

Figure 5. The complete braincase and horn cores of *Megalotragus priscus* from Erfkroon: Frontal view (A), right lateral view (B), enlarged right lateral view of the brain case (C) and a basal view (D). The scale is in centimetres.

The bullae tympanicae are not preserved, but the spaces in which they were situated are preserved and it can be deduced that they were moderately large and rounded. The foramina ovalia are quite large. From the remaining part of the frontal it is evident that the angle of the braincase to the face is small, approaching 90°. There is enough of the posterior margin of the orbits preserved to suggest that they would have been at least moderately projecting.

Table 5. Measurements (in mm) of the *Megalotragus priscus* skull and horn cores from Erfkroon. The equivalent measurements given in Von den Driesch (1976) for cattle (*Bos taurus*) are indicated in brackets.

1	Least frontal breadth (32)	143.5
2	Greatest breadth across the orbits (33)	172.8+
3	Greatest diameter of the pedicel	107.7
4	Least diameter of the pedicel	83
5	Greatest diameter of the base of the horn core (45)	108.5
6	Least diameter of the base of the horn core (46)	83
7	Greatest diameter of the horn core at mid-section	90.5
8	Least diameter of the horn core at mid-section	65.1
9	Greatest breadth of occipital condyles (26)	144*
10	Greatest breadth of the foramen magnum (28)	40.5*
11	Least breadth between the bases of the horn cores (31)	57+
12	Least breadth of the braincase (30)	142.2

+ These measurements are estimates due to incomplete preservation of the relevant parts.
* These measurements are minimum values, because of damage to the specimen.

6.3.3.2 Comparisons

It is a commonly held view that all very large-bodied alcelaphine antelope in Africa belong to one genus, *Megalotragus* Van Hoepen, 1932, and that the various fossil species of this genus are closely related (Gentry, 1978, 2010; Gentry and Gentry, 1978; Gentry *et al.*, 1995; Vrba, 1979, 1997). Previously a smaller-bodied species from Rusinga Island, Kenya, *Rusingoryx atopocranion* Pickford and Thomas, 1984, was referred to the genus *Megalotragus* (Vrba, 1997), but more recently in a revision of the morphological and phylogenetic relationships of this species it was re-assigned to *Rusingoryx* as a genus distinct from *Megalotragus*, although closely related (Pickford and Thomas, 1984; Faith *et al.*, 2011). Previously it was suggested that an East African species, *M. kattwinkeli*, is ancestral to the southern African *M. priscus*, which is considered to include two temporal forms, *M. priscus eucornutus* and *M. priscus priscus* (Gentry and Gentry, 1978). However, this hypothesis seems no longer to be supported (Gentry, 2010). There is no fossil evidence yet for the species *M. eucornutus* and *M. priscus* outside of southern Africa, although Faith *et al.* (2011) mentions the presence of large wildebeest-like alcelaphine from the Wasiriya Beds on Rusinga Island, which they refer to the genus *Megalotragus*. *M. eucornutus* is known only from Cornelia-Uitzoek, the type locality of the Cornelian Land Mammal Age (LMA) and from Cornelia-Mara, a nearby locality of similar age (Brink *et al.*, 2012; Brink, in press). It has been noted that there are grounds for maintaining the specific distinction between *M. priscus* and *M. eucornutus*, based on horn core morphology (Bender and Brink, 1992), but also on dental proportions (Brink, 2005). Thus, for the sake of clarity the distinction between names *M. eucornutus* Van Hoepen, 1932 and *M. priscus* (Broom, 1909) is maintained here.

Originally the Florisbad giant alcelaphine material was named *Bubalis helmei* by Dreyer and Lyle (1931). Van Hoepen (1932) described an isolated find of an incomplete horn core pair with intact pedicels from the farm Doornberg, on the Sand River near Kroonstad, Free State Province, as *Pelorocerus elegans*. Van Hoepen (1947) described the Mahemspan material initially as *Pelorocerus mirum*. He also redescribed the Florisbad *B. helmei* as *Lunatoceras mirum* (Van Hoepen, 1947), which was later referred to *Alcelaphus helmei* by Cooke (1952) and again to *Pelorocerus helmei* by Hoffman (1953). Subsequently Cooke (1964) referred the Florisbad material to both

Pelorocerus helmei and to *Lunatoceras mirum*. Eventually Gentry and Gentry (1978) included the Florisbad material and all other Florisian giant alcelaphine materials in *M. priscus*. This was followed by Vrba (1979, 1997), Klein (1984) and Brink (1987).

The synonymy of the genus *Megalotragus* and the species *Megalotragus priscus* is as follows:

Genus *MEGALOTRAGUS* Van Hoepen 1932
1932 *Megalotragus* Van Hoepen
1932 *Pelorocerus* Van Hoepen
1953 *Lunatoceras* Hoffman
1965 *Xenocephalus* Leakey

Type species. *Megalotragus priscus* (Broom 1909)
The type specimen (SAM 1741) is a cranial fragment with part of the left horn core preserved and it is housed in the Iziko South African Museum in Cape Town.

1909 *Bubalis priscus* Broom
1931 *Bubalis helmei* Dreyer & Lyle
1932 *Pelorocerus elegans* Van Hoepen
1947 *Lunatoceras mirum* Van Hoepen
1951 *Connochaetes grandis* Cooke & Wells

The Erfkroon specimen is very similar to the type specimen, but more complete. An almost complete set of horn cores from Florisbad, FLO 2274, is virtually identical to the Erfkroon specimen in terms of size, horn shape and horn curvature, but lacking the braincase. In FLO 2274 the frontals' suture is partially fused, as in the Erfkroon specimen and similar to the Barbary sheep, *Ammotragus lervia*. In another Florisbad specimen, FLO 2273, the horns tend to be more sharply curved, they extend further backwards at their bases before curving sideways and forwards. The bases of the horn cores lack the protuberance on the cranial surface, as seen in the Erfkroon specimen and in FLO 2274. The mutual angle between the horn bases is somewhat smaller, while the specimen is generally more gracile. This specimen is probably a female, while the Erfkroon specimen, the type specimen and FLO 2274 appear to be of males.

In addition to the above the horn core of *M. priscus* from the Ongers River near Britstown, central Karoo (Brink *et al.*, 1995), has a base that is not antero-posteriorly extended and appears to be somewhat rounded in cross section. There is no basal protuberance and, while it is difficult to estimate the degree of pedicel fusion in this specimen, it appears to have had a reasonably wide mutual angle between the horn cores. The specimen is very gracile and in size comparable to the specimen from Doornberg, C. 1711. Of all the specimens assigned to *M. priscus* the Doornberg specimen have the smallest mutual angle between the horn core bases, a condition that is considered to be plesiomorphic for *Megalotragus* (Gentry and Gentry, 1978). For this reason it is probable that the Doornberg specimen is geologically older than the other specimens of *M. priscus* under consideration here. The Ongers specimen resembles the Doornberg specimens in gracility, but has a wider mutual angle between the horn bases. The horn core bases of the Doornberg specimen are also less expanded antero-posteriorly than the Ongers River specimen and have no basal protuberances, which is also a plesiomorphic condition for *M. priscus*. Both these specimens are likely to be female.

In the type specimen of *P. mirum* (C. 2013) from Mahemspan the horn pedicels are not as extremely fused as in the Erfkroon and Florisbad specimens and the area of pedicel fusion is less elevated above the frontals (Van Hoepen, 1947). The mutual angle between the horn bases is reduced, resembling the Doornberg specimen. The basal parts of the horn cores are not as robust as in the large specimens from

Erfkroon and Florisbad. The cranial sides of the horn bases are not preserved and it cannot be established whether there were protuberances. This specimen appears to be male.

In specimen C. 2537 from Mahemspan, a frontal fragment with the basal parts of the horn cores preserved, there is a marked posterior projection of the horn bases and the horn bases are much thinner. There is a reduced mutual angle between the horn core bases, a reduced degree of pedicel fusion, while the frontals' suture appears less fused than in the Erfkroon and Florisbad specimens. C. 2537 is also more gracile than Mahemspan specimen C. 2013 and very similar to the Doornberg specimen and, consequently, is likely to be female.

Mahemspan specimen C. 2246 has an equally narrow mutual angle between the horn core bases. In contrast with the Erfkroon specimen the nuchal crest forms a convex relief, and is not inverted, and it is not as wide as in the Erfkroon specimen. In the co-type of *P. mirum* from Mahemspan, C. 2292, the horn base appears not to have a protuberance and it is not antero-posteriorly extended. The curvature of the horns is intermediate between the large forms from Florisbad and Erfkroon and the small specimens from Doornberg and Mahemspan. These specimens are probably female.

6.3.3.3 Sexual dimorphism and geographic variability in the horn cores of M. priscus

The more complete and dated materials available now for *M. priscus* allow an appreciation of sexual dimorphism and geographic variability. Although there is considerable variability in size, the horn core specimens of *M. priscus* can be separated into categories of male and female. Females are those with more gracile horn cores, with less dorso-ventral extended basal parts and with slightly shorter horn curvature. Males, on the other hand, have generally larger horn cores, with dorso-ventral expanded horn bases and with a thickening, or protuberance on the dorsal side. Male horns tend to be more horizontally positioned, as seen in the Erfkroon specimen, and to be less sharply curved. This supports the observation that the mutual angle in *M. priscus* horn cores is a sexually dimorphic character with males tending to have more downward pointing horns than females and that a greater mutual angle is associated with greater robusticity (Brink *et al.*, 1995).

The type specimen and the specimens from Florisbad and Erfkroon are considerably larger than those from Mahemspan, Doornberg and the Ongers River. In the Mahemspan specimens there is a lesser degree of horn pedicel fusion, the fused pedicels are less elevated above the frontals and the horn cores are generally more vertically inserted. Previously it has been suggested that morphological variability in *M. priscus* horns represent a temporal cline in that horns become more downward and forward pointing in the course of geological time (Cooke, 1974). This statement was based on the assumption that Mahemspan predates the Florisbad spring assemblage. However, on the evidence presented here it appears now that Mahemspan is of terminal Late Pleistocene age and that a more likely explanation for horn core variability may be that it reflects sexual dimorphism and geographic variability in populations. Although undated, the Doornberg specimen, C. 1711, named *P. elegans* by Van Hoepen (1947), is probably an early and very gracile female version of *M. priscus*. Similarly, the two forms of giant alcelaphine from Florisbad, "*P. helmei*" and "*L. mirum*", probably reflects sexual dimorphism in *M. priscus*, with the former being male and the latter female.

On the available fossil evidence *M. priscus* can be divided into two morphological entities, which may have represented two geographic variants. This is a very tentative observation and will need further testing. If the variability observed truly reflects geographic variability, it would be in parallel with the variation seen in extant populations of hartebeest, *Alcelaphus buselaphus* subspp., and tsessebe, *Damaliscus lunatus* subspp. (Kingdon, 1997).

6.4 RECONSTRUCTING THE SKULL OF *M. PRISCUS*

The horn core pair and braincase from Erfkroon, an upper jaw fragment (C. 1804) and a lower jaw (C. 2472), both from Mahemspan (Figure 5), form the basis of the reconstruction of the skull of *M. priscus* (Figures 6 and 7). Because it is possible to establish the position of the jaw articulation on the Erfkroon braincase, the Mahemspan lower jaw allows the estimation of the length of the face. Also, in conjunction with the remnant of the frontal, the mandible allows the angle of the braincase to the face to be estimated. The extreme posterior position of the horn bases and the position of the occipital condyles suggest a hanging, ox-like head position. Even if the Mahemspan materials may represent a different geographic population from the Erfkroon and Florisbad materials, this will result in only marginal distortion in the skull proportions, since the dentitions from Florisbad and Mahemspan seem to be identical in morphology and size (Brink, 2005). Because the Erfkroon specimen is used as the basis for the reconstruction, the reconstructed skull represents a male of the geographic population to which the type specimen and the Florisbad spring specimens belonged. At present nasal bones of *M. priscus* are not known in the fossil record and this reconstruction suggests only moderately inflated nasals. The nasals represent the most speculative aspect of the reconstruction.

The lower jaw of *M. priscus* is unusually elongated compared to other alcelaphines. Mesially, the diastema is enlarged to balance the increase in the posterior extension of the ramus, while the angle between the ramus and the corpus is widened to around 135°. These features are modifications to allow the mandible to fit the extremely elongated skull. A similar, but less extreme widening of the angle between the ramus and the corpus is seen it the mandible of *Rusingoryx atopocranion* (Faith *et al.*, 2011). Another parallel to this extreme morphology of the lower jaw can be found in the hartebeest and in the Barbary sheep, *Ammotragus lervia*. In both the latter species there is some degree of fusion of the horn pedicels and the horn bases are positioned posteriorly on the skull, which is reflected in the widened angle between the ramus and corpus of the mandible. In *C. taurinus* the lower jaw is also somewhat elongated, but less so than in the hartebeest, while the angle of lower jaw is not as wide. In *C. gnou* and blesbok the ramus is sub-vertical and the corpus relatively short and stout.

6.5 DISCUSSION

6.5.1 Morphological relationships of *M. priscus*

Broom emphasised the hartebeest-like horn bases of the animal (Broom, 1909), which is reflected by his choice of genus name for the specimen ("Bubalis"). Although the skull of *M. priscus* is characterised by extreme elongation and by the fusion and posterior extension of the horn pedicels, reminiscent of some of the geographic variants of *Alcelaphus buselaphus*, there are a number of characters more typical of the genus *Connochaetes.* These are the tendency for the horns to curve sideways, downward and then forward and the reduction in the prominence of nodes on the horns. The simplicity of the occlusal enamel patterns of the dentitions and the morphology of the postcrania, discussed elsewhere (Brink, 2005), are also characteristic of wildebeest.

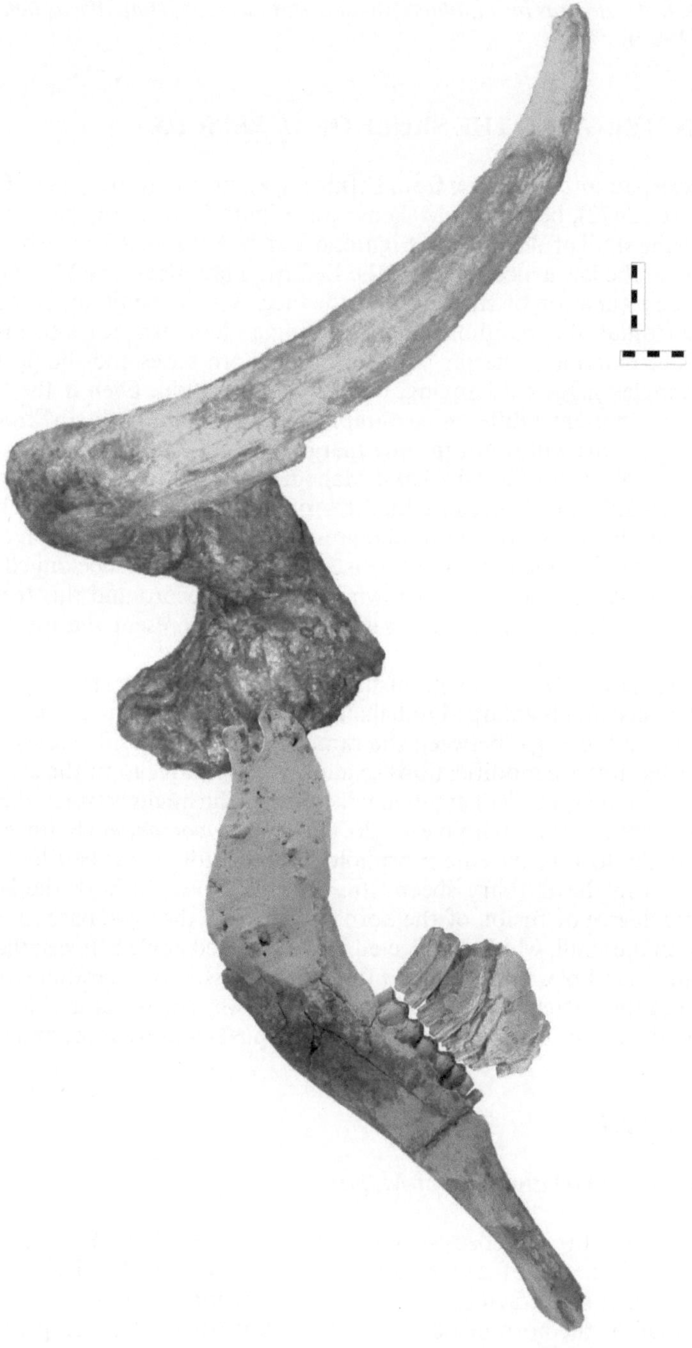

Figure 6. An arrangement in approximate natural position of the Erfkroon braincase and skull (A), a lower jaw from Mahemspan (B) and an upper jaw fragment from Mahemspan (C). This arrangement is the basis of the reconstruction of the skull of *M. priscus*, as given in Figure 7. The scale is in centimetres.

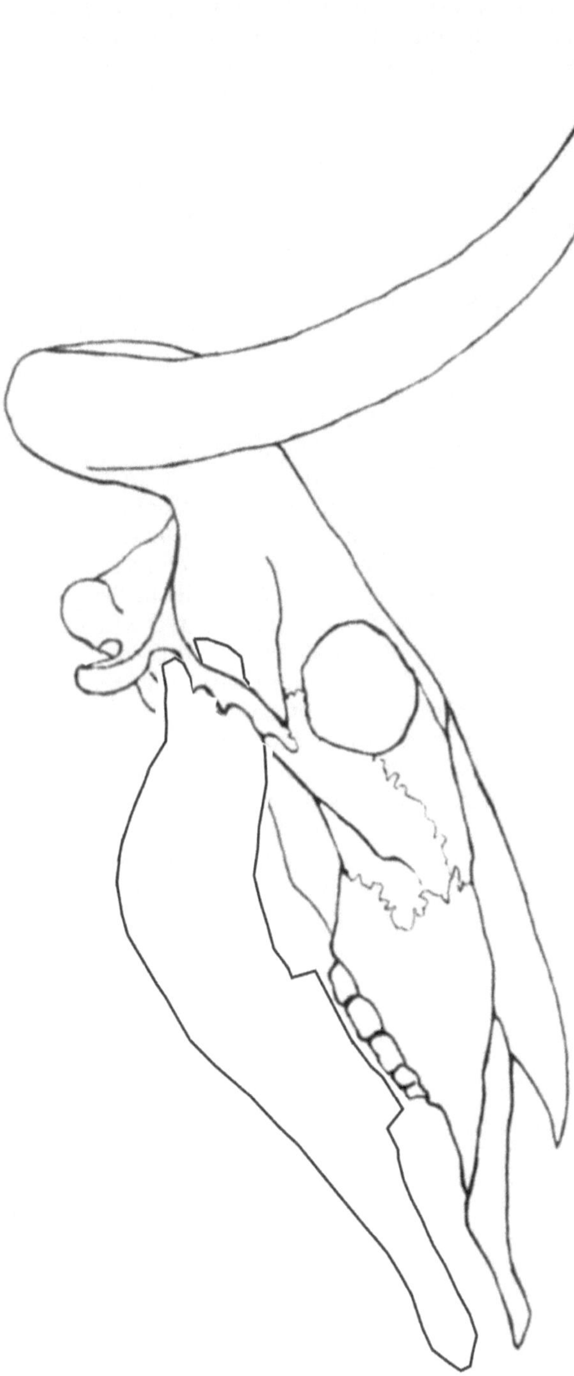

Figure 7. A reconstruction of the skull of *M. priscus*. The scale is as in Figures 6.

Previously the position of *M. priscus* within the wildebeest clade has not been apparent due to the erroneous assumption that facial lengthening, clockwise torsion in the right horn and related changes in horn base morphology are uniquely derived characters shared with the genus *Alcelaphus* (Hoffman, 1953; Wells, 1959, 1964; Klein, 1984; Vrba, 1979). It now appears that these characters reflect parallel adaptation with the hartebeest and that *M. priscus* was in fact a very large and derived form of wildebeest. It should be noted further that Vrba's phylogenetic re-analysis (Vrba, 1997) and the phylogenetic analysis of Faith *et al.* (2011) support the interpretation of the genus *Megalotragus* being closely related to the genus *Connochaetes*.

There are obvious similarities between *M. priscus* and *M. kattwinkeli,* and a number of shared derived characters unite the two species. These are elongated skulls, horns with some degree of transverse ridge development, horns inserted far behind the orbits, projecting orbits and an occipital surface that faces mainly backwards with a median vertical ridge (Gentry and Gentry, 1978; Vrba, 1997). It may be predicted that inflated nasals could be added to these characters and it will be of interest to find sufficiently well preserved nasals of *M. priscus* to test this assumption. However, there are some derived characters in *M. priscus*, which are not shared by *M. kattwinkeli*, but there are also some characters in which *M. priscus* is more plesiomorphic. Uniquely derived characters in *M. priscus* in relation to *M. kattwinkeli* are the more posterior position of the horn core insertions, the extreme state of fusion of the pedicels, very large horn size, the almost horizontal angle at which the horns curve outwards from the pedicels before they curve forward (more extreme in the Erfkroon and Florisbad specimens), the fact that the fused pedicels of the horn cores overhang the occipital and possibly large bullae tympanicae. The plesiomorphic characters of *M. priscus* in relation to *M. kattwinkeli* include a less reduced premolar row, in which the P^3 and possibly the P^2 were still present, while in *M. kattwinkeli* only the P^4 was present (Gentry and Gentry, 1978). This mosaic of characters suggests that *M. kattwinkeli* was probably not an ancestor of *M. priscus*, in spite of being closely related. *M. kattwinkeli* occurs in the fossil record until the end of the Early Pleistocene and possibly later (Faith *et al.*, 2011). *M. priscus*, first appeared in the Elandsfontein Main assemblage, which is thought to date to the end of the Early Pleistocene (Brink, in press), but it became extinct at the end of the Late Pleistocene/early Holocene (Klein, 1984). Therefore, in temporal range *M. kattwinkeli* overlaps with *M. priscus* and this temporal overlap adds to the doubt regarding the suggested ancestor-descendant relationship between *M. kattwinkeli* and *M. priscus*.

The possible ancestor-descendant relationship between *M. eucornutus* and *M. priscus*, initially mentioned by Gentry and Gentry (1978) and Vrba (1979, 1997), is not discussed further in Gentry (2010). However, this question can be further considered on the available evidence. The closely positioned horn bases, the orientation of the horn bases, the clockwise curvature and the lesser degree of fusion of the pedicels accord with *M. eucornutus* being an ancestor of *M. priscus*. However, the absence of any indication of nodes on the horns of *M. eucornutus* would be unexpected if it were ancestral to *M. priscus*. Nodes are not found on the horns of early alcelaphines, such as *Damalacra neanica* and *D. acalla* from Langebaanweg, nor in extant members of the genus *Connochaetes*, but they are present in the genera *Beatragus*, *Alcelaphus* and *Damaliscus*, which represent a different, 'non-wildebeest' branch of alcelaphine evolution. Thus, in wildebeest-like alcelaphines nodes on horns indicate a plesiomorphic condition and their presence on the horns of *M. priscus*, even though in reduced state, would have to be interpreted as an evolutionary reversal to accommodate *M. eucornutus* as the ancestor of *M. priscus*. The paucity of fossil material of *M. eucornutus* hampers proper evaluation. Apart from the horn core specimens and a few dental and postcranial elements from Cornelia-Uitzoek, very little of the body of *M. eucornutus* is known. Other than Cornelia-Uitzoek and Cornelia-Mara (Butzer, 1974; Cooke, 1974; Brink *et al.*, 2012) the only fossil locality in the interior of southern Africa that may have produced material of *M. eucornutus* is Gladysvale (Lacruz *et al.*, 2002), but so far no horn material of *M. eucornutus* has been found there. At Elandsfontein, which has produced a substantial collection of Cornelian large mammals, only *M. priscus,* and not *M. eucornutus* is recorded (Klein *et al.*, 2007). Therefore, on the existing evidence it appears unlikely that *M. eucornutus* was ancestral to *M. priscus*.

6.5.2 Behavioural implications of the skull morphology of *M. priscus*

Kingdon (1982) ascribes the increased profile of the head of the hartebeest to the importance of head signals in these animals, usually executed in slow movements. This contrasts with blue wildebeest, in which a very small proportion of the horns is visible in profile, but which has very energetic body and head movements during intraspecific encounters. Kingdon further notes that such behaviour in wildebeest may be associated with high densities when large herds form. The horn profile and the morphological characters suggesting the aggressive use of the horns in *M. priscus* are wildebeest-like. Also, the evidence for a specialised grazing niche is in accordance with the aggregation of large herds in open, Highveld-type grasslands (Brink and Lee-Thorp, 1992; Codron *et al.*, 2008), which is in contrast to the hartebeest and the genus *Damaliscus*. Hartebeest do not form herds in such large numbers as blue wildebeest. They tend to occupy ecotonal habitats rather than the more homogeneous short grass plains, favoured by *Connochaetes taurinus* in East Africa and *C. gnou* in southern Africa. The niche of the hartebeest as a roughage grazer is different from the short grass grazer niche of *C. taurinus* and *C. gnou* (Hofmann and Stewart, 1972; Codron and Brink, 2007; Brink and Lee-Thorp, 1992). In *M. priscus* the hanging head position and the forward curvature of the horns resemble *C. gnou* and the wild ancestor of cattle, *Bos primigenius*.

The picture that emerges is of a large-bodied wildebeest with some degree of territorial behaviour, where large herds formed, occupying a highly specialised grazing niche in the open grasslands of southern Africa. This niche disappeared at the end of the Late Pleistocene, evidently due to a reduction in the productivity levels of grasslands, which may be linked to increased aridity and reduced availability of soil moisture on a large geographic scale, as suggested by Brink and Lee-Thorp (1992) and supported by new evidence from the Erfkroon sedimentary record (Lyons *et al.*, 2014). The extinction of *M. priscus* coincided with the extinction of five other specialised grazing ungulates and the local extinction of a wetland faunal component in central southern Africa (Brink, in press). This coincidence reinforces the impression that aridity was a key driving factor in the end-Pleistocene extinction process.

6.6 CONCLUSION

The Erfkroon specimen and the new age estimates from Erfkroon and Mahemspan allow a revised and more detailed view of the morphological relationships of *Megalotragus priscus* and of its behaviour. It is evident that the animal was a large form of wildebeest, and not closely related to hartebeest, that it showed sexual dimorphism and probably some degree of territorial behaviour. The Erfkroon specimen allows a reconstruction of the skull of *M. priscus*, which suggests an extremely elongated skull that was held in life in a hanging, ox-like position. The head position, the forward curvature of the horns and the degree of sexual dimorphism are in accordance with the suggested territorial behaviour and add to our understanding of *M. priscus* as a large-bodied specialised grazer in the Florisian grasslands of southern Africa. Its extinction towards the end of the Pleistocene and early Holocene formed part of a southern African extinction event that included five other specialised grazers and the regional extinction of a wetland faunal component, suggesting widespread and increasing aridity in this time in the southern African subregion.

ACKNOWLEDGEMENTS

The authors thank the owners of the farms Erfkroon, Hein Bezuidenhoudt, and Orangia, Johan van der Berg, for their interest and for allowing access to their properties. We thank Lloyd Rossouw and Kris Carlson for field assistance and the Florisbad crew, Abel Dichakane, Ernest Maine, Willem Nduma, Bonny Nduma, Peter Mdala and Adam Thibeletsa for assistance in the excavation and preparation of the Erfkroon *M. priscus*. Maria Brink is thanked for reading an earlier draft of the manuscript. JSB would like to thank the French Embassy in Pretoria, South Africa, and the National Research Foundation of South Africa for financial support. The Council and Director of the National Museum are thanked for supporting research on fossil mammal evolution in the central interior of southern Africa.

REFERENCES

Bender, P.A. and Brink, J.S., 1992, A preliminary report on new large mammal fossil finds from the Cornelia-Uitzoek site. *South African Journal of Science,* **88**, pp. 512–515.

Brink, J.S., 1987, The archaeozoology of Florisbad, Orange Free State. *Memoirs of the National Museum, Bloemfontein,* **24**, pp 1–151.

Brink, J.S., 1988, The taphonomy and palaeoecology of the Florisbad spring fauna. *Palaeoecology of Africa,* **19**, pp. 169–179.

Brink, J.S., 1994, An ass, *Equus* (*Asinus*) sp., from late Quaternary mammalian assemblages of Florisbad and Vlakkraal, central Southern Africa. *South African Journal of Science,* **90**, pp. 497–500.

Brink, J.S., 2005, *The evolution of the black wildebeest,* Connochaetes gnou, *and modern large mammal faunas in central southern Africa.* DPhil dissertation, University of Stellenbosch.

Brink, J.S., in press, Faunal evidence for mid- and late Quaternary environmental change in southern Africa. In *Quaternary environmental change in southern Africa: physical and human dimensions,* edited by Knight, J. and Grab, S.W. (Cambridge: Cambridge University Press).

Brink, J.S., Herries, A.I.R., Moggi-Cecchi, J., Gowlett, J.A.J., Bousman, C.B., Hancox, J.P., Grün, R., Eisenmann, V., Adams, J.W. and Rossouw, L., 2012, First hominine remains from a ~1.0 million year old bone bed at Cornelia-Uitzoek, Free State Province, South Africa. *Journal of Human Evolution,* **63**, pp. 527–535.

Brink, J.S. and Lee-Thorp, J., 1992, The feeding niche of an extinct springbok, *Antidorcas bondi* (Antilopini, Bovidae), and its palaeoenvironmental meaning. *South African Journal of Science,* **88**, pp. 227–229.

Brink, J.S., De Bruiyn, H., Rademeyer, L.B. and Van Der Westhuizen, W.A., 1995, A new find of *Megalotragus priscus* (Alcelaphini, Bovidae) from the central Karoo, South Africa. *Palaeontologia Africana,* **32**, pp. 17–22.

Broom, R., 1909b, On a large extinct species of Bubalis. *Annals of the South African Museum*, **7**, pp. 279–280.
Butzer, K.W., 1974, Geology of the Cornelia beds. Cornelia, O.F.S., South Africa. *Memoirs of the National Museum, Bloemfontein*, **9**, pp. 7–32.
Codron, D. and Brink, J.S., 2007, Trophic ecology of two savanna grazers, blue wildebeest Connochaetes taurinus and black wildebeest Connochaetes gnou. *European Journal of Wildlife Research*, **53**, pp. 90–99.
Codron, D., Brink, J.S., Rossouw, L. and Clauss, M., 2008, The evolution of ecological specialization in southern African ungulates: competition- or physical environmental turnover? *Oikos*, **117**, pp. 344–353.
Cooke, H.B.S., 1964, Pleistocene mammal faunas of Africa, with particular reference to southern Africa. In *African ecology and human evolution*, edited by Howell, F.C. and Bourlière, F.C. (London: Methuen), pp. 65–116.
Cooke, H.B.S., 1974, The fossil mammals of Cornelia, O.F.S., South Africa. *Memoirs of the National Museum, Bloemfontein*, **9**, pp. 63–84.
Churchill, S.E., Brink, J.S., Berger, L.R., Hutchison, R.A., Rossouw, L., Stynder, D., Hancox, P.J., Brandt, D., Woodborne, S., Loock, J.C., Scott, L. and Ungar, P., 2000, Erfkroon: A new Florisian fossil locality from fluvial contexts in the western Free State. *South African Journal of Science*, **96**, pp. 161–163.
Faith, J.T., Choiniere, J.N., Tryon, C.A., Peppe, D.J., and Fox, D.L., 2011,Taxonomic status and paleoecology of *Rusingoryx atopocranion* (Mammalia, Artiodactyla), an extinct Pleistocene bovid from Rusinga Island, Kenya. *Quaternary Research*, **75**, pp. 697–707
Gentry, A.W., 1978, Bovidae. In *Evolution of African Mammals*, edited by Maglio, V. and Cooke, H.S.B. (Cambridge: Harvard University Press), pp. 540–572.
Gentry, A.W., 2010, Bovidae. In *Cenozoic mammals of Africa*, edited by Werdelin, L. and Saunders, W.J. (Berkley: University of California Press), pp. 741–796.
Gentry, A.W. and Gentry, A., 1978, Fossil Bovidae (Mammalia) of Olduvai Gorge, Tanzania. Part I. *Bulletin of the British Museum (Natural History)*, **29**, pp. 290–446.
Gentry, A.W., Gentry, A. and Mayr, H., 1995, Rediscovery of fossil antelope holotypes (Mammalia, Bovidae) collected from Olduvai Gorge, Tanzania, in 1913. *Mitteilungen der Bayerische Staatssammlung für Paläontologie und Historische Geologie*, **35**, pp. 125–135.
Grün, R., Brink, J.S., Spooner, N.A., Taylor, L., Stringer, C.B., Franciscus, R.B. and Murray, A., 1996, Direct dating of the Florisbad hominid. *Nature*, **382**, pp. 500–501.
Helgren, D.M., 1977, Geological context of the Vaal River faunas. *South African Journal of Science*, **73**, pp. 303–307.
Hoffman, A.C., 1953, The fossil alcelaphines of South Africa—genera *Peloroceras*, *Lunatoceras* and *Alcelaphus*. *Navorsinge van die Nasionale Museum, Bloemfontein*, **1**, pp. 41–56.
Hofmann, R.R and Stewart, D.R.M.,1972, Grazer or browser: a classification based on the stomach structure and feeding habits of East African ruminants. *Mammalia*, **36**, pp. 226–240.
Kingdon, J., 1982, *East African Mammals Volume III, Part C (Bovids)*. (London: Academic Press).
Kingdon, J., 1997, *The Kingdon Field Guide to African Mammals*. (London: Academic Press).
Klein, R.G., 1984, The large mammals of southern Africa: late Pliocene to recent. In *Southern African prehistory and palaeoenvironments*, edited by Klein, R.G. (Rotterdam: A.A. Balkema), pp. 107–146.

Klein, R.G., Avery, G., Cruz-Uribe, K. and Steele, T.E., 2007, The mammalian fauna associated with an archaic hominin skullcap and later Acheulean artifacts at Elandsfontein, Western Cape Province, South Africa. *Journal of Human Evolution,* **52(2)**, pp. 164–186.
Lacruz, R., Brink, J.S., Hancox, P.J., Skinner, A.R., Herries, A., Schmid, P. and Berger, L.R., 2002, Palaeontology and geological context of a Middle Pleistocene faunal assemblage from the Gladysvale Cave, South Africa. *Palaeontologia Africana,* **38**, pp. 99–114.
Lyons, R., Tooth, S. and Duller, G.A.T., 2014, Late Quaternary climatic changes revealed by luminescence dating, mineral magnetism and diffuse reflectance spectroscopy of river terrace palaeosols: a new form of geoproxy data for the southern African interior. *Quaternary Science Reviews,* **95**, pp. 43–59.
Pickford, M. and Thomas, H., 1984, An aberrant new bovid (Mammalia) in subrecent deposits from Rusinga Island, Kenya. *Proceedings: Koninklijke Nederlandse Akademie van Wetenschappen,* **87**, pp. 441–452.
Seeley, H.G., 1891, On *Bubalus Bainii. Geological Magazine,* pp. 200–202.
Tooth, S., Brandt, D., Hancox, P.J. and Mccarthy, T.S., 2004, Geological controls on alluvial river behaviour: a comparative study of three rivers on the South African Highveld. *Journal of African Earth Sciences,* **38**, pp. 9–97.
Tooth, S., Hancox, P.J., Brandt, D., McCarthy, T.S., Jacobs, Z. and Woodborne, S.M., 2013, Controls on the genesis, sedimentary architecture, and preservation potential of dryland alluvial successions in stable continental interiors: insights from the incising Modder River, South Africa. *Journal of Sedimentary Research,* **83**, pp. 541–561.
Van Hoepen, E.C.N., 1932, Voorlopige beskrywing van Vrystaatse soogdiere. *Paleontologiese navorsinge van die Nasionale Museum, Bloemfontein,* **2(5)**, pp 63–65.
Van Hoepen, E.C.N., 1947, A preliminary description of new Pleistocene mammals of South Africa.. *Paleontologiese navorsinge van die Nasionale Museum, Bloemfontein,* **2(7)**, pp. 103–106.
Von Den Driesch, A., 1976, A guide to the measurement of animal bones from archaeological sites. *Peabody Museum Bulletin,* **1**, pp. 1–136.
Vrba, E.S., 1979, Phylogenetic analysis and classification of fossil and recent Alcelaphini Mammalia: Bovidae. *Biological Journal of the Linnean Society,* **11**, pp. 207–228.
Vrba, E.S., 1997, New fossils of Alcelaphini and Caprinae (Bovidae, Mammalia) from Awash, Ethiopia, and phylogenetic analysis of Alcelaphini. *Palaeontologia Africana,* **34**, pp. 127–198.
Wells, L.H., 1959, The Quaternary giant hartebeests of South Africa. *South African Journal of Science,* **55(5)**, pp.123–128.
Wells, L.H., 1964, A large extinct antelope skull from the "younger gravels" at Sydney-on-vaal, Cape Province. *South African Journal of Science,* **60(3)**, pp. 88–91.

CHAPTER 7

Ostrich eggshell as a source of palaeoenvironmental information in the arid interior of South Africa: A case study from Wonderwerk Cave

Michaela Ecker
Research Laboratory for Archaeology and the History of Art, University of Oxford, Oxford, UK

Jennifer Botha-Brink
National Museum Bloemfontein, Bloemfontein, South Africa
Department of Zoology and Entomology, University of the Free State, Bloemfontein, South Africa

Julia A. Lee-Thorp
Research Laboratory for Archaeology and the History of Art, University of Oxford, Oxford, UK

André Piuz
Muséum d'histoire naturelle, Genève, Switzerland

Liora Kolska Horwitz
Natural History Collections, Faculty of Life Sciences, The Hebrew University, Jerusalem, Israel

ABSTRACT: The early/mid-Pleistocene sequence from Wonderwerk Cave offers a unique record with which to explore palaeoenvironmental proxies in the interior of South Africa, for a time period that is poorly represented in the regional record. This study applies a combination of stable light isotope analysis ($\delta^{18}O$ and $\delta^{13}C$), and examination of biometric parameters measured on ostrich eggshell recovered from Wonderwerk Cave spanning the period 1.96–0.78 Ma years. These complementary datasets gave concordant results that indicate a generally arid environment from ca. 1.96–1.78 Ma years, and at ca 1 Ma the onset of an even more arid environment. Ostrich diets remained largely C3, with generally modest contributions from C4 plants. The climatic changes follow those attested to in the phytolith record from the cave and larger-scale climatic changes that have been documented in southern and East Africa.

7.1 INTRODUCTION

Although the general picture of Cenozoic climate change in Africa is broadly understood (Feakins and deMenocal, 2008), the magnitude and drivers of climate change during the early and middle Pleistocene remain poorly recorded, especially on a regional scale. In recent years there has been progress in palaeoclimatic research in East Africa (Potts, 1998, 2013; Maslin and Trauth, 2009; Trauth *et al.*, 2010; Maslin *et al.*, 2014), however, in the south of the continent there are still few stratified, well-dated localities that can provide temporal resolution on factors such as rainfall, temperature, seasonality and vegetation cover.

This study aims to apply a combination of stable light isotope analysis ($\delta^{18}O$ and $\delta^{13}C$), as well as biometric parameters measured on ostrich eggshell (OES) to track changes in early to mid-Pleistocene palaeoenvironmental conditions in the arid interior of South Africa. Ostrich eggshell comprises of 96% biomineralised calcite ($CaCO_3$) and 4% organic material (primarily proteins), resulting in a dense carbonate matrix that is extremely resistant to decay processes (Brooks *et al.*, 1990; Miller *et al.*, 1992, Johnson *et al.*, 1998). It is commonly found in palaeontological and archaeological sites making it a suitable material for such studies, especially for time periods where organic remains are often limited or not preserved. We examined a diachronic series of OES from the Earlier Stone Age (ESA) sequence of Wonderwerk Cave, covering some 1 million years, from 1.96- to 0.78 Ma. Given the extensive archaeological record represented at this site (Beaumont and Vogel, 2006; Chazan *et al.*, 2008, 2012; Chazan and Horwitz, in press), such a study promises to yield invaluable information about climatic conditions and vegetation cover, especially given the rarity of records for southern Africa in this time range and the conspicuous absence of such sequences, in particular for the arid interior (Hopley *et al.*, 2007; Scott *et al.*, 2012).

7.1.1 Background to the site

Wonderwerk Cave is located in the Northern Cape Province of South Africa and has a well-dated early, middle and late Pleistocene, as well as Holocene sequence. The specimens analysed in this study are derived from excavations undertaken at the site from the 1970's through 1990's by Peter Beaumont (Beaumont, 1990; Beaumont and Vogel, 2006) and Anne and Francis Thackeray (Thackeray *et al.*, 1981; Humphreys and Thackeray, 1983) and during a small test excavation of the lowest Stratum 12 undertaken by Chazan, Horwitz and Porat in 2007 (Chazan and Horwitz, 2010). During these excavations, all archaeological remains were systematically collected and documented based on their location in the excavation grid. The excavation grid used was in square yards and the deposits were excavated in strata that were numbered sequentially, beginning with the youngest Stratum 1 at the top of the sequence to 12 at the bottom, using arbitrary 5 or 10 cm spits. Sub-phases were recognized in some strata and distinguished as a, b, c and so on.

The site is located in the summer rainfall region (Figure 1). The local vegetation is classified as Kuruman Mountain Bushveld (SVK10) (Mucina and Rutherford, 2006), with grasses following a C_4 photosynthetic pathway, whereas shrubs, trees and herbaceous plants follow the C_3 pathway. These plant groups are isotopically distinct (Smith and Epstein, 1971). Since the site is situated at the edge of the Kalahari, we expect the surrounding environment of the cave to be sensitive to climatic shifts. This is expressed in the high inter-annual variability of rainfall in the region today (current mean annual rainfall 480 mm/a; coefficient of variation 0.25; South Africa Rain Atlas).

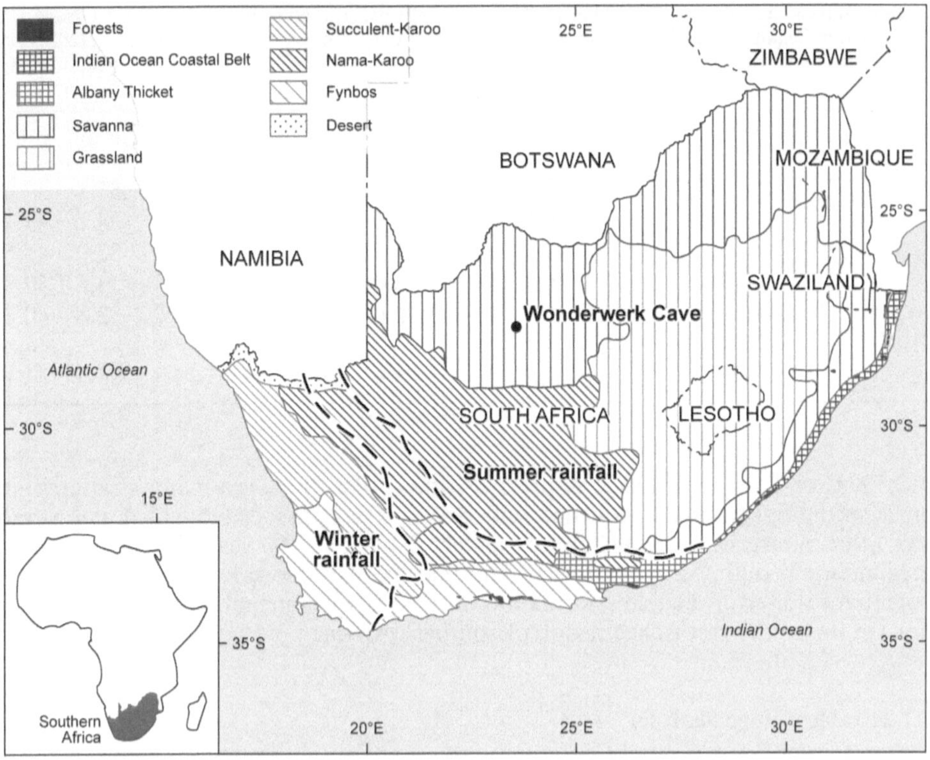

Figure 1. Location of Wonderwerk Cave in the summer rainfall region of South Africa (limit denoted by broken line), showing the main vegetation zones (after Mucina and Rutherford, 2006).

In this paper we focus on material from Excavation 1, which is currently located approximately 15 m in from the cave entrance, though the cave mouth probably extended further forward in the early Pleistocene (Matmon *et al.*, 2012). The lower sequence in Excavation 1 comprises Strata 12 to 9, which span the early to mid-Pleistocene and, as detailed in Table 1, have been dated to 1.96–0.78 Ma using palaeomagnetism and cosmogenic burial dating, corroborated by biochronology (Matmon *et al.*, 2012; Berna *et al.*, 2012). The Stratum 12 lithics have been identified as Oldowan, whereas Strata 11–9 have yielded a sequence of Acheulean industries (Chazan *et al.*, 2012; Berna *et al.*, 2012). The upper part of Stratum 9, as well as Strata 8 and 7 are as yet not well dated, although the Mode 2 lithic technology, biostratigraphy and one U-series date point to an Acheulean age older than 350,000 BP (Beaumont and Vogel, 2006). It should be noted that the U-series date clearly represents a minimum age and should be considered as such. Given the lack of refinement in dating methods and the division of some archaeological strata into multiple lithostratigraphic units (Chazan *et al.*, 2008), it is clear that some strata suffer from time-averaging. Moreover, micro-morphological investigations have demonstrated that there are clear erosional breaks within the Pleistocene sequence, such that the record represented is discontinuous (Chazan *et al.*, 2008, Goldberg *et al.*, in press).

Stratum 6 appears to contain a mixture of Acheulean and Later Stone Age artefacts, and is older than the overlying Stratum 5 which is of Late Pleistocene Age

Table 1. Basic statistics for OES stable light isotope results per stratum from Excavation 1. # U-series dates from Beaumont and Vogel (2006:Table 2). All other dates are taken from Matmon *et al.* (2012). * in the middle of Stratum 9 there is a lithostratigraphic unconformity placed at the Bruhnes-Matuyama boundary that is dated to 0.776 Ma.

Stratum	Age (Ma)	n	$\delta^{13}C$	sd	min	max	$\delta^{18}O$	sd	min	max
6	>0.349/0.350#	11	−8.4	1.0	−9.7	−6.7	40.0	3.5	35.2	47.6
7		19	−8.5	0.9	−10.4	−7.0	39.1	3.2	34.1	47.2
8		20	−8.7	0.8	−10.5	−7.7	38.8	2.9	32.9	45.5
9	0.99–0.78*	16	−8.3	1.2	−10.6	−6.6	38.7	3.7	33.6	47.8
10	0.99–1.07	16	−7.5	2.5	−10.0	−1.5	36.0	2.2	32.8	40.8
11	1.07–1.78	1	−10.0	–	–	–	35.2	–	–	–
12	1.78–1.96	5	−8.6	0.3	−8.9	−8.1	39.4	1.8	36.4	41.1

<12,500 (Lee-Thorp and Ecker, in press). At this point, there is a major stratigraphic break in the Excavation 1 sequence, with Strata 5 to 1, associated with Later Stone Age lithic industries, spanning approximately the last 12,500 years. Despite some of the limitations outlined above, it is evident that the Wonderwerk Cave record has the potential to offer invaluable information on past environmental and climatic conditions in this under-researched region of southern Africa.

7.1.2 Stable isotope analysis

Stable light isotope analysis is a well-established method for reconstructing past environments. Despite its widespread application in archaeology, few studies have explored stable isotopes in inorganic OES carbonate in southern Africa (von Schirnding, *et al.*, 1982; Johnson *et al.*, 1997; Bousman, 2005; Lee-Thorp and Ecker, in press).

Ostriches, *Struthio camelus*, are non-obligate drinkers. Today wild ostriches live in open environments and can survive without drinking by eating tender green plants and recycling metabolic water (Williams *et al.*, 1993). Their bodywater pool is therefore strongly influenced by ^{18}O enrichment in plants due to evapotranspiration. Oxygen isotope fractionation between bodywater and $CaCO_3$ is about 30‰ (following the standard fractionation from H_2O to $CaCO_3$; Sharp, 2007). Thus eggshell $\delta^{18}O$ values ($\delta^{18}O_{CaCO_3}$) strongly reflect the influence of humidity or relative humidity (RH), over and above the regional controls exerted by meteoric water $\delta^{18}O$ values ($\delta^{18}O_{mw}$). Modern studies show that OES $\delta^{18}O_{CaCO_3}$ is highly correlated with relative humidity and Mean Annual Precipitation (MAP) (Lee-Thorp and Ségalen, pers. comm. 29.10.2014).

Ostriches are, opportunistic, non-selective herbivores eating C_3, C_4 and CAM plants, with the tenderness of the plants being the overriding factor (Milton *et al.*, 1994; Williams *et al.*, 1993). The fractionation between the plant diet and eggshell $CaCO_3$ (ε^*plant-$CaCO_3$) is about 15‰ for $\delta^{13}C$ (von Schirnding *et al.*, 1982; Johnson *et al.*, 1997, 1998). Given the ostrich's requirement for tender green plants, the blood bicarbonate, and hence eggshell $CaCO_3$ isotopic composition, is not a direct reflection of the proportions of C_4 (or CAM) plants in the total local plant biomass, but rather the proportion which the ostrich finds palatable. They breed in the rainy season when there are abundant resources (Sauer and Sauer, 1966; Williams *et al.*, 1993) and eggs develop and are laid within 3–5 days. Thus, the stable light isotope signal reflects a very short period of time. To obtain a signal of the variation in ostrich diet, a sample size of at least five to 10 OES fragments per unit is recommended (Johnson *et al.*, 1998).

7.1.3 Eggshell biometry

Ostrich eggshell is composed of five different layers that vary with respect to their structure and thickness (Sauer, 1972; Sparks and Deeming, 1996; Richards *et al.*, 2000). There are three calcified layers—the mammillary layer, the palisade layer and vertical crystal layer, with an additional two non-calcified layers, the inner shell membrane and the outer shell membrane, neither of which are usually preserved in archaeological or fossil material. Running vertically through the eggshell are pore canals, which provide respiratory gas exchange (oxygen uptake, carbon dioxide ventilation) and water vapour conductance for the developing embryo. The ostrich embryo obtains its oxygen entirely by diffusion through the pores in the shell, which are branched to improve conductance and ventilation (Board, 1980). The functional conductance properties of the eggshell are established by the hen at the time of egg formation (Rahn *et al.*, 1979, 1987).

Eggshell conductance, measured as shell thickness and porosity (the latter measured by pore number, pore density and pore size), is defined as the quantity of gases (the supply of O_2 and removal of CO_2 from the embryo) that diffuse in a unit of time through the pores of an eggshell (e.g. Ar *et al.*, 1974; Rahn *et al.*, 1979; Paganelli, 1980). Conductance has been shown to be directly proportional to the area of the pores available for diffusion of water vapor and gases, and is inversely proportional to the length of the diffusion path, i.e. the length of the pore canals within the eggshell, that can be approximated by measurement of eggshell thickness (Rahn *et al.*, 1979). Thus, porous shells with a greater number of, or larger pores, and/or thinner shells, have higher gas exchange rates but risk of losing excessive quantities of water, resulting in dehydration of the embryo. In contrast, eggs with reduced porosity, i.e. those with thicker shells and/or low pore numbers and hence reduced pore densities will experience reduced conductance (Rahn *et al.*, 1979; Ar and Rahn, 1985; Ar, 1991). Clearly, an increase in eggshell thickness and/or reduction in porosity can negatively affect the developing chick, even resulting in mortality due to insufficient transpiration (e.g. Swan and Brake, 1990; Oviedo-Rondón *et al.*, 2008). Moreover, unlike other avian species, the ostrich chick has no egg tooth and hatches with the assistance of its limbs (Deeming, 1997), such that an excessive increase in eggshell strength may result in failure to break the eggshell.

In much the same way that isotopic composition of eggshell (modern and fossil), can serve as a record of the environmental conditions at the time of egg laying (Lee-Thorp and Talma, 2000; Ségalen *et al.*, 2006, Kingston, 2011), it has been shown that there is a relationship between egg conductance and nesting environment, i.e. temperature, but especially humidity (e.g. Board and Scott, 1980; Portugal *et al.*, 2010; Deeming, 2011; Stein and Badyaev, 2011; Pulikanti *et al.*, 2012). Species breeding in very humid environments have a higher degree of eggshell porosity than those breeding in very arid environments, an adaptation apparently aimed at regulating water loss (e.g. Ar and Rahn, 1985; Yom-Tov *et al.*, 1986; Thompson and Goldie, 1990). Moreover, Stein and Badyaev (2011) have shown that changes in eggshell conductance can occur rapidly. This concept has laid the basis for studies reconstructing palaeoenvironments and palaeoclimates based on the examination of conductance properties of fossil eggshells (e.g. Williams *et al.*, 1984; Deeming, 2006; Jackson *et al.*, 2008; Donaire and López-Martínez, 2009).

7.2 MATERIALS AND METHODS

7.2.1 Biometry

The Wonderwerk Cave sample of *Struthio camelus* (ostrich) eggshell fragments, from the early-mid Pleistocene layers, Strata 12 through 7 (henceforth termed 'Pleistocene'), comprised a total of 196 fragments (Table 2). There is a clear increase in OES fragment abundance over time, from ten fragments at the base of the sequence (Strata 12 and 11 combined), to 94 fragments in Stratum 7 at the top of the Pleistocene sequence. The 435 OES fragments from the topmost Holocene strata examined in this study (Strata 3–1), serve as a sub-recent proxy and represent random sub-samples taken from the hundreds of fragments found in these layers, which span the period 4800 BP to the present (Table 2). The trend of increasing OES abundance is primarily due to the larger volume of deposit excavated higher up in the sequence, but it also reflects greater exploitation of ostrich eggshell in the Holocene, i.e. a higher density of fragments. Where possible, fragments from different excavation squares and/or elevations within a stratum were sampled, in order to minimise potential inclusion of fragments from the same egg.

The OES fragments are generally well preserved though some are blackened as a result of burning or manganese encrustation. None of the fragment edges were rounded or abraded as might be expected in eggshell from high-energy palaeoenvironments (Oser and Jackson, 2014). The outer surface of many fragments exhibited fine striae, probably the result of abrasion by sediment grains and trampling by animals and hominins (Andrews and Cook, 1985; Olsen and Shipman, 1988).

For each of the 196 OES fragment from the early-mid Pleistocene strata, shell thickness was measured by means of a digital caliper, accurate to 0.01 mm. In addition, the sample of 435 fragments from the Holocene sequence (Strata 1–3) were similarly measured for comparison. All measurements were taken by one author. Measurement error calculated from repeated measurements of a random sample of 50 fragments, was <2%. Using a Mann Whitney U-test, measurements were tested for pairs of strata with statistical significance set at $P < 0.05$ using the PAST freeware (Hammer *et al.*, 2001).

Table 2. Measurement of eggshell thickness using a caliper (in mm). A statistically significant difference was found between Stratum 10 and 9. Pleistocene ages after Matmon *et al.* (2012), Holocene ages after Lee-Thorp and Ecker (in press).

Age	Stratum	N	mean	max	min	sd
Holocene BP						
last 100 yrs	1	201	1.97	2.53	1.54	0.16
190–270	2	113	1.93	2.46	1.43	0.17
2060–4800	3	121	1.91	2.26	1.49	0.13
Pleistocene Ma						
	7	99	1.81	2.27	1.31	0.16
	8	52	1.79	2.02	1.46	0.15
0.99–0.78*	9	44	1.83	2.1	1.28	0.16
0.99–1.07	10	19	1.71	2.1	1.23	0.23
1.07–1.78	11	2	1.91	1.97	1.84	
1.78–1.96	12	8	1.61	1.95	0.97	0.36

7.2.2 Histology

A sub-sample of 39 OES fragments was prepared for histological analysis following Williams *et al.* (1984). The breakdown of these samples is: 17 OES fragments from the seven early/mid-Pleistocene Strata 7–12 (Figure 2); one fragment from Stratum 6 (possibly mixed Pleistocene-Holocene); 20 OES fragments from five Holocene levels (Strata 1–5) spanning the last 12,500 years (Figure 2; Figure 7). In addition, one fragment from an extant *Struthio c. camelus* was analysed in the same way and used as a standard. Each eggshell fragment was embedded in clear Epofix Resin, to prevent the fragments from disintegrating during the thin sectioning process. The embedding was completed within a Struers CitoVac vacuum chamber to remove air bubbles from within the eggshell fragments and surrounding resin. Once the resin had set, sequential transverse sections of each egg shell fragment were cut, approximately 1.5 mm thick, using a cut-off diamond-tipped saw within a Struers Accutom-50 cut-off grinding machine. Each section was then mounted onto a frosted petrographic glass slide using the resin adhesive, Epoxy Resin LR20 and catalyst, which had been placed in the Struers CitoVac vacuum chamber to remove air bubbles. Pressure was then applied to the sections to eliminate any excess bubbles. Once set, the sections were ground to approximately 60 µm thick using a diamond-tipped grinding wheel within the Struers Accutom-50 cut-off grinding machine. The resulting thin sections were polished until smooth using a Struers LaboPol-5 polishing machine. Thin sections were viewed and photographed using a Nikon Eclipse 50i Polarizing microscope and DS-Fi1 digital camera, and then analysed using the image analysis program NIS Elements D 3.2. Micro-measurements taken on this sub-sample (after Donaire and López-Martínez, 2009) include eggshell thickness, sampled eggshell surface area, pore number, mean individual pore area, pore density, pore individual length, width, aperture size and diameter.

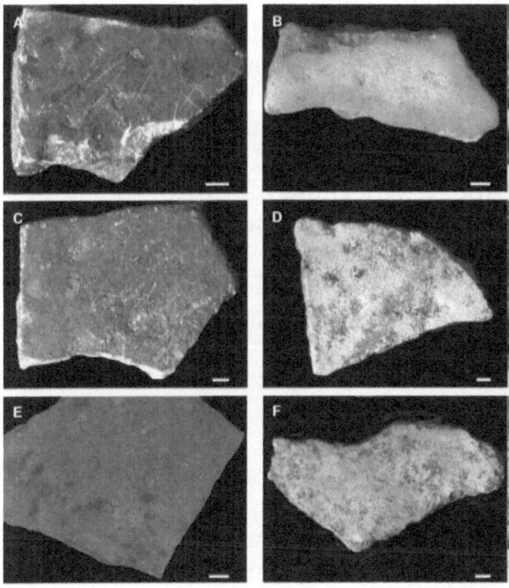

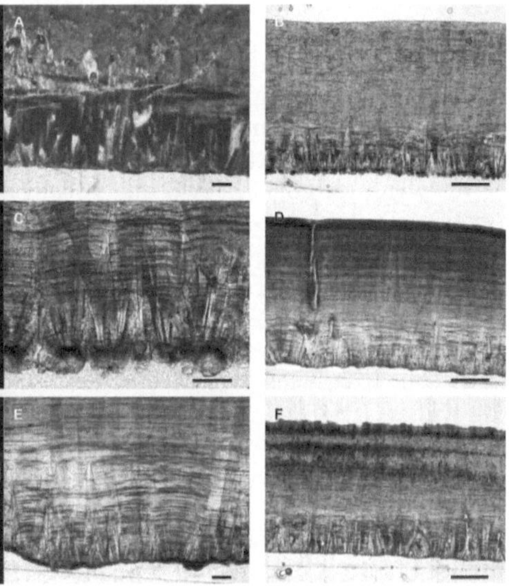

Figure 2a (left). Surface images of eggshell fragments examined (A) Str. 9 #4442, (B) Str. 9 4435, (C) Str. 10 #4397, (D) Str.0 #4383, (E) Str. 11 R29, (F) Str. 12 #84. Scale bars = 1 cm.
Figure 2b (right). Transverse sections of the egg shell microstructure (A) Str. 10 #4386, (B) Str.11 #623, (C) Str.11 #623, (D) Str. 12 #23b, (E) Str.12 #23b, (F) Str. 12 #62. A, C, E Scale bars = 100 µm, B, D, F Scale bars = 500 µm.

7.2.3 Stable isotope analysis

We measured 88 OES fragments from seven archaeological strata (Strata 12 to 6) in Excavation 1 for carbon and oxygen stable light isotopes of the inorganic fraction. Since we expect large variation between ostrich eggshell fragments, we measured >10 samples per stratum where possible.

A small corner of each fragment was removed with a diamond tipped saw blade in a handheld dremel drill. The fragment was then cleaned in an aluminium-oxide blaster and finely crushed in an agate mortar and pestle. Samples were analysed at Bradford University, using a Finnigan Gasbench II, connected to a Thermo Delta V Advantage continuous flow isotope ratio mass spectrometer. The carbonate fraction of each sample was reacted with phosphoric acid (100%) at 70°C to release CO_2, for measurement of $^{13}C/^{12}C$ and $^{18}O/^{16}O$ ratios. The reference gas was calibrated against one international (NBS 19), and three laboratory carbonate standards (MERCK $CaCO_3$, CO-1, OES), which were interspersed in all runs. The results for both isotopes are expressed as per mil (‰) in the delta (δ) notation versus the international VPDB and VSMOW standard respectively, as follows. $\delta^{13}C_{VPDB} = (R_{sample} - R_{ref})/R_{ref} \times 1000$ where $R = {}^{13}C/{}^{12}C$. $\delta^{18}O_{VSMOW} = (R_{sample} - R_{ref})/R_{ref} \times 1000$, where $R = {}^{18}O/{}^{16}O$. Analytical precision as determined from multiple replicates of the OES laboratory standard, was better than 0.1‰ for $\delta^{13}C$ and 0.2‰ for $\delta^{18}O$.

7.2.4 Scanning Electron Microscopy (SEM)

A sub-set of four OES fragments from the ESA strata were examined under a Zeiss DSM 940 A Scanning Electron Microscope (SEM) coupled with a Noran System 6 energy dispersive X-ray spectrometer (EDX). The samples examined comprised two fragments from Stratum 12 and two from Stratum 7. OES fragments with no evident surface concretions were chosen and had been cleaned with distilled water to remove adhering dust. The aims of this study were to observe the structure of OES from radial sections as well as the inner and outer surfaces of the fragments, and to investigate differences in major element composition of OES fragments from different time periods.

Samples were attached with carbon-coated tape to a stub, but the samples themselves were not coated. Two analytical positions were chosen for each sample—the outer surface of the shell and a radial section. EDX measurements were made using an accelerating voltage of 15 to 25 kV. The elements determined were C, O, Mg, Al, Si, P, S, K, Ca, Mn and Fe. Two analyses were performed at two different locations on the surface of each sample and the results averaged to obtain an individual sample mean.

7.3 RESULTS

7.3.1 Species identification

Currently the biogeographic distribution of wild ostriches is limited to Africa, where four sub-species of *Struthio camelus* are found (Dyke and Leonard, 2012). Based on an independently Ar-Ar dated volcanic tuff sequence from Laetoli (Tanzania), Harrison and Msuya (2005) place the emergence of *S. camelus* at 3.6–3.8 Ma onwards, replacing an earlier form, *S. daberasensis*. In Namibia, based on micro-mammalian chronostratigraphy, it has been proposed that this replacement occurred only ca. 2 million years ago (Senut and Pickford, 1995; Senut *et al.*, 1998; Senut, 2000; Ségalen *et al.*, 2002). In these studies, species identifications were based on differences in eggshell morphology (pore complex widths and densities) as well as thickness, with modern *S. c. camelus* having a thinner eggshell, much smaller individual pore pits (average pore diameter is 0.02–0.03 mm), and a higher pore density (100/cm^2) than fossil ratites (Pickford and Senut, 2000; Stidham, 2004; Harrison and Msuya, 2005).

Examination of eggshell thickness in the Wonderwerk Cave early/mid-Pleistocene and Holocene specimens (Table 2) indicates that the mean thickness falls within the range of extant ostriches which is given as 1.6–2.2 mm (mean 2.0 mm) by Harrison and Msuya (2005). Similarly, the shape and size of the pores, both at the surface and in cross section, of the Wonderwerk fragments most closely resemble that of *S. c. camelus* (Figures 2a and b). This was confirmed by the SEM analyses, which demonstrated that irrespective of the stratum sampled, the Wonderwerk Cave early/mid-Pleistocene OES fragments had the same physical structure and elemental composition as modern *S. c. camelus* (Figure 3). Therefore, it was concluded that all the Wonderwerk Cave eggshell fragments belong to the same ostrich species, *Struthio camelus cf. camelus*.

7.3.2 Stable isotope analysis

Basic statistics for the $\delta^{13}C$ and $\delta^{18}O$ results are given in Table 1. Boxplots showing the distribution of the results, with data points, medians and range, are shown in Figure 4, and a scatterplot of the mean values per stratum in Figure 5.

The carbon isotope data show that ostrich diet (at least in the egg-laying season) was largely dominated by C_3 plants, with a small but consistent contribution of C_4. The standard deviation for mean carbon isotope values for each of the Pleistocene strata is 2.5‰. There is a visible shift between Strata 12/11 and Stratum 10. Stratum 10 has the widest distribution of carbon isotope values, ranging from a predominantly C_3 (minimum value −10‰) to a predominantly C_4 diet (maximum value −1.5‰). A Levene's test confirmed that the data are not normally distributed in this stratum. The mean value of −7.5‰ is more negative than all strata except Stratum 11, so the extreme C_4 values seem to represent short spells within the long time span covered by this stratum. In contrast, the median $\delta^{13}C$ (as indicated in the boxplots) is similar in all strata except for Stratum 11. Stratum 10 shows high variability for $\delta^{13}C$ (as indicated by the standard deviation) and some of the most ^{13}C-enriched values in the sequence.

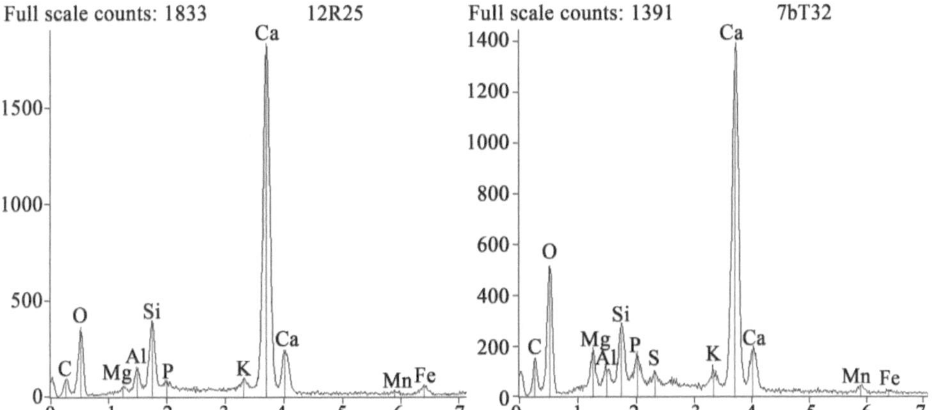

Figure 3. Two plots from the SEM/EDX showing the elemental composition on outer surface of (a) an OES fragment from Stratum 12, Square R25. (b) and an OES fragment from Stratum 7b, Square T32.

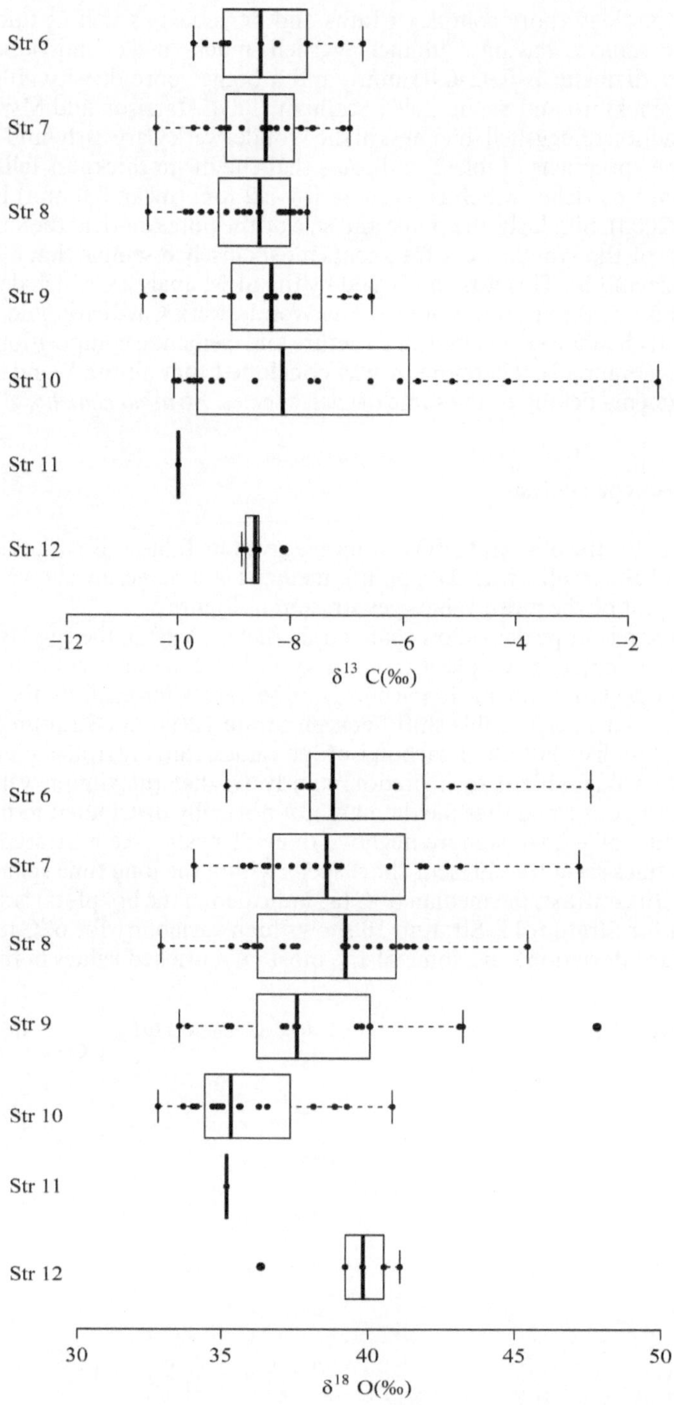

Figure 4. Boxplots showing the distribution of carbon (above) and oxygen (below) isotope data for Wonderwerk Cave Strata 12–6. The $\delta^{13}C$ values are expressed relative to VPDB, while $\delta^{18}O$ values are expressed relative to VSMOW.

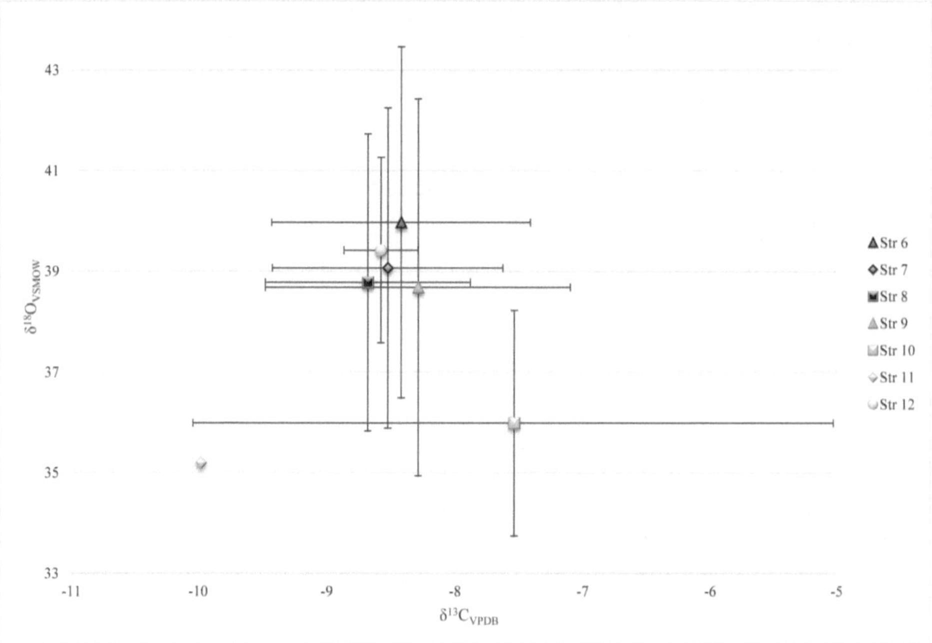

Figure 5. Scatter plot of OES isotope result carbon and oxygen mean values per stratum. Error bars represent 2 sigma standard deviations..

Overall, the oxygen isotope data for the early/mid-Pleistocene sequence indicate a generally arid setting for Wonderwerk Cave with many $\delta^{18}O$ values near or above +35‰. The $\delta^{18}O$ data are more variable than the $\delta^{13}C$ data. The lowest individual value is 32.8‰, and the maximum individual value is 47.8‰. The means range by 4.8‰.

Stratum 12 shows ^{18}O-enriched values, indicating arid conditions. However, the sample size for Stratum 12 is small (N = 5) and the results might therefore not cover the full range of variation. This is also the case for Stratum 11, where only one sample was available for analysis. This sample yielded low $\delta^{18}O$ and $\delta^{13}C$ values indicating moister conditions and C_3 plant foods. Although an isolated specimen, this result is consistent with phytolith analysis which shows a change from a more arid environment in Stratum 12 to more moist conditions with more C_3 grasses in Stratum 11 (Chazan *et al.*, 2012).

Mean values for oxygen isotopes increase from Stratum 11 upwards. As is the case for the $\delta^{13}C$ findings, Stratum 10 differs from the other strata, showing a cluster of lower $\delta^{18}O$ values indicating moister conditions. In a series of Mann-Whitney U-tests between paired strata, the $\delta^{18}O$ values of Stratum 10 were found to be significantly different from the others. In an ANOVA with a Tukey HSD test, no statistically significant differences were found between any of the other strata for either carbon or oxygen. These results suggest a highly variable environment with large shifts in moisture within the time span represented.

If we examine the isotope values for each stratum, taking into account spits which indicate depth within a stratum, then the shift from moist to arid conditions occurs not between Stratum 10 and 9, but within lower Stratum 9. Interestingly, in terms of micromorphology, there is a major unconformity within Stratum 9 between two lithostratigraphic units (Units 1 and 2), which marks a change from gray/white silt in a fine sand, to sediment rich in quartz grains in powdery carbonate (Matmon *et al.*, 2012; Goldberg *et al.*, in press). This unconformity appears to represent the Brunhes-Matuyama boundary (Matmon *et al.*, 2012; Goldberg *et al.*, in press).

It may be concluded that the carbon and oxygen isotopes data for Wonderwerk Cave OES fragments follow independent trends. This is not entirely surprising as they reflect different environmental variables, although it is often assumed that C_4 plants suggest greater aridity. Overall, ostrich diet at Wonderwerk Cave was primarily composed of C_3 plant foods, with a small contribution of C_4 plants. The proportions of C_4 plants do not follow the linear trend of aridity observed in the $\delta^{18}O$ record. Strata 12 through 10 have lower $\delta^{18}O$ values indicative of moist conditions. From Stratum 9 upwards, mean $\delta^{18}O$ values are similar, and the data indicate drier conditions (or at least more frequent dry spells). The low variability expressed in the means could be influenced by temporal smoothing, since the time depth and ranges represented by the upper strata of these Pleistocene layers remains unclear.

7.3.3 Eggshell thickness

Table 2 gives the range of eggshell thickness measured using a caliper for Strata 12–7. Significant differences were found between Strata 10 and 9 ($P < 0.035$). As illustrated in the box plots given in Figure 6, these result can primarily be attributed

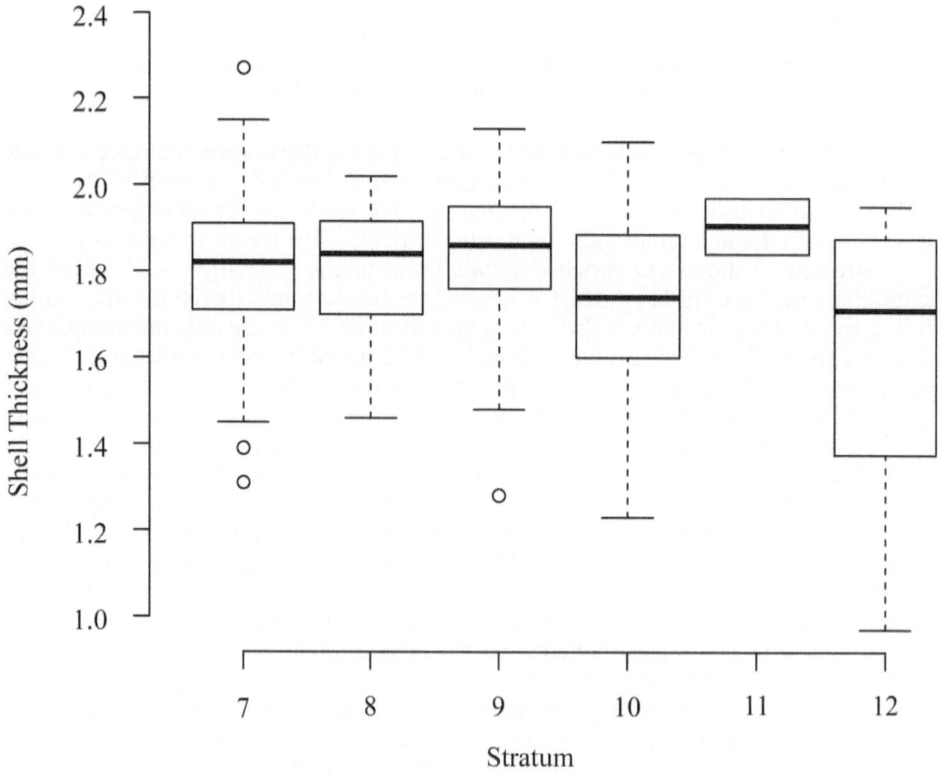

Figure 6. Box plot showing OES thickness for early/mid-Pleistocene strata.
Note the shift in the lower size range in samples from Strata 12–10 versus Strata 9–7.

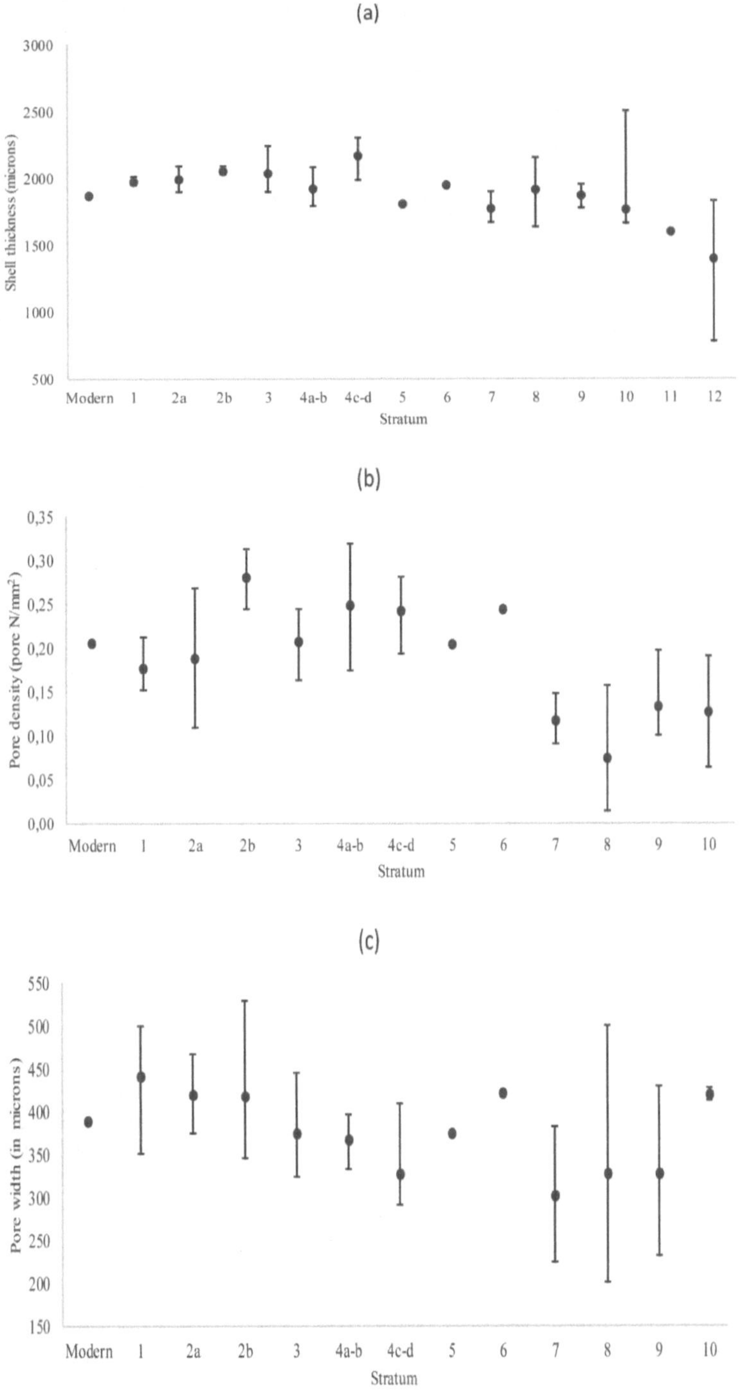

Figure 7. High-Low charts showing measurements taken on a sub-sample of OES fragments prepared for histology by stratum. Shown are mean values as well as maximum and minimum values in error bars. (a) Eggshell thickness (b) Pore density (c) Pore width.

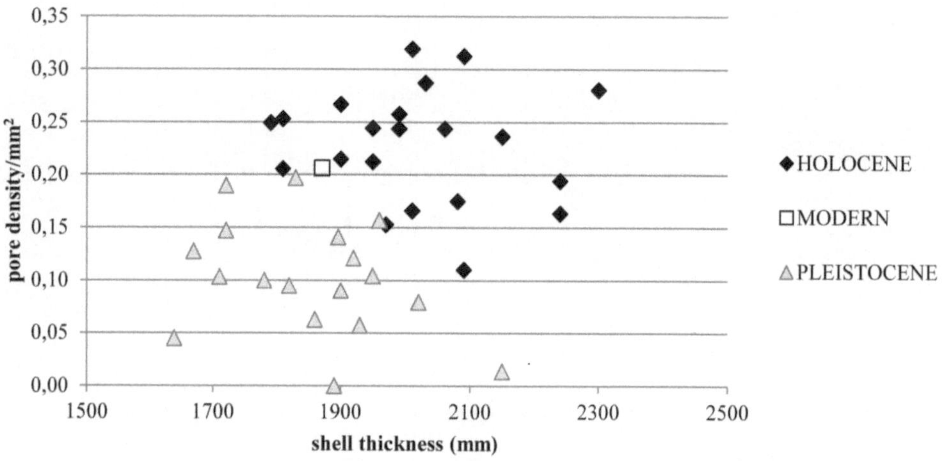

Figure 8. Scattergram of eggshell thickness against pore density for the sub-sample of OES fragments examined using histology. Note the clear and statistically significant separation of the Holocene and early/mid- Pleistocene samples indicating thinner eggshell with fewer pores in the Pleistocene sample.

to a shift in the lower-most size range. The thinner eggshell fragments present in Strata 12–10 compared to those in Strata 9–7, indicate moister conditions in the earlier layers.

7.3.4 Pore density and thickness

The small sub-samples of OES fragments examined for pore density and size were also measured for thickness. Figures 7a–c summarise the results and demonstrate the general trends. The results show an increase in eggshell thickness from the earliest to the latest strata and indicate a concomitant increase in pore density (i.e. the number of pores per mm^2) accompanied by an increase in pore size (width). Since the sample sizes are extremely small, this trend could not be tested for statistical significance, but supports the conclusions based on measurements of shell thickness for the larger samples described above. Given the small size of the samples, data on pore density and shell thickness for all the early/mid-Pleistocene strata were combined as were those of the Holocene. Significant differences were found in both parameters between periods ($P < 0.000$; Figure 8), with the Holocene sample having thicker eggshell, but wider as well as a greater number of pores than the Pleistocene fragments, indicative of drier/warmer conditions in the Holocene.

7.4 DISCUSSION

Based on the information presented here, it is evident that only one ratite species, *Struthio camelus,* was represented in the Wonderwerk Cave early/mid- Pleistocene and Holocene sequence. Notably, ostrich eggshell fragments (and skeletal remains) are extremely rare in published Plio-Pleistocene faunal lists from southern Africa pre-dating the Middle Stone Age. Exceptions are Swartkrans Member 5, where nine fragments of eggshell (as well as an ostrich phalanx) were identified and attributed to *Struthio* sp. (Watson, 1983) and Kromdraai B where a few ostrich eggshell fragments and bones were found (Brain, 1981), while McKee (1993) noted the presence of eggshell in the "Dart deposits" D-A at Taung, though no specific identification was given. In this respect Wonderwerk Cave more closely resembles East African mid-Pleistocene sites, such as Olduvai (BK) Bed II (Hay, 1976) where large quantities of ostrich eggshell were discovered.

A phase of highly variable climate has been documented for the period ca. 1.9–1.7 Ma which is the time-span covering Strata 12 to 11 at Wonderwerk Cave. It has been associated with the development of the Walker circulation (Cerling *et al.*, 1988; Hopley *et al.*, 2007; Maslin *et al.*, 2014) and was characterised by expansion of the savannah biome (Hopley *et al.*, 2007) resulting in a shift in large mammal communities to more grazing species (Bobe and Behrensmeyer, 2004; Lee-Thorp *et al.*, 2007). Our carbon and oxygen stable isotope data for Strata 12 at Wonderwerk Cave, indicate a generally arid, C_3 environment with a small component of C4 plants. Overall, these results are corroborated by the findings for grass phytoliths (Chazan *et al.*, 2012), which indicate a warm Savanna/Nama Karoo grassland in Stratum 12 with a slight shift by the top of Stratum 11—which becomes more pronounced in the overlying layers, towards an even more arid environment supporting a modern Succulent Karoo biome. Similarly, the data on eggshell thickness for Strata 12–10, follows a shift of increasing aridity for the egg producing periods.

It has been proposed that ca. 1 Ma, there was a phase of climate instability in Africa (deMenocal, 1995; Maslin and Christensen, 2007; Trauth *et al.*, 2009; Potts, 2013; Maslin *et al.*, 2014). In East Africa at this time there is evidence for wet phases (Trauth *et al.*, 2010) as well as aridity (deMenocal, 2004), and East African soil carbonate data shows a C_4 spike in soil carbonate $\delta^{13}C$ around 1 Ma and more positive values thereafter (Levin, 2013). The records for southern Africa at this time include geological data from a dated section in the Mamatwan Mine, located ca. 40 km northwest of Wonderwerk Cave in the Northern Cape Province, South Africa, that corroborate this change. At this locality, Matmon *et al.* (2014) have documented moist conditions in the southern Kalahari ca. 1.5–1 Ma based on the presence of a stable though shallow, low-energy water body in the region. This water body then disappeared due to rapid sediment accumulation close to the onset of the early-middle Pleistocene Transition ca. 1 Ma. Our stable isotope data suggests fluctuating conditions in Stratum 10 (ca. 1 Ma) with C_3 dominance as well as extreme C_4 spells in the vegetation, in an overall moist setting. As this is the best constrained early/mid-Pleistocene stratum, covering less than 100 ka years, the stable isotope analysis is able to pick up a more complex local picture of high variability. The significant change in eggshell thickness and pore density between Strata 10 and 9 supports this assertion.

All subsequent Pleistocene strata at Wonderwerk Cave represent a mixed C_3/C_4, warm environment. This is supported by stable isotope values for bovids which demonstrate elevated C_4 grass consumption, in conjunction with more limited faunal diversity during the period 1.0–0.6 Ma, indicative of a trend towards semi-arid conditions in the interior of southern Africa (Codron *et al.*, 2008; Brink *et al.*, 2012).

Examining these results in a broader context, there is a statistically significant difference in OES associated early/mid-Pleistocene deposits compared to those associated with late Holocene deposits. The thicker eggshell in the late Holocene is associated with an increase in pore density indicating more arid conditions in the more recent periods. Holocene OES isotopes from Wonderwerk Cave show small, but persistent amounts of C_4 in the ostrich diet, similar amounts to the Pleistocene strata, without any extremes. In contrast, the $\delta^{18}O$ record or the Holocene indicates a sequence of mostly arid, but highly variable, moisture levels through the sequence. We were able to identify moist peaks at the beginning of the Holocene and in the mid-Holocene but with increasing aridity in the uppermost Strata 3 to 1 (Lee-Thorp and Ecker, in press), which is broadly consistent with the palynological results for the site (van Zinderen Bakker, 1982; Scott *et al.*, 2012) and the biometric data presented here.

7.5 CONCLUSION

The results of the analyses presented, clearly demonstrate diachronic changes in biometric parameters of Pleistocene ostrich eggshell at Wonderwerk Cave. The most salient trend, supported by direct measurement of OES fragments as well as measurements of histologically prepared samples of OES, is an increase in the thickness of eggshell. Within the early/mid-Pleistocene sequence, there is a statistically significant increase in eggshell thickness between Stratum 10 and 9. In a palaeoclimatic context, this indicates that the thinner eggshell with fewer and smaller pores, which characterises Strata 12–10, was laid under relatively moister climatic conditions. Following this, the climate shifted to a more arid regime which continued into the Holocene and probably became increasingly arid over time. Of particular note is that this scenario agrees with the palaeoclimatic reconstruction obtained from carbon and oxygen isotopes, measured on OES fragments from the same strata at Wonderwerk Cave.

The findings presented here show a large degree of concordance in results obtained using two independent methods for reconstructing palaeoenvironment that are based on different principles, and which were applied to the same OES sample. We propose that, taken together, they can provide complementary measures of palaeoclimatic conditions.

All proxies show variable shifts between early/mid-Pleistocene Strata 9 and 10 (ca. 1 Ma) at Wonderwerk Cave. This coincides with a proposed global cooling phase at the initiation of the 100 ky glacial-interglacial cycles that resulted in highly fluctuating climatic conditions. In conclusion, the early/mid-Pleistocene sequence from Wonderwerk Cave has proved to be a repository of invaluable information on palaeoenvironmental conditions in the interior of South Africa for a time period which has been poorly represented in the regional record.

ACKNOWLEDGEMENTS

We would like to extend our warm thanks to: Michael Chazan and the Wonderwerk Cave team for their assistance and support; Louis Scott for kindly providing Figure 1; Andy Gledhill (University of Bradford) for assistance with the stable isotope measurements at the Stable Isotope Laboratory, Division of Archaeological, Geographical and Environmental Sciences at Bradford University, UK; James Brink (Director of the Florisbad Quaternary Research Center, National Museum, Bloemfontein, South Africa) for hosting LKH and ME while undertaking parts of this study; and two reviewers for their comments.

ME received funding from the German Academic Exchange Service (DAAD), the School of Archaeology and the Boise Trust Fund, University of Oxford; JBB acknowledges the National Research Foundation (Grant No. 91602) for financial support.

REFERENCES

Andrews, P. and Cook, J., 1985, Natural modifications to bones in a temperate setting. *Man,* **20**, pp. 675–691.

Ar, A., 1991, Egg water movement during incubation. In *Avian Incubation*, edited by Tullet, S.G. (London: Butterworth-Heinemann), pp. 157–174.

Ar, A. and Rahn, H., 1985, Pores in avian eggshells: Gas conductance, gas exchange and embryonic growth rate. *Respiratory Physiology,* **61**, pp. 1–20.

Ar, A., Paganelli, C.V., Reeves, R.B., Greene D.G. and Rahn H., 1974, The avian egg: Water vapor conductance, shell thickness and functional pore area. *The Condor,* **76**, pp. 153–158.

Beaumont, P.B., 1990, Wonderwerk Cave. In *Guide to the Archaeological Sites in the Northern Cape,* edited by Beaumont, P.B. and Morris, D. (Kimberley: McGregor Museum) pp. 101–134.

Beaumont, P.B. and Vogel, J.C., 2006, On a timescale for the past million years of human history in central South Africa. *South African Journal of Science,* **102**, pp. 217–228.

Berna, F., Goldberg, P., Horwitz, L.K., Brink, J., Holt S., Bamford, M. and Chazan, M., 2012, Microstratigraphic evidence of in situ fire in the Acheulean strata of Wonderwerk Cave, Northern Cape province, South Africa. *Proceedings of the National Academy of Sciences,* **109**, pp. 1215–1220.

Board, R.G., 1980, The avian eggshell—a resistance network. *Journal of Applied Bacteriology,* **48**, pp. 303–313.

Board, R.G. and Scott, V.D., 1980, Porosity of the avian eggshell. *American Zoologist,* **20**, pp. 339–349.

Bobe, R. and Behrensmeyer, A.K., 2004, The expansion of grassland ecosystems in Africa in relation to mammalian evolution and the origin of the genus *Homo*. *Palaeogeography, Palaeoclimatology, Palaeoecology,* **207(3)**, pp. 399–420.

Bousman, C.B., 2005, Coping with risk: Later Stone Age technological strategies at Blydefontein Rock Shelter, South Africa. *Journal of Anthropological Archaeology,* **24**, pp. 193–226.

Brain, C.K., 1981, *The Hunters or the Hunted? An Introduction to African Cave Taphonomy*. (Chicago: University of Chicago Press).

Brink, J.S., Herries, A.I.R., Moggi-Cecchi, J., Gowlett, J.A.J., Bousman, C.B., Hancox, J.P., Grün, R., Eisenmann, V., Adams, J.W. and Rossouw, L., 2012, First hominine remains from a ~1.0 million year old bone bed at Cornelia-Uitzoek, Free State Province, South Africa. *Journal of Human Evolution,* **63**, pp. 527–535.

Brooks, A.S., Hare, P.E., Kokis, J.E., Miller, G.H., Ernst, R.D. and Wendorf, F., 1990, Dating Pleistocene archaeological sites by protein diagenesis in ostrich eggshell. *Science,* **248**, pp. 60–64.

Cerling, T.E., Bowman, J.R. and O'Neil, J.R., 1988, An isotopic study of a fluvial-lacustrine sequence: The Plio-Pleistocene Koobi Fora sequence, East Africa. *Palaeogeography, Palaeoclimatology, Palaeoecology,* **63(4)**, pp. 335–356.

Chazan, M. and Horwitz, L.K., in press, *An Overview of Recent Research at Wonderwerk Cave, South Africa.* Proceedings of the 13th Congress of the PAA/SAfA, Dakar, Senegal 2010.

Chazan, M. and Horwitz, L.K., 2010, *Wonderwerk Cave and Related Research Projects in the Northern Cape Province.* Progress Report submitted to SAHRA, August 2010.

Chazan, M., Avery, D.M, Bamford, M.K., Berna, F., Brink, J., Holt, S., Fernandez-Jalvo, Y., Goldberg, P., Matmon, A., Porat, N., Ron, H., Rossouw, L., Scott, L. and Horwitz, L.K., 2012, The Oldowan horizon in Wonderwerk Cave (South Africa):

Archaeological, geological, paleontological and paleoclimatic evidence. *Journal of Human Evolution,* **63**, pp. 859–866.

Chazan, M., Ron, H., Matmon, A., Porat, N., Goldberg, P., Yates, R., Avery, M., Sumner, A. and Horwitz, L.K., 2008, Radiometric dating of the Earlier Stone Age sequence in Excavation 1 at Wonderwerk Cave, South Africa: Preliminary results. *Journal of Human Evolution*, **55**, pp. 1–11.

Codron, D., Brink, J.S., Rossouw, L. and Clauss, M., 2008, The evolution of ecological specialization in southern African ungulates: Competition- or physical environmental turnover? *Oikos,* **117**, pp. 344–353.

Deeming, D.C., 2011, Importance of nest type on the regulation of humidity in bird nests. *Avian Biology Research,* **4(1)**, pp. 23–31.

Deeming, D.C., 2006, Ultrastructural and functional morphology of eggshells supports the idea that dinosaur eggs were incubated buried in a substrate. *Palaeontology*, **49**, pp. 171–185.

Deeming, D.C., 1997, *Ratite Egg Incubation—A Practical Guide,* (Oxford: Ratite Conference, Oxford Print Centre).

deMenocal, P.B., 2004, African climate change and faunal evolution during the Pliocene–Pleistocene. *Earth and Planetary Science Letters,* **220(1)**, pp. 3–24.

deMenocal, P.B., 1995, Plio-Pleistocene African climate. *Science,* **270**, pp. 53–59.

Donaire, M. and López-Martínez, N., 2009, Porosity of Late Paleocene *Ornitholithus* eggshells (Tremp Fm, south-central Pyrenees, Spain): Palaeoclimatic implications. *Palaeogeography, Palaeoclimatology, Palaeoecology,* **279 (3–4)**, pp. 147–159.

Dyke, G.J. and Leonard, L.M., 2012, Palaeognathae. In *eLS* (Chichester: John Wiley & Sons), http://www.els.net [doi:10.1002/9780470015902.a0001550.pub3].

Feakins, S.J. and deMenocal, P.B., 2008, Golbal and African regional climate during the Cenozoic. In *Cenozoic Mammals of Africa,* edited by Werdelin, L. and Sanders, W.J. (University of California Press), pp. 45–55.

Goldberg, P., Berna, F. and Chazan, M., in press, Deposition and Diagenesis in the Earlier Stone Age of Wonderwerk Cave, Excavation 1, South Africa. *African Archaeological Review*.

Hammer, Ø., Harper, D.A.T. and Ryan P. D., 2001, PAST: Paleontological statistics software package for education and data analysis. *Palaeontologica Electronica,* **4(1)**, http://palaeo-electronica.org/2001_1/past/issue1_01.htm.

Harrison, T. and Msuya, C.P., 2005, Fossil struthionid eggshells from Laetoli, Tanzania: Taxonomic and biostratigraphic significance. *Journal of African Earth Sciences,* **41**, pp. 303–315.

Hay, R.L., 1976, *Geology of the Olduvai Gorge: A Study of Sedimentation in a Semiarid Basin*. (Berkeley: University of California Press).

Hopley, P.J., Weedon, G.P., Marshall, J.D., Herries, A.I.R., Latham, A.G. and Kuykendall, K.L., 2007, High- and low-latitude orbital forcing of early hominin habitats in South Africa. *Earth and Planetary Science Letters,* **256**, pp. 419–432.

Humphreys, A.J. and Thackeray, A.I., 1983, *Ghaap and Gariep: Later Stone Age Studies in the Northern Cape,* (Cape Town: South African Archaeological Society).

Jackson, F.D., Varricchio, D.J., Jackson, R.A., Vila, B. and Chiappe, L.M., 2008, Comparison of water vapor conductance in a titanosaur egg from the Upper Cretaceous of Argentina and a *Megaloolithus siruguei* egg from Spain. *Paleobiology,* **34**, pp. 229–246.

Johnson, B.J., Fogel, M.L. and Miller, G.H., 1998, Stable isotopes in modern ostrich eggshell: A calibration for paleoenvironmental applications in semi-arid regions of southern Africa. *Geochimica et Cosmochimica Acta,* **62(14)**, pp. 2451–2461.

Johnson, B.J., Miller, G.H., Beaumont, P.B. and Fogel, M.L., 1997, The determination of late Quaternary paleoenvironments at Equus Cave, South Africa, using stable isotopes and amino acid racemization in ostrich eggshell. *Palaeogeography Palaeoclimatology Palaeoecology,* **136**, pp. 121–137.

Kingston, J.D., 2011, The Laetoli hominins and associated fauna. In *Paleontology and Geology of Laetoli: Human Evolution in Context. Volume 2: Fossil Hominins and the Associated Fauna* edited by Harrison, T. (Dordrecht: Springer), pp. 293–328.

Lee-Thorp, J.A. and Ecker, M., in press, Holocene climate and environmental changes from stable isotopes in ostrich egg shell at Wonderwerk Cave, South Africa. *African Archaeological Review*.

Lee-Thorp, J.A. and Talma, E.S., 2000, Stable light isotopes and past environments in the Southern African Quaternary and Pliocene. In *The Cenozoic of Southern Africa,* edited by Partridge, T.C. and Maud, R. (Oxford: Oxford University Press), pp. 236–251.

Lee-Thorp, J.A., Sponheimer, M. and Luyt, J.C., 2007, Tracking changing environments using stable carbon isotopes in fossil tooth enamel: An example from the South African hominin sites. *Journal of Human Evolution,* **53**, pp. 595–601.

Levin, N.E., 2013, *Compilation of East African soil carbonate stable isotope data.* Integrated Earth Data Application. http://dx.doi.org/10.1594/IEDA/100231.

Maslin, M.A. and Christensen, B., 2007, Tectonics, orbital forcing, global climate change, and human evolution in Africa: Introduction to the African paleoclimate special volume. *Journal of Human Evolution,* **53**, pp. 443–464.

Maslin, M.A. and Trauth, M.H., 2009, Plio-Pleistocene east African pulsed climate variability and its influence on early human evolution. In *The First Humans—Origins of the Genus Homo,* edited by Grine, F.E., Leakey, R.E. and Fleagle, J.G. (Vertebrate Paleobiology and Paleoanthropology Series: Springer Verlag), pp. 151–158.

Maslin, M.A., Brierley, C.M., Milner, A.M., Shultz,S., Trauth, M.H. and Wilson, K.E., 2014, East African climate pulses and early human evolution. *Quaternary Science Reviews,* **101**, pp. 1–17.

Matmon, A., Hidy A.J., Vainer S., Crouvi, O., Fink D., Erel Y., ASTER TEAM, Horwitz, L.K. and Chazan, M., 2014, *New chronology for the southern Kalahari Group sediments: Implications for sediment cycle dynamics and hominin occupation.* Poster presented at the Annual Meeting of The Geological Society of America, 19–22 October, Vancouver, BC Canada.

Matmon, A., Ron, H., Chazan, M., Porat, N. and Horwitz, L.K., 2012, Reconstructing the history of sediment deposition in caves: A case study from Wonderwerk Cave. South Africa. *Geological Society of America Bulletin,* **124(3–4)**, pp. 611–625.

McKee, J.K., 1993, Faunal dating of the Taung hominid fossil deposit. *Journal of Human Evolution*, **25**, pp. 363–376.

Miller, G.H., Beaumont, P.B., Jull, A.J.T. and Johnson, B., 1992, Pleistocene geochronology and paleothermometry from protein diagenesis in ostrich eggshells: Implications for the evolution of modern humans. *Philosophical Transactions of the Royal Society, London B,* **337**, pp. 149–157.

Milton, S.J., Dean, W.R.J. and Siegfried, W.R., 1994, Food Selection by Ostrich in Southern Africa. *The Journal of Wildlife Management*, **58(2)**, pp. 234–248.

Mucina, L. and Rutherford, M.C., 2006, *The vegetation of South Africa, Lesotho and Swaziland,* (South African National Biodiversity Institute).

Olsen, S.L. and Shipman, P., 1988, Surface modification on bone: Trampling versus butchery. *Journal of Archaeological Science,* 15, pp. 535–553.

Oser, S.E. and Jackson, F.D., 2014, Sediment and eggshell interactions: Using abrasion to assess transport in fossil eggshell accumulations. *Historical Biology,* **26(2)**, pp. 165–172.

Oviedo-Rondón, E.O., Small, J., Wineland, M.J., Christensen, V.L., Mozdziak, P.S., Koci, M.D., Funderburk, S.V.L., Ort, D.T. and Mann, K.M., 2008, Broiler embryo bone development is influenced by incubator temperature, oxygen concentration and eggshell conductance at the plateau stage in oxygen. *British Poultry Science,* **49(6)**, pp. 666–676.

Paganelli, C.V., 1980, The physics of gas exchange across the avian egg. *American Zoology,* **20**, pp. 329–349.

Pickford, M. and Senut, B., 2000, Geology and paleobiology of the central and southern Namib Desert, Southwestern Africa, vol. 1: Geology and History of Study. *Geological Survey of Namibia, Memoir,* **18**, pp. 1–155.

Portugal, S.J., Maurer, G. and Cassey, P., 2010, Eggshell permeability: A standard technique for determining interspecific rates of water vapor conductance. *Physiological and Biochemical Zoology,* **83**, pp. 1023–1031.

Potts, R., 2013, Hominin evolution in settings of strong environmental variability. *Quaternary Science Reviews,* **73**, pp. 1–13.

Potts, R., 1998, Environmental hypotheses of hominin evolution. *Yearbook of Physical Anthropology,* **41**, pp. 93–136.

Pulikanti, R., Peebles, E.D., Zhai, W. and Gerard, P.D., 2012, Determination of embryonic temperature profiles and eggshell water vapor conductance constants in incubating Ross × Ross 708 broiler hatching eggs using temperature. *Poultry Science,* **91(9)**, pp. 2183–2188.

Rahn, H., Ar, A. and Paganelli, C.V., 1979, How bird eggs breathe. *Scientific American,* **240**, pp. 46–55.

Rahn, H., Paganelli, C.V. and Ar, A., 1987, Pores and gas exchange of avian eggs: A review. *Journal of Experimental Zoology,* **1**, pp. 165–172.

Richards, P.D.G., Richards, P.A. and Lee, M.E., 2000, Ultrastructural characteristics of ostrich eggshell: Outer shell membrane and the calcified layers. *Journal of the South African Veterinary Association,* **71(2)**, pp. 97–102.

Sauer, E.G.F., 1972, Ratite eggshells and phylogenetic questions. *Bonner Zoologische Beiträge,* **1**, pp. 3–48.

Sauer, E. G. F. and Sauer, E.M., 1966, Social behaviour of the South African ostrich *Struthio camelus australis*. *Ostrich: Journal of African Ornithology,* **37**, pp. 183–191.

Scott, L., Neumann, F.H., Brook, G.A., Bousman, C.B., Norström, E. and Metwally, A.A., 2012, Terrestrial fossil-pollen evidence of climate change during the last 26 thousand years in Southern Africa. *Quaternary Science Reviews,* **32**, pp. 100–118.

Ségalen, L., Renard, M., Lee-Thorp, J.A., Emmanuel, L., Le Callonnec, L., de Rafélis, M., Senut, B., Pickford, M. and Melice, J.-L., 2006, Neogene climate change and emergence of C4 grasses in the Namib, southwestern Africa, as reflected in ratite 13C and 18O. *Earth and Planetary Science Letters,* **244**, pp. 725–734.

Ségalen, L., Renard, M., Pickford, M., Senut, B., Cojan, I., Le Callonnec, L. and Rognon, P., 2002, Environmental and climatic evolution of the Namib Desert since the Middle Miocene: The contribution of carbon isotope ratios in ratite eggshells. *Comptes Rendus Geoscience,* **334**, pp. 917–924.

Senut, B., 2000, Fossil ratite eggshells: A useful tool for Cainozoic biostratigraphy in Namibia. *Communications of the Geological Survey of Namibia,* **12**, pp. 367–373.

Senut, B. and Pickford, M., 1995, Fossil eggs and Cenozoic continental biostratigraphy of Namibia. *Palaeontologia Africana,* **32**, pp. 33–37.

Senut, B., Dauphin, Y. and Pickford, M., 1998, Nouveaux restes aviens du Neogene de la Sperrgebiet (Namibie): Complement a la biostratigraphie avienne des eolianites du desert Namib. *Comptes Rendus de l'Academie des Sciences, Sciences de la Terre et des Planets,* **327**, pp. 639–644.

Sharp, Z., 2007, *Principles of Stable Isotope Geochemistry,* (Upper Saddle River, NJ: Pearson Education).

Smith, B.N. and Epstein, S., 1971, Two categories of 13C/12C ratios for higher plants. *Plant physiology,* **47(3)**, pp. 380–384.

Sparks, N.H.C. and Deeming, D.C., 1996, Ostrich eggshell ultrastructure—a study using electron microscopy and x-ray diffraction. In *Improving our Understanding of Ratites in a Farming Environment,* edited by Deeming, D.C. (Oxford: Proceedings Ratite Conference), pp. 164–165.

Stein, L.R. and Badyaev, A.V., 2011, Evolution of eggshell structure during rapid range expansion in a passerine bird. *Functional Ecology,* **25(6)**, pp. 1215–1222.

Stidham, T.A., 2004, Extinct ostrich eggshell (Aves: Struthionidae) from the Pliocene Chiwondo Beds, Malawi: Implications for the potential biostratigraphic correlation of African Neogene deposits. *Journal of Human Evolution,* **46**, pp. 489–496.

South African Rain Atlas, online resource http://134.76.173.220/rainfall/index.html, accessed 20/08/2013.

Swann, G.S. and Brake, J., 1990, Effect of dry-bulb temperature, relative humidity, and eggshell conductance during the first three days of incubation on egg weight loss and chick weight. *Poultry Science,* **69**, pp. 535–544.

Thackeray, A.I., Thackeray, J.F., Beaumont, P.B. and Vogel, J.C., 1981, Dated rock engravings from Wonderwerk Cave, South Africa. *Science,* **214(4516)**, pp. 64–67.

Thompson, M.B. and Goldie, K.N., 1990, Conductance and structure of eggs of Adelie Penguins, *Pygoscelis adeliae*, and its implications for incubation. *The Condor,* **92**, pp. 304–312.

Trauth, M.H., Larrasoana, J.C. and Mudelsee, M., 2009, Trends, rhythms and events in Plio-Pleistocene African climate. *Quaternary Science Reviews,* **28**, pp. 399–411.

Trauth, M.H., Maslin, M.A., Deino, A.L., Junginger, A., Lesoloyiae, M., Odada, E.O., Olago, D.O., Olaka, L.A., Strecker, M.R. and Tiedemann, R., 2010, Human evolution in a variable environment: The amplifier lakes of Eastern Africa. *Quaternary Science Reviews*, **29**, pp. 2981–2988.

van Zinderen Bakker, E.M., 1982, Pollen analytical studies of the Wonderwerk Cave, South Africa. *Pollen et Spores,* **24**, pp. 235–250.

von Schirnding, Y., Van der Merwe, N.J. and Vogel, J.C., 1982, Influence of diet and age on carbon isotope ratios in ostrich eggshell. *Archaeometry,* **24**, pp. 3–20.

Watson, V., 1993, Composition of the Swartkrans bone accumulations, in terms of skeletal parts and animals represented. In *Swartkrans: A Cave's Chronicle of Early Man* edited by C.K. Transvaal Museum, Pretoria, **8**, pp. 35–73.

Williams, D.L.G., Seymour, R.S. and Kerourio, P., 1984, Structure of fossil dinosaur eggshell from the Aix basin, France. *Palaeogeography, Palaeoclimatology, Palaeoecology,* **45**, pp. 23–37.

Williams, J.B., Siegfried, W.R., Milton, S.J., Adams, N.J., Dean, W.R.J., du Plessis, M.A., Jackson, S. and Nagy, K.A., 1993, Field metabolism, water requirements, and foraging behaviour of wild ostriches in the Namib. *Ecology*, **74**, pp. 390–404.

Yom-Tov, Y., Wilson, R. and Ar, A., 1986, Water loss from Jackass Penguin *Speniscus demersus* eggs during natural incubation. *Ibis,* **128**, pp. 1–8.

CHAPTER 8

First chronological, palaeoenvironmental, and archaeological data from the Baden-Baden fossil spring complex in the western Free State, South Africa

Andri C. van Aardt
Department of Plant Sciences, University of the Free State, Bloemfontein, South Africa

C. Britt Bousman
Anthropology Department, Texas State University, San Marcos, Texas
GAES, University of the Witwatersrand, Johannesburg, South Africa

James S. Brink
Florisbad Quaternary Research Department, National Museum, Bloemfontein, South Africa
Centre for Environmental Management, University of the Free State, Bloemfontein, South Africa

George A. Brook
Department of Geography, University of Georgia, Athens, USA

Zenobia Jacobs
Centre for Archaeological Science, School of Earth and Environmental Sciences, University of Wollongong, Wollongong, Australia

Pieter J. du Preez
Department of Plant Sciences, University of the Free State, Bloemfontein, South Africa

Lloyd Rossouw
Department of Plant Sciences, University of the Free State, Bloemfontein, South Africa
Archaeology Department, Museum, Bloemfontein, South Africa

Louis Scott
Department of Plant Sciences, University of the Free State, Bloemfontein, South Africa

ABSTRACT: The Baden-Baden spring mound is one of the extremely scarce archaeological sites in the dry central and western interior of South Africa, where fossil fauna and also palaeobotanical material are preserved. This is the first and preliminary summary of ongoing palaeoenvironmental research at this spring mound complex, which is situated 70 km northwest of Bloemfontein near Dealesville. Topographic mapping, radiocarbon and OSL dating complement the archaeological, faunal and palynological records from Baden-Baden and compare it to other spring, pan and alluvial sites in the region like Florisbad, Deelpan and Erfkroon. OSL and radiocarbon dating places the available sequences within the last ~160 ka. Holocene archaeological and faunal remains were recovered from several excavations on the east side of the primary mound. These materials provide unique insights into prehistoric human adaptations in the grassveld. Pollen, extracted from a peat mound and buried organic layers beneath sand accumulations, suggests cooler, moist conditions during the late Pleistocene and drier conditions in the Holocene. These palaeoenvironmental proxy indicators offer the potential for better understanding of long-term climate and vegetation changes in the western Free State.

8.1 INTRODUCTION

Baden-Baden is a complex of spring mounds located 70 km northwest of Bloemfontein in the western Free State Province near a large pan known as Annaspan (Bousman and Brink, 2008, 2012) (Figure 1). The site is locally well known due to a pre-Boer War bath house built over a large mineral spring on the east side of the largest mound. The bath house is historically interesting because of the inscriptions written by bathers on the outside walls. Historical records indicate that Dr. Brownlow from Europe believed

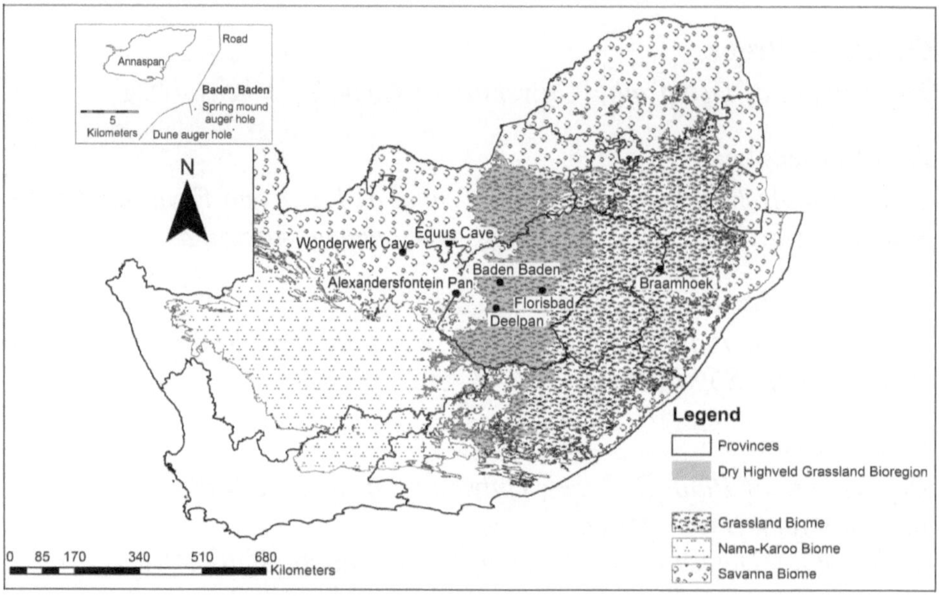

Figure 1. Locality map of Baden-Baden in its bioregion and three major surrounding biomes. Insert: Annaspan and the positions of auger holes in the main Baden-Baden spring mound and the dune south of the site.

the spring water, which is rich in NaCl and $MgSO_4$, had healing properties and established a hospital near the site ca. 1905 (N. Nel, pers. comm., letter from Dr. Albert Wessels). The palaeoenvironmental and archaeological potential of this site was first recognized by Louis Scott in 1987.

Annaspan (Figure 1) is one of the largest pans in South Africa measuring ~7.5 km in its maximum dimension and covering approximately 1850 hectares. Pans are very common in the western Free State and frequently occur in ancient abandoned stream channels (Marshall, 1987, 1988). They form by wind erosion that removes unconsolidated sediments down to bedrock (Butzer and Oswald, in press). This wind erosion creates large aeolian dunes on the leeward sides of the pans. Frequently, when bedrock allows, springs emerge on the slopes of pans and sandy mounds form around these springs. Rarely these sand accumulations grow to form large spring mounds such as found at Baden-Baden and Florisbad (Grobler and Loock, 1988; Visser and Joubert, 1991; Kuman *et al.*, 1999).

The primary spring mound at Baden-Baden is a sandy hillock that stands eight meters above the surrounding landscape (Figures 2 and 3). Two short streams drain the eastern and western side of the large mound and flow into Annaspan. On the eastern drainage, a number of active springs and seeps emanate from a complex series of small mounds. The western drainage starts at a different set of large and small springs where a bathing pool, now abandoned, was built. Baden-Baden has *in situ* faunal remains and artefacts dating to the Later Stone Age and sporadic surface Middle Stone Age occurrences. It is similar to the Florisbad spring mound about 30 km to the east and well-known for its fossil human cranium and faunal remains (Grün *et al.*, 1996; Kuman *et al.*, 1999; Brink, 2005), but the primary mound contains fewer organic layers than Florisbad (Bousman and Brink, 2008, 2012). The deposits at Baden-Baden are important because of their archaeological, faunal and plant microfossil records relating to human occupations and palaeoenvironments of the central interior of South Africa.

Several research teams have worked at Baden-Baden. In 2003, Louis Scott and his student, Bokang Theko, sampled for pollen analysis (Theko *et al.*, 2003). Later that same year and in 2006 a team from Texas State University and the National Museum, headed by James Brink and Britt Bousman, undertook topographic mapping, archaeological excavations, and geological trenching. In 2005 and 2006 a team from the University of Georgia under George Brook's direction, cored the main sandy spring mound and the sand dune to the south, and also collected samples from Annaspan. The archaeological, faunal, palynological and dating results reveal a complex set of Holocene and Pleistocene deposits with preserved Holocene human occupations. Three blocks (North, Central and South) were excavated archaeologically along the eastern edge of the spring mound. Here we present a brief description of the site and report the combined initial results that can serve as a basis for ongoing research and to assess the potential of the site for making contributions to palaeoenvironmental studies in the central South African interior. There seems to be renewed interest in this often neglected region with activities focusing at sites near Baden-Baden like Erfkroon, Florisbad and Deelpan and further to the west in the Southern Kalahari at Equus Cave, Wonderwerk Cave and Kathu Pan (Scott, 1987; Beaumont, 2004; Gil Romera *et al.*, 2014; Scott and Thackeray, 2014; Butzer and Oswald, in press). Together with these efforts, future Baden-Baden studies promise to expand the palaeoenvironmental information in the region where previous research, such as at Alexandersfontein, indicated marked environmental changes, including the existence of a palaeolake in that pan basin during the late Quaternary (Butzer *et al.*, 1973).

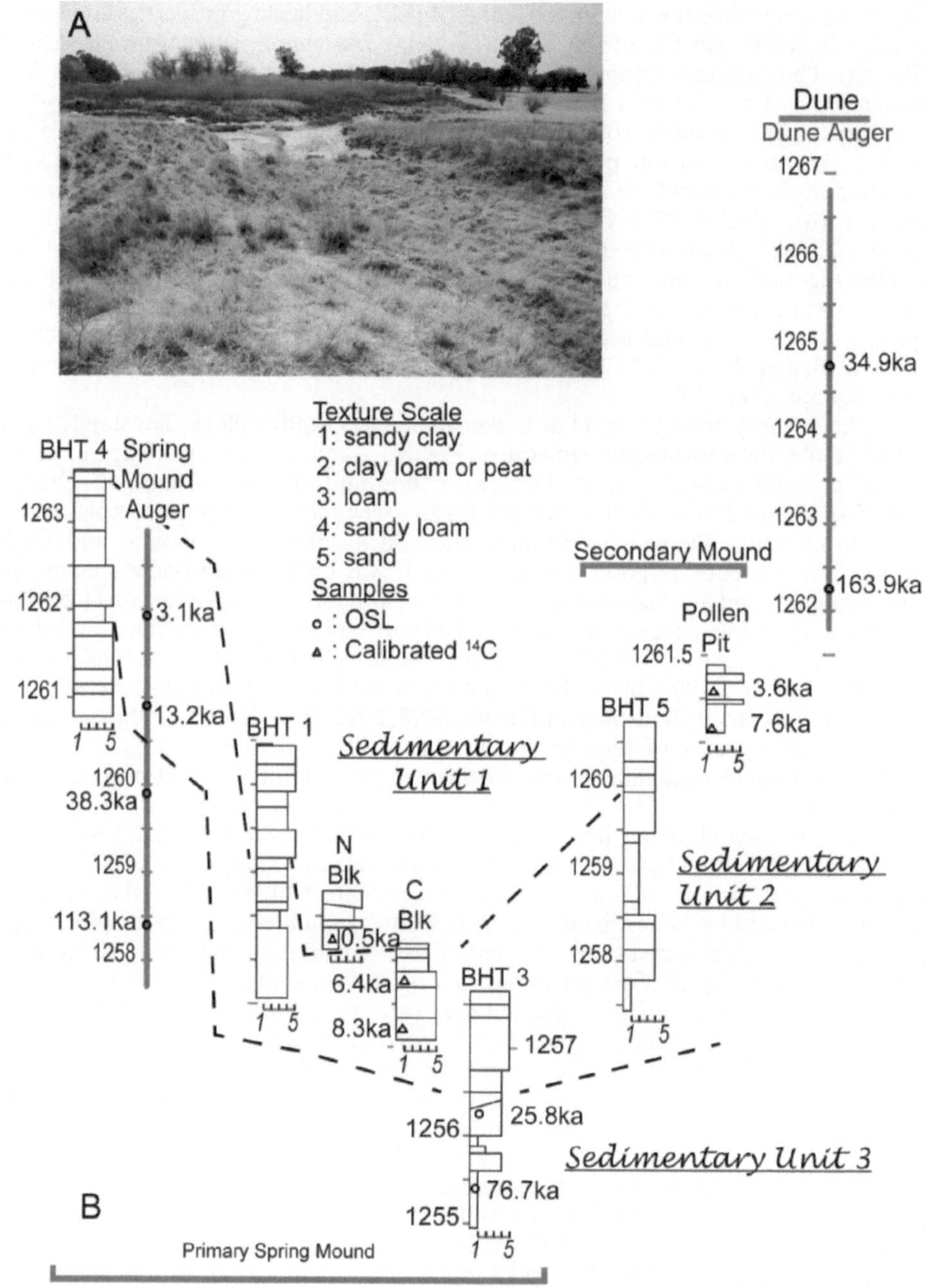

Figure 2. A, The *Phragmites australis* reed community which dominates the moister parts of the secondary spring mounds with some planted exotic trees. B, Profile cross-section showing elevations for described backhoe trenches, excavation units, augers and the Pollen Pit. Backhoe trench and excavation sediment textures and colours are correlated by sedimentary units (according to Tables 1–7). Augers and the Pollen Pit sections illustrated by elevation but not correlated into sedimentary units. OSL and calibrated radiocarbon ages and sample locations plotted by elevations on appropriate profiles.

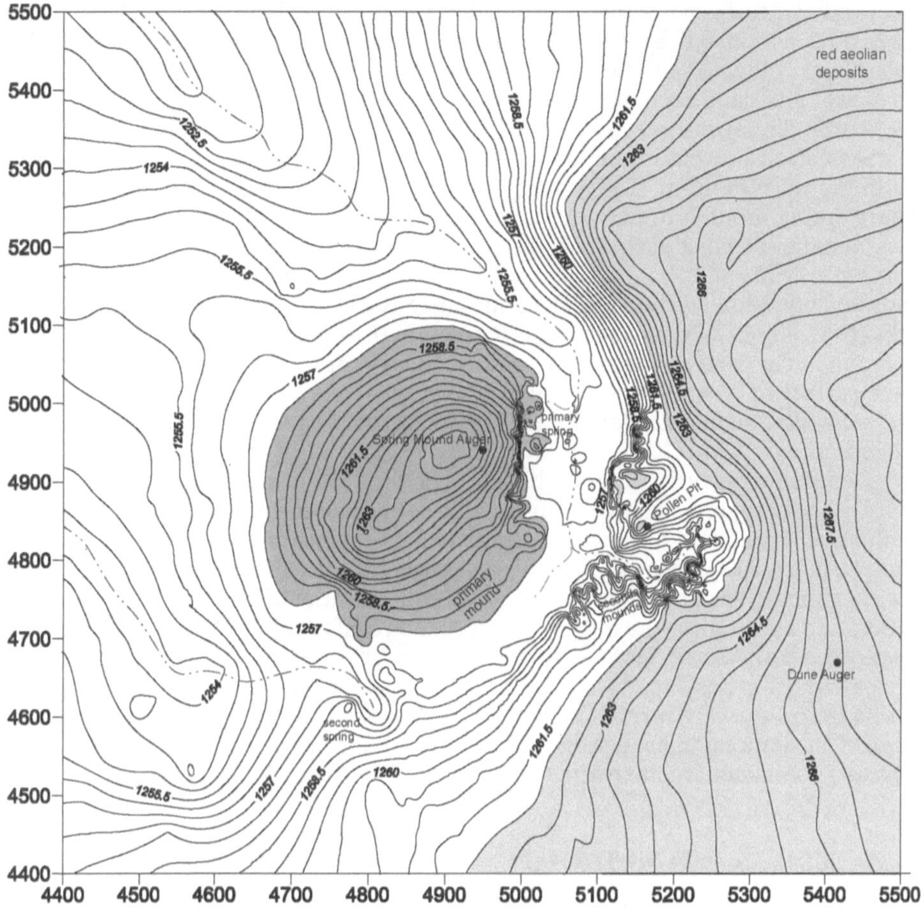

Figure 3. Topographic map of the Baden-Baden site showing the main mound in the centre (shaded) the secondary mounds (centre right) and the adjacent dune (shaded, far right).

8.2 PRESENT DAY ENVIRONMENT

The study area falls within the Grassland Biome of South Africa (Mucina and Rutherford, 2006) (Figure 1). The altitude in the area is ~1 300 m amsl (above mean sea level), sloping towards the west (Holmes *et al.*, 2008). The interior of South Africa experiences temperatures below freezing during the winter months, and frost at the site is a common phenomenon (Schulze, 1972; Bousman, personal observation). This is a summer-rainfall area (MAP 380–530 mm) with high summer temperatures (Geldenhuys, 1982; Schulze, 1997). Evaporation increases towards the west of the Province—estimates are that 91% of South Africa's MAP is returned to the atmosphere (Schulze, 1997).

Baden-Baden is in the western portion of the Grassland Biome within the Dry Highveld Grassland Bioregion (Mucina and Rutherford, 2006, pp. 32–51). The mound is positioned at the boundary between the Vaal-Vet Sandy Grasslands and the Western Free State Clay Grasslands (Mucina and Rutherford, 2006). The Western Free State Clay Grasslands are a species-poor dry grassland community growing on thin clayey soils overlying Ecca shales that form the lower-lying flat plains. The Vaal-Vet Sandy Grasslands, dominated by *Themeda triandra*, occurs in areas of higher elevation on aeolian and colluvial sandy hills and ridges overlying Ecca shales. At Baden-Baden this type of grassland occurs on aeolian dune deposits eroded from Annaspan. Vegetation surrounding Annaspan is classified as Highveld Salt Pan Vegetation.

Several local vegetation communities occur at Baden-Baden on the spring mounds and in the valley-bottom. The primary spring mound and the drier parts of the secondary spring mounds are covered by a *Cynodon dactylon—Eragrostis lehmanniana* grass community which include the karroid shrub *Chrysocoma ciliata* and the grasses *Digitaria eriantha* and *Aristida congesta*. The elevated mounds have deep well-drained sandy soils that are leached of salts and minerals. Burrowing animals such as aardvarks (*Orycteropus afer*) and suricates (*Suricata suricatta*) caused some disturbance on the mounds. On parts of the primary and secondary spring mounds where water seeps to the surface or where the water table is relatively close to the soil surface a *Phragmites australis* community dominates (Figure 2A) accompanied by aquatic forbs like *Ranunculus multifidus*.

Depending on variations in salt concentrations, soil texture, surface water availability, leaching and the degree of disturbance a mosaic of three different plant communities occurs at the base of the spring mounds in the valley-bottom wetland. For example, in patches where soil pH and salt concentrations are high a *Limonium dregeanum* community occurs. Areas with high salt concentration that result from seep water evaporation are relatively hostile to most wetland plants except for the halophytic rush *Juncus rigidus* and some *Sporobolus virginicus*. Large dry parts of the valley floor have bare patches or are covered by reddish sands that support a community dominated by *Suaeda fruticosa* dwarf shrubs and fewer *Salsola aphyla*. Around deep water along the fringe of the impoundment a sedge community with *Schoenoplectus triqueter* (*Scirpus triqueter*) dominates together with some stands of bulrush (*Typha capensis*).

8.3 MATERIALS AND METHODS

Topographic surveying started during 2003 when elevation measurements were collected in order to construct a detailed map of the site (Figure 3). A local arbitrary XYZ grid was established using a Sokkia SET 600 Total Station with a Sokkia data collector. Over 1855 surface elevation points were measured including a trig marker with a known elevation approximately 2 km to the southeast of the primary mound. The trig marker elevation was used to calibrate the arbitrary heights to elevations above mean sea level. Elevations of all excavations and backhoe trenches were also recorded and used to construct a topographic map in Surfer 8.0.

8.3.1 Archaeological excavations

A series of 1 × 1 meter excavation units were dug at the base and on the edge of the primary mound at Baden-Baden. Most were grouped in northern, central and southern blocks, but a few isolated units were excavated also. All artefacts, bone, sediment samples and other materials were plotted with a Sokkia Total Station and locations recorded with a data collector. All sediment was wet sieved in 5 mm screen and recorded by arbitrary 10 cm units and natural stratigraphy. Sediment, pollen and dating samples were collected from each block. Seven backhoe trenches were excavated at various locations to provide geological observations in areas not sampled by archaeological excavations. The artefacts and fauna from the archaeological excavations were cleaned, analysed and curated at the Florisbad Quaternary Research Station.

The sediments in the archaeological excavations and backhoe trenches were recorded using a combination of pedogenic and geologic methods. Sediment textures were determined in the field using Olsen's (1981, pp. 23–24) methods. Sediment colour was determined on moist sediment using a Munsell Soil Color Chart. Soil structure and other features classification follows the definitions in the *Soil Survey Manual* and *Keys to Soil Taxonomy* (Soil Survey Staff, 1993, 2010), and sedimentary structures follow those outlined in *Depositional Sedimentary Environments* (Reineck and Singh, 1975) and *The North American Stratigraphic Code* (North American Commission on Stratigraphic Nomenclature, 2005).

8.3.2 Dating

To provide a provisional chronology for the site's deposits, samples were collected from different backhoe trenches, excavation units, cores and pollen profiles for radiocarbon (^{14}C) and Optically Stimulated Luminescence (OSL) dating at locations shown in Figure 4. Four ^{14}C ages have been obtained from the North and Central Blocks from Quaternary Dating Research Unit (QUADRU) at the CSIR in Pretoria, South Africa, in 2003. A further two samples of organic spring deposits from a pit (the Pollen Pit) in one of the secondary spring mounds, were submitted and dated by Beta Analytic. The pit was dug separately for exploratory pollen analysis by LS and not recorded by the same criteria of sediment description as given above.

The first two sediment samples for OSL dating (OSL#4 and OSL#6) were processed at QUADRU. These samples came from a backhoe trench adjacent to the archaeological and palaeontological excavation from which samples for ^{14}C dating were collected. OSL#4 was collected from Zone 6, ~143 cm below the present ground surface, and OSL#6 from Zone 10, ~222 cm below the present ground surface (Figure 4). The samples were taken by inserting a metal pipe into the section wall to prevent the sample from being exposed to light. Both ends of the pipe were sealed and the sample tube was labelled.

Auger samples for further OSL dating were processed at the Luminescence Laboratory of the University of Georgia (UGA). In July 2005 the spring mound at Baden-Baden (28° 34.402 S; 25° 49.822 E) was augered using a sand bucket auger to 5.2 m depth. Samples of sediment and samples for OSL dating were collected at 0.97, 2, 3 and 4.5 m depth (BB-1 to BB-4). Further sampling was undertaken in July 2006 on a relict dune ridge at the edge of a corn field (28° 34.623 S; 25° 50.016 E). The dune was augered to 5 m depth and samples for OSL dating were collected in a bucket auger at 1, 2, 3, 4, and 4.55 m depth (Figure 4).

When samples were taken, the auger was twisted deep into the sand to force new sand upwards through the bucket. At the surface a piece of PVC pipe or a tin can with a pierced base was pushed into the top of the sand in the bucket after cleaning away a few centimeters of sand. The auger and pipe were turned upside down; the pipe was removed and capped at both ends, wrapped in thick, opaque black plastic, sealed with duct tape and labeled. In the laboratory, the potentially light-exposed portions at both ends of the sampling tubes were removed.

In both laboratories, all sample preparation and luminescence measurements were carried out in subdued red light. Raw samples were treated with 10% HCl and 20% H_2O_2 to remove carbonate and organic matter. After drying, the samples were sieved to select the grain size in the size ranges of 90–125 μm, 125–180 μm and 180–250 μm. Heavy liquids of densities 2.62 and 2.75 g/cm^3 separated the grain fraction to obtain

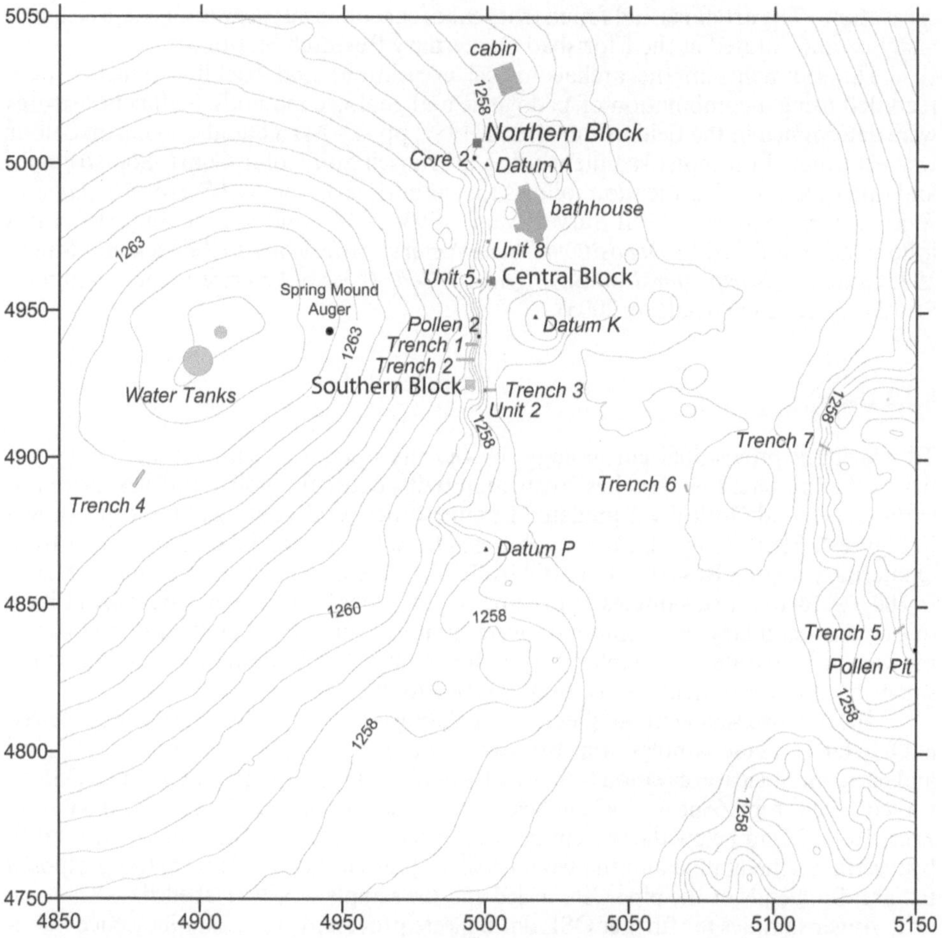

Figure 4. Topographic map of the study area adjacent to the main mound at Baden-Baden showing the positions of the auger hole for OSL dating and the excavation areas of the North, Central and South Blocks.

quartz and feldspar. The quartz grains were treated with 40% HF for 60 min to remove the outer layer irradiated by alpha particles and remaining feldspars. The grains were then treated with 1 mol/L HCl for 10 min to remove fluorides created during the HF etching. Pure quartz fractions with grains in the size range 125–180 µm (UGA) or 180–212 µm (QUADRU) were acquired finally.

The OSL measurements were carried out in both laboratories, using an automated Risø TL/OSL-DA-15 reader (Markey *et al.*, 1997). Light stimulation of quartz mineral extracts was undertaken with an excitation unit containing blue light-emitting diodes ($\lambda = 470 \pm 30$ nm) (Bøtter-Jensen *et al.*, 1999). Detection optics comprised two Hoya 2.5 mm thick U340 filters and a 3 mm thick stimulation Schott GG420 filter coupled to an EMI 9635 QA photomultiplier tube. Laboratory irradiation was carried out using calibrated $^{90}Sr/^{90}Y$ sources mounted within the readers.

The equivalent dose (D_e), the numerator in the OSL age equation, of quartz was determined by the SAR protocol (Murray and Wintle, 2000). The purity of quartz was checked by IRSL at 50°C (UGA) or the OSL-IR depletion ratio test (QUADRU) and the results showed that none of the samples contained feldspar in the quartz fraction. The OSL dating recuperation test and dose recovery test showed that the SAR protocol was reliable for the samples examined in this study. Data were analyzed using the ANALYST program of Duller (1999). It is possible that some sediments may be insufficiently and/or unevenly bleached prior to burial. Whether the samples were fully bleached prior to burial can be detected by measuring aliquots that contain a small number of grains and examining the distribution among D_e values obtained from many individual aliquots from the same sample. Overdispersion (OD) values ranging from 5 ± 2% to 55 ± 10% were obtained and were spread symmetrically around a common value; OD is the amount of scatter left after all sources of measurement uncertainty are taken into account. The individual D_e values were all statistically consistent with each other and showed no clear evidence for partial bleaching and were combined, to obtain a single value of D_e for age determination, using the central age model of Galbraith *et al.* (1999). Overdispersion values much greater than 20% (at two sigma limits) for auger samples UGA06OSL-331 and 332 indicate mixing or grains of various ages or partial bleaching of grains. Under conditions of partial bleaching a minimum age model can be applied but in this case we believe the samples contain a mixture of grains of different age, some of the mixing possible due to incorporation of younger grains that fell into the auger hole during the augering process, and were incorporated into older samples despite the precautions taken. Because of this, central, rather than minimum ages were calculated for these samples. However, the very high over-dispersion values for these two samples question their accuracy and here we use them only as approximate estimates of age, despite the ages being in correct stratigraphic order.

The environmental dose rate, the denominator in the OSL age equation, is created by the radioactive elements existing in grains of the sample and the surrounding sediments, with a small contribution from cosmic rays. For all the samples measured in both laboratories, thick source Daybreak alpha counting systems were used to estimate U and Th for the dose rate calculation. K contents were measured by ICP90, using the sodium peroxide fusion technique at the SGS Laboratory in Canada for the samples from UGA, and by X-ray fluorescence for the samples from QUADRU. All measurements were converted to alpha, beta and gamma dose rates according to the conversion factors of Aitken (1985, 1998). The dose rate from cosmic rays was calculated based on sample burial depth and the altitude of the section (Prescott and Hutton, 1994). Water content of the sediment samples examined here must have changed drastically over time after they were buried. This is because conditions varied between more arid to less arid resulting in alternating periods of decreased and increased surface and groundwater flow. Therefore, estimation of the water content was based on what is known about the history of the deposits and on as-collected values of moisture content which were all less than 5% of dry sample weight. Since it is not possible to accurately determine the mean water content during the sediment burial period, a slightly higher value for water content (5 ± 2.5%) than that measured in the samples was assumed in age calculations for the sediments measured at UGA; the current measured water contents of 17 ± 5 and 9 ± 3% were used for QUADRU samples OSL#4 and OSL#6, respectively.

8.3.3 Palynology

Pollen was extracted from deposits using standard methods, which included digestion in HCl (10%) for carbonates or KOH (5%) for peaty samples, and mineral separation using $ZnCl_2$ solution (specific gravity 2) and mounted in glycerine jelly for microscope analysis.

8.3.4 Phytoliths

Grass short-cell phytoliths were extracted from samples BB-1, BB-2, BB-3 and BB-4, which were taken at the spring mound auger site. Standard laboratory steps included deflocculation, the elimination of carbonates using HCl in low concentration (10%), removal of clays by means of sedimentation and mineral separation with a heavy liquid solution of sodium polytungstate (specific density = 2.3). Fractions were mounted on microscope slides in glycerin jelly and scanned under a Nikon 50i polarizing microscope at x400 magnification. Provisional observations focussed on the identification of non-lobate grass short-cell phytoliths as indicators of C_3 and C_4 palaeoenvironmental conditions (Rossouw, 2009).

8.4 RESULTS

The topographic analysis of the site shows that the crest of the primary mound forms a bi-lobed oval and this is likely due to the presence of two large mounds that have merged into a single mound (Figures 3 and 4). Two streams drain the numerous springs. The larger drainage is on the eastern side of the primary mound where a bathhouse was constructed at a large spring eye on the eastern flank of the primary mound. The eastern stream flows north then turns northwest toward Annaspan. A dam was constructed at that point. The eastern stream also extends upstream to the southwest where it is joined by a minor tributary. A number of smaller secondary mounds are also evident on the eastern stream. Some minor mounds flank the primary mound on its southeast side and others occur on an eastern tributary drainage. These smaller mounds adjacent to the tributary drainage appear to be coalescing into a new mound complex and they merge with a large dune deposit to the east and south. The number and distribution of smaller mounds illustrates the complex nature of mound genesis. A second major spring is southwest of the primary mound. This is drained by a smaller stream that flows to the northwest toward Annaspan. A bathing pool was also constructed on the south side of the primary mound at the second major spring but it no longer holds water and is not in use.

8.4.1 Geology

The archaeological excavations, geological profiles, pollen profiles and geological cores for dating are on the eastern and southern side of the mound (Figure 4). Previously the slopes on the eastern side of the primary mound were removed creating a steep cut bank. This cut bank provided a vertical view of the mound's lower slope sediments and exposed archaeological materials eroding from the deposits. The surface archaeological materials were used to determine the location of the archaeological excavations. In addition to the excavation units, a number of backhoe trenches were excavated to gain a better understanding of the depositional history of the site. A description of the geological profiles demonstrates the nature of sediment accumulation in the mound. The main sediment units and relative elevation of various sequences are illustrated in Figure 2B.

8.4.2 Auger holes

Major changes in sediment characteristics revealed by augering (Table 1, Figure 2B) showed that the spring mound was almost pure fine sand to ~4 m with cemented sand

Table 1. Sediment descriptions for the spring mound and dune auger sites.

Sample	Depth (cm)	Description
SPRING MOUND		
BB-1	97	Greyish brown (10YR 5/2) fine sand with rare cemented sand 'nodules'
BB-2	200	Light yellowish brown (2.5YR 6/4) fine sand
BB-3	300	Strong brown (7.5YR 5/8) fine sand with scarce cemented sand 'nodules'
BB-4	450	Light brownish grey (2.5Y 6/2) very fine sand, silt and clay with yellowish red mottles (5 YR 5/6). The sediment is entirely cemented and was 'green' when first exposed to the air.
DUNE		
BBDR-1	100	Strong brown (7.5YR 4/6) fine sand cemented by $CaCO_3$
BBDR-2	200	Red (2.5YR 4/6) fine sand cemented by $CaCO_3$
BBDR-3	300	Red (2.5YR 5/6) fine sand
BBDR-4	400	Yellowish red (5YR 5/8) fine sand
BBDR-4.55	455	Strong brown (7.5YR 5/8) fine sand and silt partially cemented by $CaCO_3$. Slight 'green' colour when first exposed to the air
BBDR-5 m	500	Light yellowish brown (10YR 6/4) very fine sand, silt and clay. 'Green' colour when first exposed to the air

'nodules' between 75 and 90 cm depth and again between 300 cm and 380 cm with the nodules larger and more abundant at greater depth. Between 400 cm and 520 cm depth the sandy sediment was light brownish grey (2.5Y 6/2) with yellowish red mottles (5YR 5/6). Dark grey organic-rich horizons were apparent at 410 and 470 cm. At 505 cm depth the sediments consisted of very fine sand with abundant silt and clay all cemented by $CaCO_3$. The sand 'nodules' were clearly produced by evaporation of water from soil water containing $CaCO_3$ in solution. In the upper part of the profile this is probably not related to a ground water table, rather to evaporation as rainfall percolates downward through the sand. However, below 3 m the 'nodules' and mottling are clearly related to the capillary zone above a water table, perhaps 1–2 m below, and evaporation in this zone precipitated the $CaCO_3$ that formed the 'nodules'. When first brought to the surface, sediments deeper than 450 cm were a green colour that rapidly fades to light brownish grey after exposure to the atmosphere (see Visser and Joubert, 1991, p. 126).

The sand dune south of the mound consists of brown, red and then yellowish red fine sand with some silt and clay to 400 cm depth. Below this depth, at 455 cm, the fine sand is strong brown in colour with higher silt content. At 500 cm there is cemented, light yellowish brown very fine sand with silt and clay (Table 1). Augering was terminated at 500 cm. The sediments from 455 and 500 cm were 'green' when first exposed to the atmosphere but this colour quickly faded after exposure to the atmosphere. At both the spring mound and the dune the green colour and finer texture of the cemented sediments at depth suggests wetter conditions while the overlying, generally loose aeolian sands record a much drier period.

8.4.3 Backhoe trenches

Trench 4 was excavated near the crest of the primary spring mound. Backhoe trenches allowed the sediments to be observed in place and a range of other observations could be made that are not possible in auger samples (Figure 2B). Twelve soil horizons were identified in three sedimentary units (Table 2). An A-C pedon was recorded in the upper 14 cm of sandy aeolian sediment. This sat unconformably on a buried A horizon that formed the top of the second sedimentary unit. Sedimentary Unit 2 had a 2 Ab-2B1-2B2-2B3-2B4 pedon spanning 161 cm with thin $CaCO_3$ films on ped faces in yellowish-red to brownish yellow sands. Sedimentary Unit 3 is marked by a marked increase in $CaCO_3$ accumulations as large nodules grading down to a rubified weathered horizon. A pale olive coarse sand horizon was found below and it is tentatively grouped with Sedimentary Unit 3 but it could also be part of an older sedimentary unit. All of these deposits are well sorted sandy loams to sand and are differentiated by the degree of pedogenesis, soil carbonate accumulation and weathering.

Trench 1 (Table 3) was excavated on the eastern edge of the spring mound. Trench 2 was nearby and sampled similar deposits so it is not described. These deposits provide the broader sedimentary context for the Central Block Excavations. Trench 1 was excavated from the top of the cut bank lip formed by the removal of sediment on the east side of the mound. The sediments from Trench 1 are divided into 13 zones and two sedimentary units. At the top of the trench grasses grow on sands. This (O-C horizon sequence) sand layer appears to be a very recent aeolian deposit that caps the entire mound, and sits unconformably on a buried A horizon inceptisol formed between 23–35 cm below the surface and sitting on a C horizon. An unconformity separates the first from the second sedimentary unit, which is characterized by an A-B-Bg sequence with gleyed sediments in the lower meter of deposits.

Trench 3 (Table 4) was excavated on the floor of the drainage adjacent to Excavation Unit 2. The sediments are correlated to Trench 1 following the soil horizon and sediment unit designations. The sediments begin at the top of Trench 3 with a truncated gleyed B horizon followed by additional B horizons consisting of light yellowish brown to greyish brown back to light yellowish brown sand shifting to a light olive grey sandy loam at the bottom of this sedimentary unit. These B horizons sit unconformably on a dark grey sandy loam A horizon recorded as Zone 5. This grades down into a series of greenish grey sandy loam B horizons which form Sediment Unit 3. Below this are two B horizons consisting of grey and greenish grey clay loam to sandy loam soil horizons. The final soil horizon is a greenish grey sandy clay.

Trench 5 was excavated in one of the larger secondary mounds on the east side of the site near the so-called Pollen Pit. It was placed in an attempt to record the sediments for the Pollen Pit in more detail but it is not entirely clear to what degree the sediments vary from one spot to the other. In any case, this profile does provide an indication of the sediment sequence and variability in these smaller mounds (Table 5). Two depositional units are visible. The upper depositional unit is capped by a yellowish brown sandy loam that may represent fairly recent sediments. The lower sedimentary unit is mostly gleyed with varying amounts of organic matter.

The Pollen Pit was 78 cm deep and revealed mainly black "peat" with dark sandy layers at 10–20 cm and 40–45 cm. Rootlets were observed down to the bottom.

Table 2. Profile description for Trench 4.

Zone (horizon)	Depth (cm)	Description
1 (A)	0–2	Strong brown (7.5YR 4/6) friable sandy loam, common rootlets, sod layer, very abrupt smooth lower boundary.
2 (C)	2–14	Brown (7.5YR 5/4) friable sand with fine cross bedding, aeolian, few rootlets, abrupt smooth lower boundary.
		—unconformity—
3 (2Ab)	14–30	Dark brown (7.5YR 4/4) slightly firm sandy loam, weak coarse subangular blocky structure, few roots and rootlets, clear smooth lower boundary.
4 (2Bb1)	30–70	Yellowish red (7.5YR 4/6) slightly firm sandy loam, weak medium subangular blocky structure, $CaCo_3$ films on ped faces and root pores, insect burrows filled with grey and green sand, gradual smooth lower boundary.
5 (2Bb2)	70–110	Strong brown (7.5YR 5/6) sandy loam, weak coarse subangular blocky structure, few small roots, $CaCo_3$ films on ped faces and root pores, in lower 10 cm few very firm yellowish red (5YR 5/8) iron oxidized clasts (≤5 cm) that seem to have formed in place, abrupt wavy lower boundary.
6 (2Bb3)	110–156	Yellow (10YR 7/6) firm sand with few yellowish red (5YR 5/6) mottles, coarser sand than above, medium moderate subangular to angular blocky structure, very small Mg flecks, few rootlets, clear smooth lower boundary.
7 (2Bb4)	156–175	Brownish yellow (10YR 6/6) friable sandy loam, few insect burrows, very few roots, very abrupt highly irregular sloping (toward south) boundary.
		—unconformity—
8 (3Bb1)	175–190	Very pale brown to yellow (10YR 7/4 to 7/6) slightly firm fine sand with 20% slightly firm $CaCO_3$ pale brown (10YR 8/3) nodules formed in globular to vertical patterns, small Mg flecks, irregular clear to abrupt lower boundary.
9 (3Bb2)	190–228	Pale yellow (2.5YR 5/6) sand with brown (7.5YR 5/6) sand filling insect burrows, Mg films, 5–7% very firm light brownish grey (10YR 6/2- in the interior) $CaCo_3$ nodules up to 15 cm long and 4 cm thick that form discontinuous nodules, clear smooth lower boundary.
10 (3Bb3)	228–245	Slightly firm strong brown (7.5YR 5/6) sand with weak medium subangular blocky structure, pale olive (5Y 6/3) sandy loam filling insect burrows extending down from above zone, very abrupt and irregular lower boundary.
11 (3Bb4)	245–256	Pinkish white to reddish yellow (7.5YR 8/2 to 6/8) very firm sand $CaCo_3$ crust with Mg concretions, very abrupt irregular lower boundary.
		—unconformity—
12 (3Bb)	256–282+	Pale olive (5Y 6/3) coarse sand with strong brown (7.5YR 5/6) mottles and filled insect burrows, abundant Mg flecks, 20–30% insect burrows, 15% larger very firm $CaCo_3$ nodules with Mg flecks in nodules, lower boundary not observed.

Table 3. Trench 1 profile descriptions. Sediments from Trench 2 are similar.

Zone (horizon)	Depth (cm)	Description
1 (O)	0–2	Grass and sod layer, very abrupt lower boundary.
2 (C)	2–23	Reddish yellow (7.5YR 6/6) very friable fine medium sand, common rootlets decreasing slightly down profile, no structure, few reddish yellow (7.5YR 5.5/6) sand infilled burrows, southerly sloping very abrupt lower boundary with occasional insect burrows on boundary.
		—unconformity—
3 (Ab1)	23–35	Brown (7.5YR 5/4) friable sand, few roots, abrupt smooth lower boundary.
4 (Cb1)	35–54	Light brown (7.5YR 6/4) slightly firm to friable sand, common rootlets, no structure, abrupt wavy lower boundary turbated by insect burrows.
		—unconformity—
5 (Ab2)	54–72	Yellow brown (10YR 5/4) slightly firm sandy loam, common rootlets, dark yellow brown (10YR 4/4) clay films and bodies along root pores, clear smooth lower boundary.
6 (Bb2)	72–98	Brown (10YR 5/3) friable to slightly firm sandy loam with slightly more silt than above, common rootlets, few observable insect burrows, very weak coarse subangular blocky structure, clear smooth lower boundary.
7 (C2)	98–130	Pale brown (10YR 6/3) friable sand, common rootlets, abrupt wavy lower boundary.
		—unconformity—
8 (2Ab1)	130–142	Very firm yellowish brown (10YR 5/6) sandy loam with weak coarse subangular blocky structure, few roots and rootlets, stone artefacts in zone adjacent to trench, clear smooth lower boundary.

8.4.4 Archaeological excavations

In the Central Block, six soil horizons were evident in the west wall of Unit 4 and 7 (Table 6). During the excavation informal labels were used and those are presented in parentheses at the end of the descriptions. These were deposited and altered by spring activity and have been correlated to the sedimentary units in Trenches 1 and 3 described above. The uppermost 6 cm of sediment is a light brown sandy loam with recent historic artefacts. Below this is a series of organic-rich A horizons separated by C horizons. The A horizons range from greyish brown to black sands and sandy loams while the C horizons are pale yellow to light grey sands and sandy loams. Zone 5 is a spring deposit that extruded through the A horizon that forms Zone 6 and covered it with sand. There is an unconformity that has truncated the Zone 4 A horizon but the sediments above and below this unconformity are very similar and not different enough to place into separate sedimentary units.

Sediments in the North Block are divided into two sedimentary units (Table 7); however both are correlated to Sedimentary Unit 1. The uppermost consists of a sandy cap of recent deposits with historic artefacts sitting unconformably on a buried A

Table 4. Trench 3 profile descriptions. Soil horizon designations correlated to Trench 1.

Zone	Depth (cm)	Description
1 (2Bg4)	0–15	Light yellowish brown (2.5Y 6/4) firm structureless sand, few medium faint reddish yellow (7.5YR 7/6) mottles, few distinct fine dark greyish brown (10YR 4.2) vertical stained root pores, clear smooth lower boundary.
2 (2B5)	15–30	Greyish brown (10YR 5/2) firm sand, few distinct fine dark greyish brown (10YR 4/2) vertical stained root pores, clear smooth lower boundary.
3 (2B6)	30–90	Light yellowish brown (2.5Y 6/3) friable massive moist sand with fine common distinct dary greyish brown (10YR 4/2 vertical and horizontal stained root pores, few faint fine light yellowish brown (2.5Y 6/4) mottles, gradual smooth lower boundary.
4 (2B7)	90–115	light olive grey (5Y 6/2) moist friable sandy loam, few fine faint yellow (2.5Y 7/6) mottles, few vertical very dark greyish brown (10YR 3/2) mottles along root pores, abrupt smooth lower boundary.
		—unconformity—
5 (3Ab)	115–125/134	Dark grey (N 4/0) sandy loam with few faint medium greenish grey (5G 6/1) mottles, within this zone three very dark greyish (N 3/0) 1–2 cm thick bands, lower boundary is clear smooth and sloping toward center of drainage.
6 (3Bb1)	124/134–165	Greenish grey (5G 6/1) sandy loam with few fine distinct bluish grey (5B 6/1) mottles with few fine distinct grey (N 5/0) mottles around root pores, OSL 4 sample at 143 cm (25.6 ± 1.4 ka), clear smooth lower boundary.
7 (3Bb2)	165–177	Greenish grey (5G 5/1) sandy clay with few medium distinct grey (N 5/0) mottles, abrupt smooth sloping lower boundary.
		—unconformity—
8 (3Btb3)	177–185	Grey (N 5/0) clay loam, with fine weak subangular blocky structure, clear sloping lower boundary.
9 (3Bb4)	185–205	Greenish grey (5G 5/1) sandy loam, clear smooth lower boundary.
		—unconformity—
10 (3Bt5)	205–270+	Greenish grey (5BG 5/1) sandy clay, moderate medium prismatic structure with dark brown stains around root pores, OSL 6 sampled at 222 cm (76.7 ± 6.2 ka), lower boundary not observed.

horizon. This Ab horizon is very dark grey clay loam and contained abundant mammal bones (see **Faunal and stone tool analysis**) and a few prehistoric ceramic sherds and fractured dolerite cobbles. No chipped stone artefacts were present which make the assignment of cultural affiliation difficult. The Ab horizon is the second sedimentary unit and it represents a marsh deposit.

Table 5. Trench 5 profile descriptions.

Zone	Depth (cm)	Description
1 (A/B)	0–45	Yellowish brown (10YR 5/4) sandy loam, abundant roots and rootlets that decrease in frequency down profile, clear smooth lower boundary.
2 (B)	45–60	Brown (10YR 5/3) loose sandy loam, common distinct coarse yellowish red (5YR 4/6) mottles, small faint greenish grey (5GY 5/1) mottles, clear wavy lower boundary.
3 (Ab1)	60–80	Dark grey (10YR 4/1) sandy loam with thin 2–3 cm very pale brown (10YR 7/4) sand bedsets of alternating very pale brown (10YR 7/4) sand, common roots and rootlets, abrupt smooth lower boundary.
4 (Ab2)	80–127	Black (10YR 2/1 sandy loam alternating with dark grey (10YR 5/1) sandy loam thin (2–10 cm thick) beds, few rootlets, abrupt irregular lower boundary,
		—unconformity—
5 (2Bg1)	127–138	Light olive grey (5Y 6/2) clay loam, few rootlets, dark grey (10YR 4/1) stains around root pores, clay films in root pores, clear smooth lower boundary.
6 (2Bg2)	138–205	Greenish grey (5G 5/1) clay loam, many vertical roots still decaying, grey (N5/0) root pores, clear smooth lower boundary.
7 (2Bg3)	205–220	Greyish green (5G 5/2) sandy clay loam, common rootlets, grey staining around root pores as above, clear smooth lower boundary.
8 (2Bg4)	220–230	Gr-10eenish grey (5GY 5.5/1) sandy loam, common rootlets, clear smooth lower boundary,
9 (2Bg5)	230–260	Light grey (10YR 7/2) sandy loam, common rootlets, water flowing at lower boundary, clear smooth lower boundary.
10 (2Bg6)	260–295	Greenish grey (5GY 5/1) sandy loam, abundant rootlets, gradual smooth lower boundary.
11 (2Bg7)	295–330+	Very dark grey (N 3/0) clayey sand, lower boundary not observed.

8.4.5 Chronological information

The age relationships of various sequences are summarized in Figure 2B. Radiocarbon samples were collected at different times by a number of researchers. Three radiocarbon laboratories processed the samples: Pretoria, Beta Analytic and Groningen. Samples analysed at the Pretoria lab used traditional Beta count methods while the other two labs used AMS techniques. Radiocarbon ages in ^{14}C yr BP (^{14}C BP) were calibrated at the 2σ probability level to calendar years BP (cal BP) using CALIB 7.0 (Stuiver and Reimer, 1993) and the Southern Hemisphere (SHcal13) atmospheric calibration curve of Hogg *et al.* (2013) (Table 8). All $\delta^{13}C$ values for the dated organic material ranged from −21.7 ‰ to −27.5 ‰ indicating that C_3 plant material was dated, most likely aquatic plant material growing in the marshes associated with spring outflow in the

Table 6. Profile descriptions for Central Block Excavations in Unit 7 and 4. Soil horizon designations correlated to Trench 1 and 3.

Zone	Depth (cm)	Description
1 (A1)	0–6	Light brown to brown mottled (10YR 6/4) loose sandy loam, common historic artefacts, abrupt lower boundary, slope wash.
		—unconformity—
2 (2Ab)	6–16	Friable very dark greyish brown to dark greyish brown (10YR 3/2 to 4/2) sandy loam with very dark grey (10YR 3/1) sandy lenses with brown (10YR 4/3) mottles, abrupt lower boundary, "Upper Black Sand"
3 (2Cb)	16–32	Friable light grey (2.5Y 7/2) sandy loam, abrupt lower boundary, "White Sand".
		—unconformity—
4 (2Ab)	32–50	Slightly firm black (N2.5 /0) sand, abundant lithic artefacts and bone, very abrupt wavy lower boundary, unconformity, "Lower Black Sand".
5 (2Cb)	50–85	Friable pale yellow to light brownish grey (2.5Y 7/3 to 2.5Y 6/2) sand, common dark reed fragments, spring eye deposits extend down into zone 6 in Unit 4 forming a circular vertical spring vent deposit, very few lithic artefacts or bones, "Green Sand".
6 (2Ab)	85–110	Slightly firm greyish brown (2.5Y 5/2) sand, lower boundary not observed.

past. Radiocarbon ages ranged from 490 ± 40 ^{14}C BP (545–340 cal BP) to 7570 ± 40 ^{14}C BP (8412–8205 cal BP).

Organic material recovered from levels at 30–35 cm and 73–78 cm depth in the Pollen Pit excavated in a secondary peat spring mound included plant fragments as well as fine organic sediment, both of which were radiocarbon dated. In both cases, the plant material was significantly younger than the organic sediment (Table 8). In the 30–35 cm level the plant material dated to 656–557 cal BP and the organic sediment to 861–1638 cal BP; in the 73–78 cm level the plant material gave an age of 3684–3475 cal BP and the organic sediment 7672–7575 cal BP. As rootlets have been observed in the pit section, a likely explanation for this difference is that the plant material dated consisted of these roots that had penetrated the older deposit, thus providing a younger age. A second possibility is that the fine organic sediment is old organic material transported to the site, thus providing an age that would be too old. Particularly in dry climates organic material can be preserved in the environment for thousands of years and if this material is introduced into younger deposits it can give misleading ages for the deposit (e.g. Martin and Johnson, 1995). As 7 of the 9 ages in Table 8 are on organic sediments this would present a problem in ascribing an age to the pollen recovered from these sediments. In fact, we believe the plant material was most likely roots and that the organic sediment ages are reliable but further work is needed to prove this.

OSL ages show that the spring mound sediment sequence dates from recent to 113 ± 13 ka, and the dune sequence from recent to 164 ± 15 ka (Table 9). The old

Table 7. Profile descriptions for North Block Excavations in Unit 11, west wall profile.

Zone	Depth (cm)	Description
1 (A)	0–15/20	Greyish brown (10YR 5/2) loose sand lenses separating thin lenses of finely chopped plant matter (1–2 mm thick), extremely common rootlets, fine moderate platty structure, abrupt highly irregular lower boundary,
2 (A/B)	15/20–34	Yellowish brown (10YR 5/4) friable sandy loam, common rootlets and insect burrows, much scattered decomposed organic matter, abrupt wavy to irregular lower boundary.
3 (B)	34–42	Pale brown (10YR 6/3) sand with common very dark greyish brown (10YR 3.2) irregular lenses of slightly firm sandy infilled insect burrows, abrupt smooth to irregular lower boundary.
		—unconformity—
4 (Ab)	42–61	Very dark grey (10YR 3/1) firm clay loam, medium moderate subangular blocky structure, vertical pedogenic cracks filled with yellowish brown (10YR 5/4) sand, bone common in upper 6–7 cm, lower boundary not observed.

Table 8. Radiocarbon ages dates.

Lab no.	$\delta^{13}C$‰	Age ^{14}C BP	Age cal BP (2σ)	Provenience
Pta-9195	–21.7	490 ± 40	545–340	North Block, Unit 11, west wall, Black Clay Loam, 56–60 cm
Pta-9199	–23.8	5600 ± 90	6601–6128	Central Block, Unit 8, west wall, Lower Black Sand, 67–70 cm
GrA-25206	–24.4	5630 ± 45	6465–6291	Central Block, Unit 7, west wall, Upper Black Sand, 48–52 cm
Gr-A 27637	–24.46	7570 ± 40	8412–8205	Central Block, Unit 10, Upper Dark Green Sand, 89–93 cm
Pta-8840	–22.6	2420 ± 70	2718–2209	Central Block, near Unit 9, Black sand ("peat"), 71–79 cm (Theko *et al.* 2003)
Beta-386716	–27.5	680 ± 30	656–557	Pollen Pit, 30–35 cm, plant material
Beta-387920	–24.5	1870 ± 30	1861–1638	Pollen Pit, 30–35 cm. organic sediment
Beta-386717	–27.0	3380 ± 30	3684–3475	Pollen Pit, 73–78 cm, plant material
Beta-387921	–26.0	6810 ± 30	7672–7575	Pollen Pit, 73–78 cm, organic sediment

Table 9. Equivalent dose (De) and dose rate information, together with OSL ages for the spring mound and dune sediments at Baden-Baden.

Laboratory number*	Sample ID	Zone and depth (cm)	Grain Size µ	No. of ali-quots	De (Gy)	OD (%)	U (ppm)	Th (ppm)	K (%)	Water content (%)	Cosmic dose rate (Gy/ka)	Dose rate (Gy/ka)	Age (ka)
UGA06OSL-331	BB-1	97	150–180	13	4.2±0.46	36±7	1.40±0.22	1.94±0.79	0.74±0.1	5±2.5	0.22±0.02	1.36±0.13	3.09±0.45
UGA06OSL-332	BB-2	200	150–180	15	17.64±2.56	55±10	1.60±0.25	2.03±0.87	0.70±0.1	5±2.5	0.19±0.02	1.34±0.13	13.15±2.30
UGA06OSL-330	BB-3	300	150–180	16	51.86±2.48	17±3	1.47±0.11	2.78±0.41	0.71±0.1	5±2.5	0.17±0.02	1.36±0.11	38.28±3.61
UGA06OSL-333	BB-4	450	150–180	19	189.32±10.52	25±4	1.41±0.41	4.75±1.41	0.95±0.1	5±2.5	0.14±0.02	1.67±0.17	113.11±13.27
UGA09OSL-655	BBDR-2	200	180–250	19	51.02±2.18	16±3	1.49±0.22	4.02±0.78	0.72±0.1	5±2.5	0.19±0.02	1.46±0.13	34.92±3.38
UGA09OSL-656	BBDR-4.55	455	180–250	18	221.29±3.88	5±2	0.92±0.20	3.75±0.74	0.81±0.1	5±2.5	0.14±0.02	1.35±0.12	163.95±15.22
QUADRU-OSL#4	OSL#4	Z6 143	180–212	24	38.48 ± 0.59	6±1	1.52±0.23	5.09±0.74	0.84±0.1	17±5	0.20±0.02	1.49±0.07	25.82 ± 1.35
QUADRU-OSL#6	OSL#6	Z10 222	180–212	48	172.33±11.96	19±6	1.64±0.28	6.81±0.90	1.47±0.01	9±3	1.49±0.07	2.25±0.08	76.70 ± 6.18

* UGA06OSL 330 to 333from primary spring mound, UGA09OSL 655 to 656 samples from the dune, and QUADRU OSL #4 and #6 from Trench 3.
UGA samples were dated at the University of Georgia Luminescence Dating Laboratory in Athens, Georgia, U.S.A.; QUADRU samples were dated at the Quaternary Dating Research Unit (QUADRU) at the CSIR in Pretoria, South Africa. All ages were determined on quartz sand.

ages are not surprising given dates from nearby Florisbad spring of up to 279 ± 47 ka (Grün *et al.*, 1996). Trench 3 provided two OSL ages of 26 ± 1.4 ka at 143 cm depth and 77 ± 6 ka at 222 cm depth. All of the OSL ages appear reliable although the older ages have large uncertainties that might suggest some mixing of sands of different ages at some point during their burial history.

The age data for the spring mound and dune are not sufficiently detailed to allow major inferences about past conditions; however, some tentative conclusions are possible. The mound ages indicate that there was about 1 m of sand accumulation between 13.15 ± 2.30 ka and 3.09 ± 0.45 ka or about 100 cm in 10 ka and a further 1 m of accumulation between 3.09 ± 0.45 ka and present (100 cm in 3 ka). In contrast, there appears to have been only 1 m of accumulation between 38.28 ± 2.3 ka and 13.15 ± 2.30 ka, or 100 cm in 25 ka. The much slower accumulation in this older period might indicate much wetter conditions in the Baden-Baden area at this time. In addition, the dune age of 34.92 ± 3.38 ka at 200 cm corresponds closely with the 38.28 ± 2.3 ka age from the spring mound and both could be evidence of an aeolian period that preceded a long interval of wetter conditions. The age of 25.82 ± 1.35 ka for sample OSL#4 from 143 cm in Zone 6 in Trench 3 also suggests active spring flow and possibly wetter conditions in the ca. 38–13 ka interval (Table 9).

In fact Burrough *et al.* (2009) report three high stands of Palaeolake Makgadikgadi between ca. 39 ka and 17 ka (17.1 ± 1.6 ka, 26.8 ± 1.2 ka, and 38.7 ± 1.8 ka), while Brook *et al.* (2013) obtained an organic age for Stromatolite MART1-14 from Etosha Pan, of 34.2–32.9 cal ka, for a major wet phase in northern Namibia when the lake filling Etosha Pan was at least 8 m above the present floor. Isotope and pollen evidence from a stalagmite in Wonderwerk Cave near Kuruman in the Northern Cape suggested wetter conditions than today at ca. 33 ka, from 23 to 17 ka (Brook *et al.*, 2010). At Soutpan, Peats III and IV record two periods of increased discharge at the Florisbad spring site at ca. 23.2 ka and 6.3 ka (Butzer, 1984a). At Deelpan, Liebenbergspan and Alexandersfontein Pan in the western Free State and Northern Cape, high lakes and spring activity are indicated from ca. 19.2–16.5 ka, and possibly from 25.1–16.5 ka (Butzer, 1984a, b). Alexandersfontein is today only an evaporation pan but according to Butzer *et al.* (1973) it contained a 44 km^2 lake with an average depth of 8 m around 19 ka if ages on carbonate are reliable. Using water balance calculations and assuming a temperature depression of 6°C during the LGM, Butzer *et al.* (1973) estimated that rainfall must have been about twice that of today to support such a large lake. OSL ages for sediments at Witpan in the Northern Cape confirm a wet LGM climate in the South African summer rainfall zone (Telfer and Thomas, 2007). Pan sediments deposited when water occupied the pan indicate two major lake periods: one at ca. 32 ka, the other during the LGM at 20 ka (Telfer *et al.*, 2009). There is thus strong evidence that the period 38–13 ka was wetter than today in the Western Free State.

More rapid deposition of sand on the spring mound between 13.15 ± 2.30 ka and 3.09 ± 0.45 ka, and then between 3.09 ± 0.45 ka and present, suggests drier conditions overall at Baden-Baden during the Holocene. This agrees with evidence of lunette development at several pans in the Western Free State at 12–10 ka, 5.5–3 ka, 2–1 ka, and 0.3–0.07 ka (Holmes *et al.*, 2008) and with evidence of drier conditions in the early Holocene at Wonderwerk Cave, near Kuruman in the Northern Cape (Brook *et al.*, 2010). Burrough *et al.* (2009) report a major high-stand of Palaeolake Makgadikgadi centred on 8.5 ± 0.2 ka, which is in the interval between periods of pan lunette development in the Western Free State at 12–10 ka and 5.5–3 ka reported by Holmes *et al.* (2008). In fact, 4 of 7 radiocarbon ages on organic sediment (not plant material) for the Baden-Baden archaeological sites and Pollen Pit fall in the range ca. 8.4–6.1 ka

(8412–8205, 7672–7575, 6465–6291 and 6601–6128 cal BP) suggesting increased spring flow in the Western Free State around the time when Palaeolake Makgadikgadi was at a high level (8.5 ± 0.2 ka). More confidence will only be gained with more dating of the secondary spring mounds and when evidence from more sequences in the wider region becomes available to test this hypothesis.

The age of 76.70 ± 6.18 ka for sample OSL#6 at 222 cm depth in Zone 10 in Trench 3 is evidence of active spring flow and possibly wetter conditions during MIS 4. Burrough *et al.* (2009) report a high lake stand of Palaeolake Makgadikgadi in Botswana at 64.2 ± 2.0 ka possibly synchronous with wet conditions in the Free State. However, at 205 cm depth there is an unconformity recording either cessation of spring flow because of reduced groundwater levels or a change in the location of the spring outlet. As flow had resumed prior to 25.82 ± 1.35 ka, former lower ground water levels seem more likely. In fact, Riedel *et al.* (2009) report that fossil gastropods in the now mainly dry Boteti River indicate that it was permanently flowing through the western Makgadikgadi Pans at ca. 46 ka, indicating only slightly wetter conditions than today as Lake Palaeo-Makgadikgadi must have been comparatively small at this time as the Boteti River flowed for long distances across the exposed bed of this former lake. Riedel *et al.* (2009) indicate that the hard water effect could not be estimated for the shell and suggest that the true age could be a few thousand years younger. Nevertheless, this approximate time is when the hiatus between Zones 9 and 10 in Trench 3 occurred, supporting an interpretation of relatively dry conditions.

Perhaps the most significant results of the OSL dating are the apparent very old ages for the cemented green, silt/clay-rich fine sands at the base of each auger hole. At the spring mound, the age at 450 cm is 113.11 ± 13.27 ka; at the dune, the age at 455 cm is 164 ± 15.2 ka (Table 8). Because of the large uncertainties in both of these ages, it is unclear in which Marine Isotope Stage (MIS) the respective sediments belong. The spring mound age suggests that the wet period recorded by the clays is MIS 5. In fact, Burrough *et al.* (2009) report high levels of Palaeolake Makgadikgadi at 104.6 ± 3.1 ka and 131 ± 11 ka, either of which might correlate with our spring mound age of 113.11 ± 13.27 ka. In contrast, the dune age suggests wetter conditions during MIS 6 and given that MIS 2 and MIS 4 may have also been wet it would not be surprising if MIS 6 was wet also. Given the height of the clays at the dune site, it is likely that there was abundant spring flow and a high ground water table at this time and possibly also a sizeable lake in Annaspan.

The accumulation of sediments on the east side of the primary mound is complex. All of the deposits are covered with a recent sand layer containing historic artefacts. The marsh deposit in the Northern Block that contains abundant mammal bone is the youngest dated deposit. It is also covered by a very recent sand deposit with historic artefacts.

8.4.6 Faunal and stone tool analysis

The Northern Block, also known as the Bone Bed, was discovered during the last week of the 2003season. A single 1 × 1 m unit (Unit 11) was excavated then and eight additional units (not all a full 1 × 1 m) were excavated in 2006. The only artefacts recovered in the Northern Block were 20 stone cobbles and four fibre tempered plain sherds (Table 10). A single radiocarbon date (Pta-9195) produced an age of 490 ± 40 ^{14}C BP (ca. 496 cal BP). The most common species in the Bone Bed assemblage is black wildebeest, with some hartebeest, springbok, impala and warthog, but there is no evidence for plains zebra (Table 11). We know from historical records and from late Holocene

Table 10. Prehistoric artefacts recovered from excavation blocks at Baden-Baden.

	Northern block	Central block	Southern block	Total
Flakes	0	113	288	401
Adzes	0	6	1	7
Edge Modified Flakes	0	19	0	19
Ceramic sherds	4	0	0	4
Grindstones	0	2	0	2
Bored Stone	0	1	0	1
Hammerstones	0	0	1	1
Cores	0	11	1	12
Cobbles	20	23	13	56
Total	24	175	304	503

Table 11. List of identifiable vertebrate remains by excavation block, as recovered in the 2003 and 2006 field seasons. The counts are the number of identified specimens (NISP).

		Northern block	Central block	Total
REPTILIA				
	Tortoise	–	1	1
AVES				
	Indet.	1	–	1
MAMMALIA				
Carnivora				
	Indet.	1	–	1
Perissodactyla				
	Equus quagga subsp. (plains zebra)	–	1	1
Artiodactyla				
	Phacochoerus aethiopicus/ africanus (warthog)	8	3	11
	Alcelaphus buselaphus (hartebeest)	7	–	7
	Damaliscus pygargus (blesbok)	2	–	2
	Connochaetes gnou (black wildebeest)	68	6	74
	Antidorcas marsupialis (springbok)	11	–	11
	Aepycerus melampus (impala)	3	–	3
	Ovis aries (sheep)	11	–	11
	Bovidae indet.			
	Large-medium	88	11	99
	Small-medium	10	1	11
	Small	2	–	2
	Total	212	23	235

Note: Bone preservation in the South Block was poor and only nine bone specimens were recorded, but these were not identifiable and are not included in this table.

sites, such as Deelpan (Scott and Brink, 1992) that plains zebra was a common faunal element in the central interior. Given the abundance of black wildebeest in the Bone Bed and the known ecological association between plains zebra and wildebeest, the absence of plains zebra is more likely explained by human prey selection and is probably not a reflection of palaeoenvironmental conditions. The size of the black wildebeest suggests a late Holocene age, which accords with the radiocarbon date and with the presence of domestic sheep in the assemblage. Most of the Northern Block Bone Bed specimens had been intensively processed and were highly fragmented in the process of extracting marrow. The artefacts suggest a very narrow range of activities primarily focused on the fracturing of bone and cooking.

The faunal remains from the Northern Block demonstrate intensive processing activities by LSA hunter-gatherers. The degree to which bones were broken, so that even phalanges were split, suggests marrow-extracting or fat rendering activities. Since chipped stone tools are lacking, there is little evidence for this being the location of the initial stage of butchering and none to suggest this was the exact locale where the kills were made, although both were probably very close. The narrow focus on black wildebeest suggests that this species was systematically hunted in an organized manner. The preference for black wildebeest as prey is in contrast to the evidence for Middle Stone Age hunting behaviour in a similar setting at Florisbad, where there is a focus on medium-sized antelope, mainly blesbok and springbok and their extinct relatives (Brink 1987; Brink and Henderson, 2001). The presence of domestic sheep in hunter-gatherer subsistence is not unusual and has been recorded previously in the Riet River occupations (Brink *et al.*, 1992).

Of particular interest is the presence of impala in the Bone Bed (Figure 5) assemblage, since there is uncertainty on its occurrence in historic times in the central

Figure 5. A view of the dense clustering of bone specimens in Unit 11 in Northern Block.

interior (Skinner and Smithers, 1990). Impala is usually associated with savanna grassland, where there is some degree of tree cover. However, the modern vegetation around Baden-Baden is a dry, open grassland (see above), and would not provide suitable habitat for impala. It is noteworthy that the bone assemblages from Deelpan do not contain impala (Scott and Brink, 1992; Brink, 2005; Butzer and Oswald, in press), and its presence in the Bone Bed assemblage may suggest a slight shift in the local vegetation during recent times with possibly localised patches of tree cover (see pollen results). The predominance of black wildebeest in the Bone Bed assemblage would not suggest tree cover on a wide scale, since the species requires open habitat (Brink, 2005).

In the Central Block seven 1 × 1 m units were excavated in 2003 and three single isolated units (Units 2, 5 and 8) were placed nearby. LSA stone artefacts, stone cobbles, and faunal remains were recovered. The Central Block and Unit 8 produced material dated to the early-mid Holocene. These occupations are interesting because they are not microlithic Wilton components. The most distinctive tools are adzes and a single bored stone (Table 10). The lack of microlithic tools is unusual as this material overlaps chronologically with Wilton occupations, however, similar artefacts have been found at Voigstpost and at Florisbad dating to the same approximate period. It is possible that this presents a terminal Lockshoek or some other non-microlithic tradition (Horowitz *et al.*, 1978; Kuman *et al.*, 1999). The predominance of black wildebeest in the faunal remains from the central block confirms the focus on this species as the preferred prey also earlier during the Holocene.

The bones in the Central Block were generally very poorly preserved compared to the Northern Block bones, which is probably due to the proximity of the spring drainage area. At Florisbad poorly preserved bones and stone artefacts were found in a similar setting close to the drainage of the spring. Bone preservation was markedly better toward the centre of the mound than on the eastern side with the spring and drainage, suggesting that the quality of bone preservation will improve into the mound and away from the spring drainage area. As in the Northern Block, the species composition in the Central Block is dominated by black wildebeest, with warthog also frequent, however, the evidence of intensive processing is absent in the Central Block.

The Southern Block also had very poor bone preservation probably because the occupation was in a fine friable sandy deposit. Most of the artefacts, made on hornfels, were flakes but a hammerstone, adze and core were recovered in the excavation. Immediately below the cut bank where the Southern Block artefacts were exposed we recovered a cryptocrystalline straight-backed bladelet that must have eroded out of the deposit. No absolute dates have been obtained on this occupation but the bladelet and the general degree of pedogenic development suggests a late Holocene age and a Wilton occupation.

8.4.7 Palynological potential and preliminary results

Pollen samples were taken from the 70 cm of organic material in the North Block bone bed; no pollen was recovered. The lack of pollen could be due to exposure and drying out but these deposits have potential for phytolith analysis. The secondary spring mounds, which showed the above-mentioned discrepancy between fractions of organic material in one mound, have good potential for further palynological research. The 89 cm sequence from the Pollen Pit, which consists of peat with some sandy horizons at 15 and 42 cm, contained rich pollen assemblages mainly including Poaceae (Figure 6).

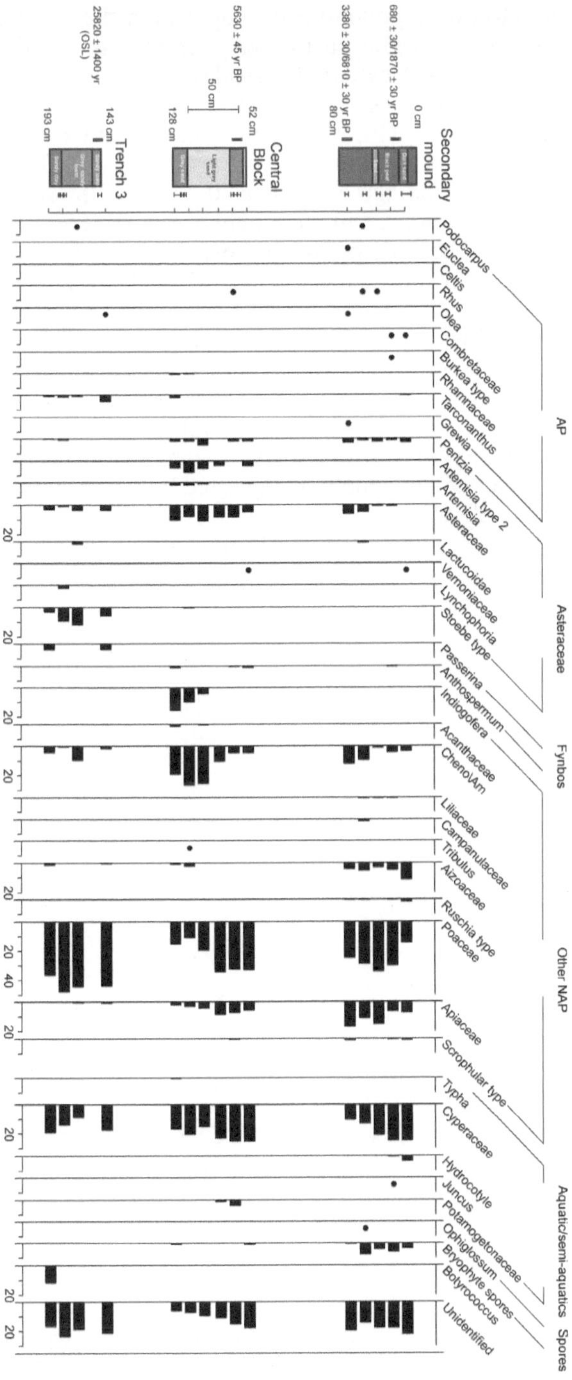

Figure 6. Pollen diagram summarizing the main pollen types in organic layers of the Pollen Pit in a secondary mound (top), Unit 7 in the archaeological excavation in the Central Block (centre), and from Trench 3 at Baden-Baden (bottom). Dots indicate presence where percentages were too low to be illustrated effectively as bars.

Middle to late Holocene polleniferous deposits in the Central Block are dark organic sands or peat that formed part of an energized spring flow. Similar to the Pollen Pit, the results suggest essentially treeless vegetation with very low numbers of woodland pollen, possibly transported a long distance by wind from the neighbouring Savanna Biome (Figure 1), but this should be supported by more analyses at higher resolution. Two sections are available, viz., the upper section, P1, of Theko *et al.* (2003) with an age of 2420 ± 70 ^{14}C BP (Pta-8840, ca. 2452 cal BP, Table 7) corresponding to an area immediately above Unit 9 (Bousman *et al.*, n.d. interim report; details of pollen not shown here), and the deeper adjacent excavation Unit 7 with a radiocarbon date of 5630 ± 45 ^{14}C BP (GrA-25206, ca. 6362 cal BP). Pollen composition in the former consists of a grass-rich assemblage that overlies grey sand with lower organic contents which are richer in Chenopodiaceae and Amaranthaceae pollen (Cheno/Ams). The pollen in the latter section (Unit 7) yielded results with a strong presence of Asteraceae pollen; indicative of a grassy karroid veld ca. 6362 cal BP (Figure 6). The high presence of Cheno/Ams in the lower levels suggests drier conditions. Further work will be needed to establish the nature of conditions ca. 8.5 ka, which as suggested by the geological results (above), might have been relatively wet.

A sequence (Pollen 2, Figure 3) some 15 m south of the spring structure of the Central Block with less organic material in the sands contained much lower pollen concentrations (Theko *et al.*, 2003). The Last Glacial period is represented in Trench 3

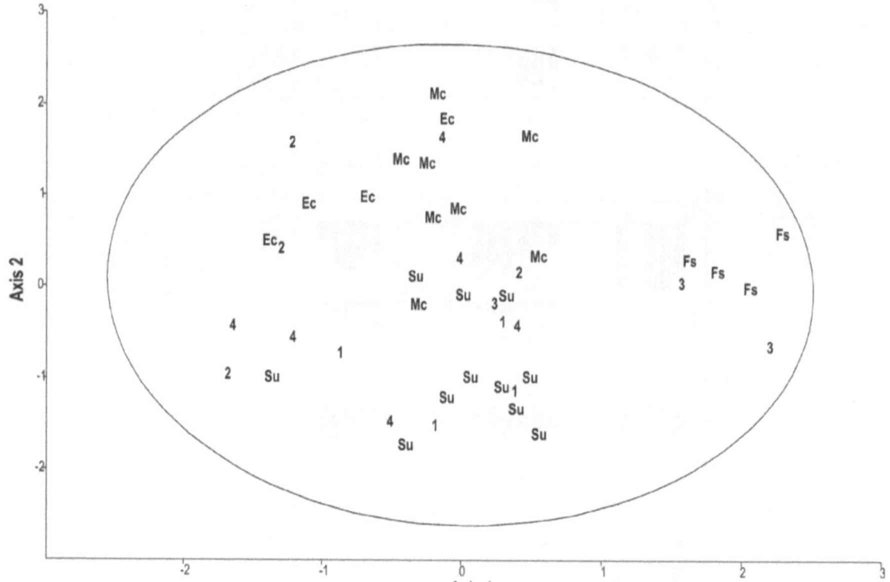

Figure 7. A two-dimensional matrix of data generated by a Detrended Correspondence Analysis based on the morphometrical dimensions of non-lobate morphotypes recorded in samples BB-1–BB-4, compared to the morphometrical dimensions of non-lobate morphotypes produced by modern C_4 and C_3 grasses. 1 = BB-1; 2 = BB-2; 3 = BB-3; 4 = BB-4; Mc = C_4 mesophytic *Microchloa*; Ec = C_4 mesophytic or xerophytic *Eragrostis*; Su = C4 xerophytic *Stipagrostis* and Fs = C_3 mesophytic *Festuca*. Eigenvalues for Axis 1 and Axis 2 are 0.02792 and 0.002662, respectively.

Table 12. Morphometrical analysis of non-lobate short-cell morphotypes from BB-1, BB-2, BB-3 and BB-4 based on the assessment of three different measurements. The non-lobate category includes diagnostic short-cell morphotypes commonly recognized as cubical, round, oblong, saddle-shaped or elongated. NECKX = minimum width of the anterior aspect of the silica body; AX = maximum width of the anterior aspect of the silica body; AY = maximum length of the anterior aspect of the silica body. Values for NECKX, AX and AY are expressed in microns (μm).

Sample ID	Slide #	Ref #	Morphotype	NECKX	AX	AY	Ratio x/y
BB-1	BB 378	249	Saddle	5.47	4.84	6.04	0.801
		252	Saddle	6.36	5.89	8.41	0.700
		253	Saddle	7.29	7.64	6.81	1.122
		245	Saddle	5.96	6.12	8.2	0.746
BB-2	BB 379	255	Saddle	8.46	8.86	5.79	1.530
		256	Saddle	4.95	6.58	5.16	1.275
		257	Saddle	5.51	6.03	8.22	0.734
		259	Saddle	5.36	6.18	4.91	1.259
BB-3	BB 380	267	Oblong	6.8	7.33	9.29	0.789
		269	Oblong	5.02	4.97	10.15	0.490
		269	Oblong	5.43	4.49	12.09	0.371
BB-4	BB 382	279	Saddle	3.24	4.11	4.91	0.837
		280	Saddle	5.45	5.47	7.72	0.709
		284	Saddle	5.91	6.13	5.11	1.200
		286	Saddle	4.72	4.13	4.82	0.857
		288	Saddle	4.26	4.65	5.82	0.799
		289	Saddle	5.93	6.22	4.47	1.391

with organic lenses estimated to be ca. 25 ka old (OSL dating) that contain pollen types such as *Passerina* and *Stoebe*, indicating that fynbos elements occurred, which are probably associated with cooler grassy conditions. Data from this age are not available for Florisbad although this site contains limited pollen evidence from a previous cooler period (possibly MIS 6 or 8) in levels that have yet to be dated effectively (van Zinderen Bakker, 1989; Scott and Rossouw, 2005). They seem to have been deposited under wetter conditions as indicated by Ericaceae pollen grains.

8.4.8 Grass short-cell phytolith potential and preliminary results

Based on the OSL ages provided for the spring mound auger site, as well as a multivariate analysis (DCA) of non-lobate morphotypes recorded in the samples (Figure 7, Table 12 and Plates 1–4), the phytolith data indicate an overall prevalence of warm, C_4 grassy conditions at 3.09 ± 0.45 ka (sample BB-1), 13.2 ± 2.3 ka (sample BB-2) and again at 113.1 ± 13.3 (sample BB-4). This inference is based on the presence of medially indented, as well as medially convex saddle-shaped morphotypes (see measurements in Table 12), which are exclusively produced by chloridoid and aristidoid grasses adapted to warm and moderately wet to arid conditions (Rossouw, 2009). Conversely, a predominance of C_3-affiliated cubical and oblong—shaped silica bodies at the cost of saddle-shaped morphotypes at 38.3 ± 3.6 (sample BB-3) concur with the inference for regionally cooler and wetter conditions around 39 ka and 34 ka as mentioned earlier in the paper.

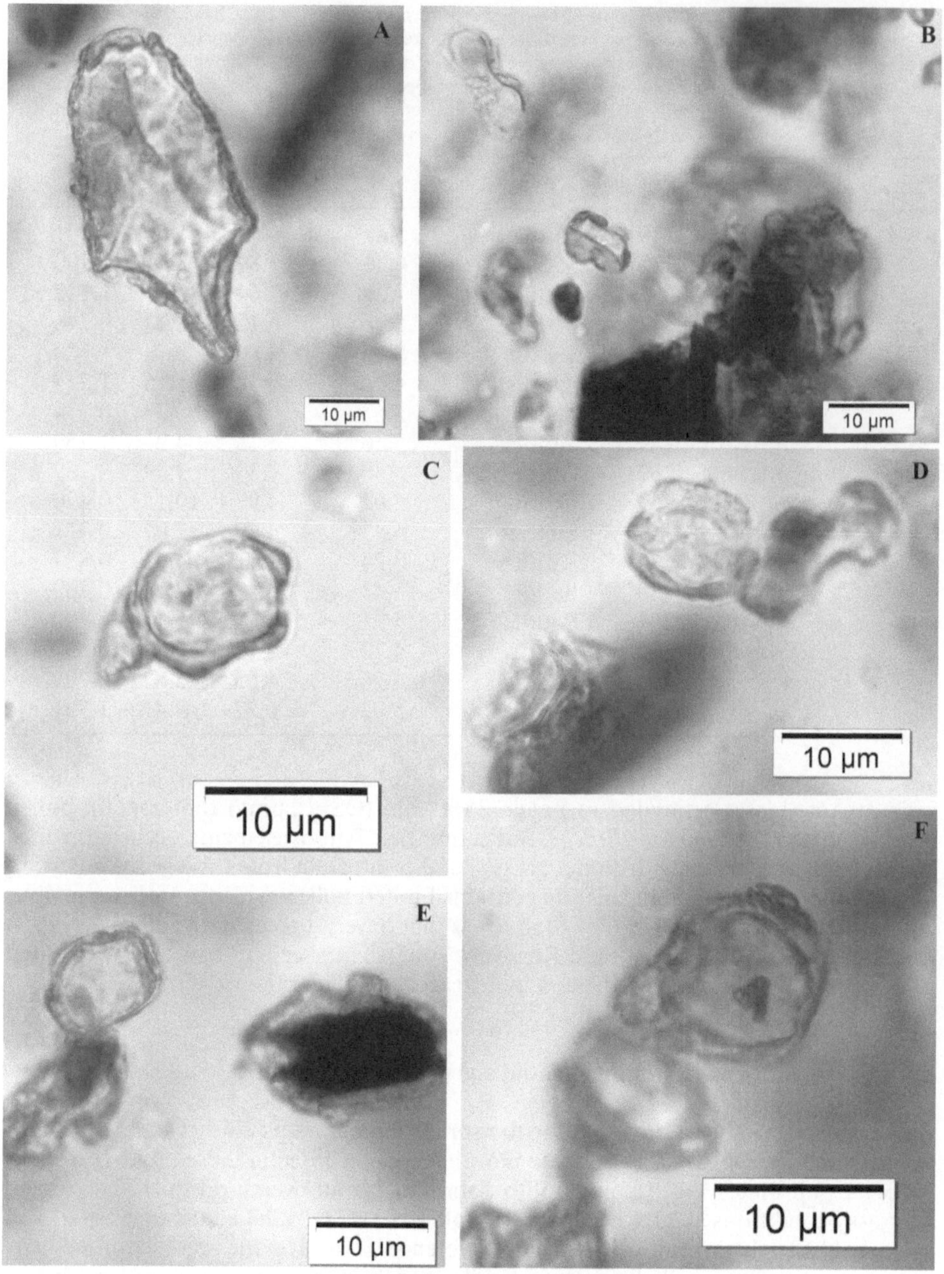

Plate 1. Spring Mound Auger Site sample BB-1. (A) cuneiform bulliform silica body; (B) short-necked bilobate, anterior and side view; (C) medially indented saddle morphotype, anterior view; (D) medially convex saddle morphotype, posterior view; (E) medially convex saddle morphotype, anterior view; (F) medially indented saddle morphotype, anterior view.

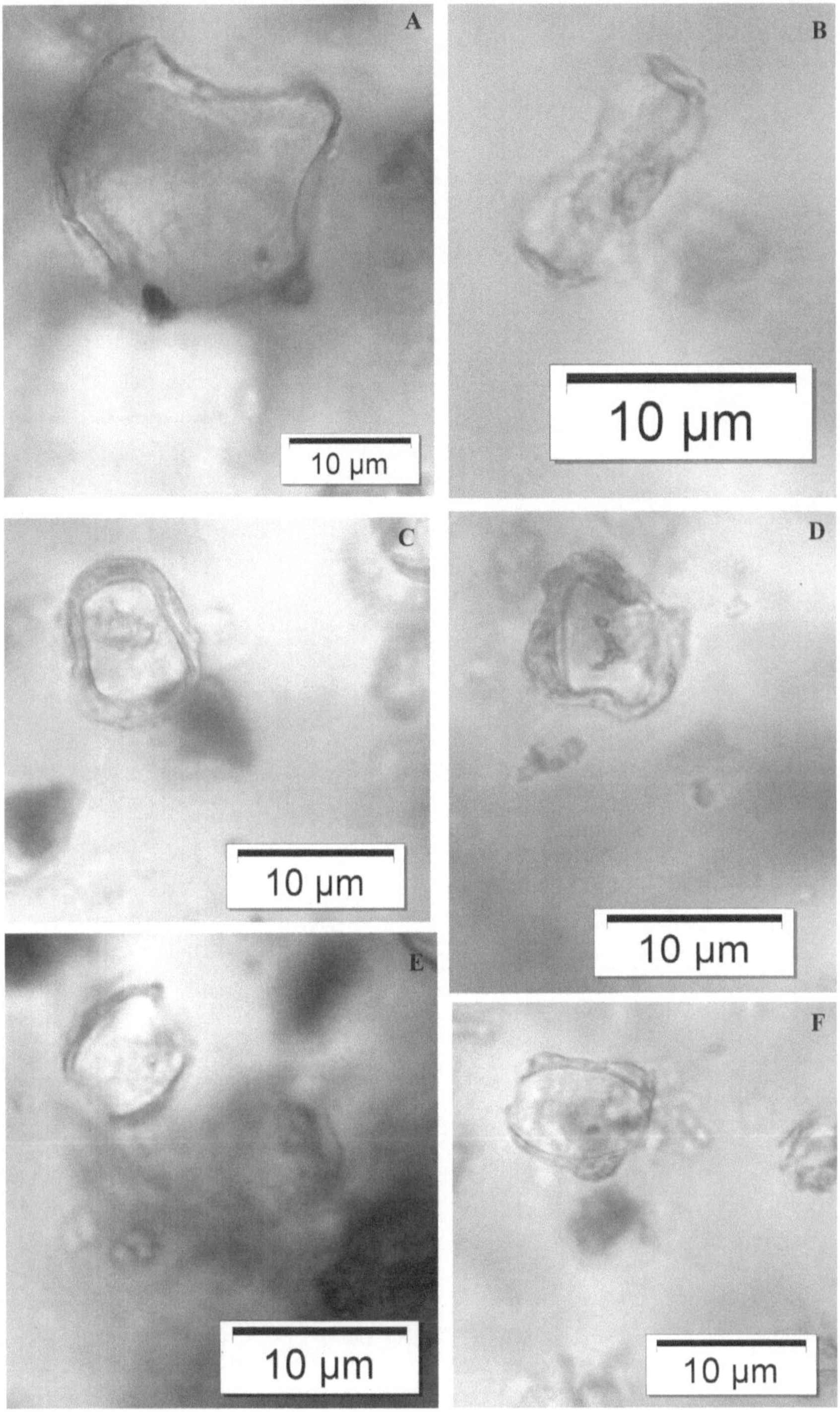

Plate 2. Spring Mound Auger Site sample BB-2. (A) cuneiform bulliform silica body; (B) short-necked bilobate, anterior view; (C–F) medially indented saddle morphotype, anterior view.

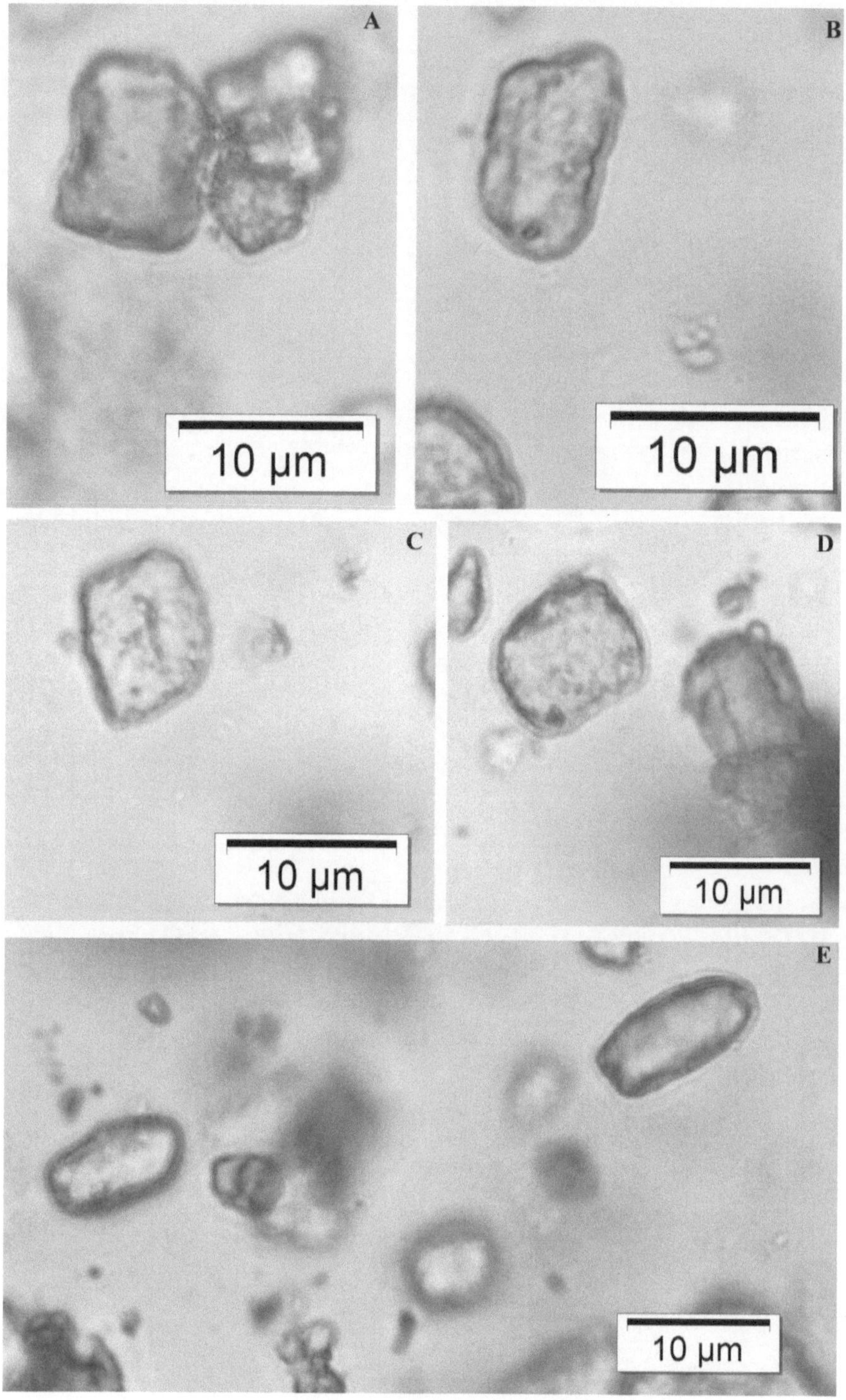

Plate 3. Mound Auger Site sample BB-3. (A–D) trapezoid silica body, anterior and side view; (E) oblong silica body, anterior view.

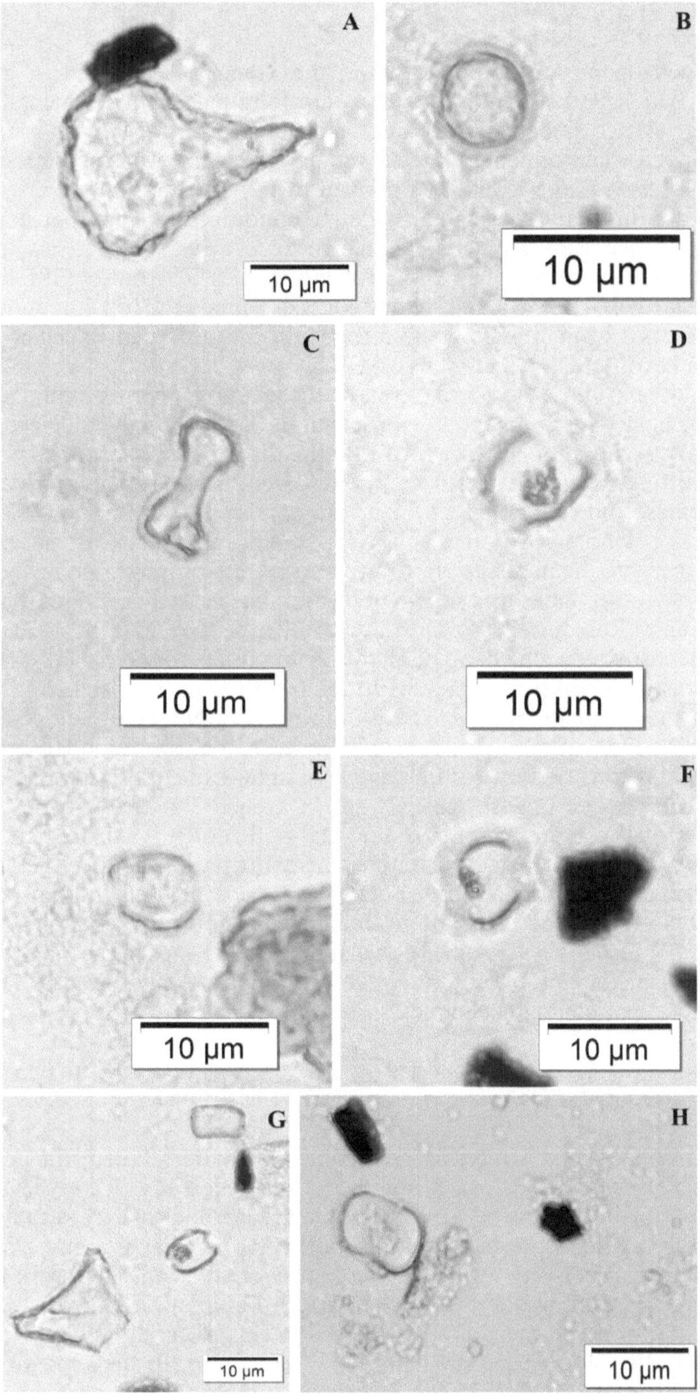

Plate 4. Spring Mound Auger Site sample BB-4. (A) cuneiform bulliform silica body; (B) globular silica body with rugose surface (C) short-necked bilobate, anterior view; (D–H) medially indented saddle morphotype, anterior view.

8.5 CONCLUSION

The OSL ages for the spring mound, dune and Trench 3 sediments are a clear indication that Baden-Baden could provide a very long palaeoenvironmental record for central South Africa associated with human occupations. In both auger holes, augering was stopped when cemented, green, silt/clay-rich sand was encountered, and yet deeper, and older, samples could be obtained from both sites. A future research program linking spring activity at Baden-Baden to conditions in Annaspan could provide one of the longest records of climate change for central South Africa, particularly if it was supplemented by pollen and phytolith studies. The oldest OSL age we have obtained so far is 164 ± 15 ka and we have only examined the top 5 m of sediments at the site. There is a strong possibility that the sediments at Baden-Baden could provide information beyond MIS 6 and even to MIS 8.

The middle to late Holocene deposits of the secondary spring mounds show good potential for palaeoenvironmental reconstruction based on palynological research in the secondary spring mounds provided that the discrepancy in dating between plant material and fine organic fractions can be resolved. Microscopic charcoal data (not shown here) also indicate potential for investigating past burning in the area. In the Central Block palynological potential is marginal since some levels do not contain pollen. It will therefore be necessary in future also to rely on phytolith analysis in these deposits as phytoliths have been shown to be useful in the semiarid Free State environments (Scott and Rossouw, 2005). Overall the pollen sequences from the Baden-Baden Holocene deposits indicate changing plant communities that reflect alternating variations in moisture conditions—parallel to the results from Florisbad and Deelpan in the western Free State (Scott, 1988; Scott and Brink, 1992; Scott and Nyakale, 2002; Butzer and Oswald, in press) that can be integrated to improve the Holocene palaeoenvironmental reconstruction of the region. A significant finding at Baden-Baden is that it provided pollen of last "glacial" age in Trench 3 at the base of the primary mound. The levels containing fynbos and grass pollen of this age are limited to a few horizons. Nevertheless it makes an important contribution about vegetation conditions in the relatively dry central Free State because coeval levels from the other major site in the region, Florisbad, are absent. Further palynological research and dating at Baden-Baden, promises to fill some spatial and temporal gaps in the long-term palaeoenvironmental history of the interior region of South Africa by providing a more complete picture of changing climates.

ACKNOWLEDGEMENTS

We would like to thank Mr Nellis Nel, owner of Baden-Baden, for access to the site, logistical assistance, historical notes, and his generosity. We especially wish to thank Sarel Greyling for his help in during the archaeological excavations. The town of Dealesville provided the backhoe and both Hanlie Greyling and "Witkop" are thanked for their generosity to the archaeological crew. The pollen work is based on the research supported by the National Research Foundation. Any opinion, finding and conclusion or recommendation expressed in this material is that of the authors and the NRF does not accept any liability in this regard. Brook's research was supported by NSF grant NSF-0725090. Bousman and Brink's research was funded by a grant from the Leakey Foundation, Texas State University, the National Research Foundation and the National Museum. Bokang Theko, Petrus Chakane, Frank Neumann, Abel Dichakane, Peter Mdala, Willem Nduma, Lebohang Nyenye, Adam

Thibeletsa, Koos Mzondi, Zoe Henderson, Sharon Holt, Duma Mbolekwa, Bonny Nduma, Peter Ntulini, Isaac Thapo, Gary Trower and helpers from the National Museum, Bloemfontein, assisted with fieldwork. Thanks to Tania R. Marshall for copies of articles.

REFERENCES

Aitken, M.J., 1985, *Thermoluminescence Dating*, (Orlando: Academic Press).

Aitken, M.J., 1998, *An Introduction to Optical Dating,* (Oxford: Oxford University Press).

Beaumont, P.B., 2004, Wonderwerk Cave. In *The Northern Cape: Some Key Sites*, edited by Morris, D. and Beaumont, P.B. (Kimberley: Archaeology, McGregor Museum), pp. 31–36.

Bøtter-Jensen, L., Duller, G.A.T., Murray, A.S. and Banerjee, D., 1999, Blue light emitting diodes for optical stimulation of quartz in retrospective dosimetry and dating. *Radiation Protection Dosimetry,* **84**, pp. 335–340.

Bousman, C.B, Brink, J.S. and Weinstein, A., n.d., L. S. B. Leakey Foundation Grant Interim Report: A Search for Modern Human Behavior at Baden-Baden, South Africa (Unpublished report).

Bousman, C.B. and Brink, J.S., 2008, Holocene Geoarchaeology and Archaeology at Baden-Baden; Symposium: New views on ancient Africa. Papers in honour of C. Garth Sampson. Presentation, Society for American Archaeology, Vancouver.

Bousman, C.B. and Brink, J.S., 2012, Baden-Baden—*A Smithfield Antelope Processing Site in the western Free State, South Africa.* Presentation. Society of Africanist Archaeologists, Toronto, June 2012.

Brink, J.S., 1987, The archaeozoology of Florisbad, Orange Free State. *Memoirs of the National Museum, Bloemfontein,* **24**, pp. 1–151.

Brink, J.S., 2005, The evolution of the black wildebeest (Connochaetes gnou) and modern large mammal faunas of central southern Africa. DPhil dissertation, University of Stellenbosch.

Brink, J.S. and Henderson, Z.L., 2001, A high-resolution last interglacial MSA horizon at Florisbad in the context of other open-air occurences in the central interior of southern Africa: an interim statement. In *Middle Palaeolithic and Middle Stone Age settlement systems*, edited by Conard, N. (Tuebingen: Kerns-Verlag), pp. 1–20.

Brink, J.S., Dreyer, J.J.B. and Loubser, J., 1992, Rescue excavations at Pramberg, Jacobsdal, south-western Orange Free State. *Southern African Field Archaeology,* **1**, pp. 54–60.

Brook, G.A., Scott, L., Railsback, L.B. and Goddard, E.A., 2010, A 35 ka pollen and isotope record of environmental change along the southern margin of the Kalahari from a stalagmite and animal dung deposits in Wonderwerk Cave, South Africa. *Journal of Arid Lands,* **74**, pp. 870–884.

Brook, G.A., Cherkinsky, A., Railsback, L.B., Marais, E. and Hipondoka, M.H.T., 2013, Radiocarbon dating of organic residue and carbonate in stromatolites from Etosha Pan, Namibia: the radiocarbon reservoir effect, correction of published carbonate ages, and evidence of a >8 m deep lake during the Late Pleistocene. *Radiocarbon,* **55**, pp. 1156–1163.

Burrough, S.L., Thomas, D.S.G. and Bailey, R.M., 2009, Mega-Lake in the Kalahari: A Late Pleistocene record of the Palaeolake Makgadikgadi system. *Quaternary Science Reviews,* **28**, pp. 1392–1411.

Butzer, K.W., Fock, G.J., Stuckenrath, R. and Zilch, A., 1973, Palaeohydrology of Late Pleistocene Lake, Alexanderfontein, Kimberley, South Africa. *Nature,* **243**, pp. 328–330.
Butzer, K.W., 1984a, Late Quaternary environments in South Africa. In *Late Cainozoic Palaeoclimates of the Southern Hemisphere,* edited by Vogel, J.C. (Rotterdam: Balkema), pp. 235–264.
Butzer, K.W., 1984b, Archaeogeology and Quaternary environment in the interior of southern Africa. In *Southern African Prehistory and Palaeoenvironments,* edited by Klein, R.G. (Rotterdam: Balkema), pp. 1–64.
Butzer, K.W. and Oswald, J.F., in press, Dry Lakes or Pans of the western Free State, South Africa: Environmental History and possible Human Impact at Deelpan. *Palaeoecology of Africa,* **33**.
Duller, G.A.T., 1999, *Luminescence Analyst computer programme V2.18*. Department of Geography and Environmental Sciences, University of Wales, Aberystwyth.
Galbraith, R.F., Roberts, R.G., Laslett, G.M., Yoshida, H. and Olley, J.M., 1999, Optical dating of single and multiple grains of quartz from Jinmium rock shelter, northern Australia, part 1, Experimental design and statistical models. *Archaeometry,* **41**, pp. 339–364.
Geldenhuys, J.N., 1982, Classification of the pans of the western Orange Free State according to vegetation structure, with reference to avifaunal communities. *South African Journal of Wildlife Research,* **12**, pp. 55–62.
Gil-Romera, G., Neumann, F.H., Scott, L., Sevilla-Callejo, M. and Fernández-Jalvo, Y., 2014, Pollen taphonomy from hyaena scats and coprolites: preservation and quantitative differences. *Journal of Archaeological Science,* **46**, pp. 89–95.
Grobler, N.J., and Loock, J.C., 1988, Morphological development of the Florisbad deposit. *Palaeoecology of Africa,* **19**, pp. 163–168.
Grün, R., Brink, J.S., Spoons, N.A., Taylor, L., Stringer, C.B., Franciscus, R.G. and Murray, A.S., 1996, Direct dating of Florisbad hominid. *Nature,* **382**, pp. 500–501.
Hogg, A.G., Hua, Q., Blackwell, P.G., Buck, C.E., Guildersno, T.P., Heaton, T.J., Niu, M., Palmer, J.G., Reimer, P.J.; Reimer, R.W., Turney, C.S.M. and Zimmerman, S., 2013, SHCal13 Southern Hemisphere Calibration, 0–50,000 years cal BP. *Radiocarbon,* **55(4)**, pp. 1889–1903.
Holmes, P.J., Bateman, M.B., Thomas, D.S.G., Telfer, M.W., Barker, C.H. and Lawson, M.P., 2008, A Holocene-late Pleistocene aeolian record from lunette dunes of the western Free State panfield, South Africa. *The Holocene,* **18(8)**, pp. 1193–1205.
Horowitz, A, Sampson, C.G., Scott, L. and Vogel, J. C., 1978, Analysis of the Voigtspost Site, O. F. S., South Africa. *South African Archaeological Bulletin,* **33**, pp. 152–159.
Kuman, K.M., Inbar, I. and Clark, R.J., 1999, Palaeoenvironments and cultural sequence of the Florisbad Middle Stone Age hominid site, South Africa. *Journal of Archaeological Science,* **26**, pp. 1409–1425.
Markey, B.G., Bøtter-Jensen, L. and Duller, G.A.T., 1997, A new flexible system for measuring thermally and optically stimulated luminescence. *Radiation Measurements,* **27**, pp. 83–90.
Marshall, T.R., 1987, Morphotectonic analysis of the Wesslesbron panveld. *South African Journal of Geology,* **90(3)**, pp. 209–218.
Marshall, T.R., 1988, The origin of the pans of the western Orange Free State, a morpholtectonic study of the palaeo-Kimberley River. *Palaeoecology of Africa,* **19**, pp. 97–108.
Martin, C.W. and Johnson, W.C., 1995, Variation in radiocarbon ages of soil organic matter fractions from late Quaternary buried soils. *Quaternary Research,* **43**, pp. 232–237.

Mucina, L. and Rutherford, M.C., 2006, *The vegetation of South Africa, Lesotho and Swaziland* (Pretoria: South African National Biodiversity Institute, Strelitzia 19).

Murray, A.S. and Wintle, A.G., 2000, Luminescence dating of quartz using improved single-aliquot regenerative-dose protocol. *Radiation Measurement,* **32**, pp. 57–73.

North American Commission on Stratigraphic Nomenclature, 2005, North American Stratigraphic Code. *The American Association of Petroleum Geologists Bulletin,* **89, 11**, pp 1547–1591.

Olson, G.W., 1981, *Soils and environment, a guide to soils surveys and their applications,* (New York: Chapman and Hall).

Prescott, J.R. and Hutton, J.T., 1994, Cosmic ray contributions to dose rates for luminescence and ESR dating: large depths and long-term time variations. *Radiation Measurements,* **23**, pp. 497–500.

Reineck, H.-E. and Singh, I.B., 1975, *Depositional sedimentary environments, with reference to terrigenous clastics,* (Berlin: Springer-Verlag).

Riedel, F., von Rintelen, T., Erhardt, S. and Kossler, A. 2009, A fossil Potadoma (Gastropoda: Pachychilidae) from Pleistocene central Kalahari fluvio-lacustrine sediments. *Hydrobiologia,* **636**, pp. 493–498.

Rossouw, L., 2009, The application of fossil grass-phytolith analysis in the reconstruction of late Cainozoic environments in the South African interior. PhD Dissertation. University of the Free State. Bloemfontein.

Schulze, B.R., 1972, South Africa. In *Climates of Africa* edited by Griffiths, J.F. in *World Survey of Climatology Volume 10,* edited by Landsberg, H.E. (Amsterdam: Elsevier Publishing Company), pp. 501–586.

Schulze, R.E., 1997, Climate. In *Vegetation of Southern Africa,* edited by Cowling, R.M., Richardson, D.M. and Pierce S.M. (Cambridge University Press), pp. 21–42

Scott, L., 1987, Pollen Analysis of Hyena Coprolites and Sediments fron Equus Cave, Taung, Southern Kalahari (South Africa). *Quaternary Research*, **28**, pp. 144–156.

Scott, L., 1988, Holocene environmental change at western Orange Free State pans, South Africa, inferred from pollen analysis. *Palaeoecology of Africa and the surrounding Islands,* **19**, pp. 109–118.

Scott, L. and Brink, J.S., 1992, Quaternary palaeoenvironments of pans in central South Africa: Palynological and palaeontological evidence. *South African Geographer,* **19(1/2)**, pp. 22–34.

Scott, L. and Nyakale, M., 2002, Pollen indications of Holocene palaeoenvironments at Florisbad spring in the central Free State, South Africa. *The Holocene,* **12(4)**, pp. 497–503.

Scott, L. and Rossouw, L., 2005, Reassessment of Botanical Evidence for Palaeoenvironments at Florisbad, South Africa. *South African Archaeological Bulletin,* **60(182)**, pp. 96–102.

Scott, L. and Thackeray, J.F., 2014, Palynology of Holocene deposits in Excavation 1 at Wonderwerk Cave, Northern Cape (South Africa). *African Archaeological Review* (in press).

Skinner, J. and Smithers, R.H., 1990, *The mammals of the southern African subregion.* (Pretoria: University of Pretoria, 2nd edition).

Stuiver, M. and Reimer, P.J., 1993, Extended ^{14}C database and revised CALIB radiocarbon calibration program. *Radiocarbon,* **35**, pp. 215–230.

Telfer, M.W. and Thomas, D.S.G., 2007, Late Quaternary linear dune accumulation and chronostratigraphy of the southwestern Kalahari: implications for aeolian palaeoclimatic reconstructions and predictions of future dynamics. *Quaternary Science Reviews,* **26**, pp. 2617–2630.

Telfer, M.W., Thomas, D.S.G., Parker, A.G., Walkington, H. and Finch, A.A., 2009, The palaeoenvironmental potential form a southeastern Kalahari pan floor. *Palaeogeography, Palaeoclimatology, Palaeoecology,* **273**, pp. 50–60.
Theko, B., Scott, L. and Bousman, C.B., 2003, *Preliminary Report on the Palynology of Baden-Baden Spring Deposits in the Free State, South Africa.* SASQUA Conference, University of the Witwatersrand, Johannesburg.
Van Zinderen Bakker, E.M., 1989, Middle Stone Age palaeoenvironments at Florisbad (South Africa). *Palaeoecology of Africa and the surrounding Islands,* **20**, pp. 133–154.
Visser, J.N.J. and Joubert, A., 1991, Cyclicity on the Late Pleistocene to Holocene spring and lacustrine deposits at Florisbad, Orange Free State. *South African Journal of Geology,* **94**, pp. 123–131.

CHAPTER 9

Charcoal from pre-Holocene Stratum 5, Wonderwerk Cave, South Africa

Marion K. Bamford

Evolutionary Studies Institute and School of Geosciences, University of the Witwatersrand, Johannesburg, South Africa

ABSTRACT: Charcoal from the archaeological layer, Stratum 5 of Wonderwerk Cave in the Northern Cape Province of South Africa is described and identified. The stratum has been dated by several methods and is pre-Holocene in age, up to about 15 ka. About three quarters of the 134 pieces are identifiable and belong to eight types (six species and two to generic level) of woody plants: *Ozoroa paniculosa* and *Searsia* (*Rhus*) *lancea* (Anacardiaceae), *Ehretia* sp. (Boraginaceae), *Commiphora* sp. (Burseraceae), *Dombeya rotundifolia* (Pentapetalaceae), *Olinia ventosa* (Oliniaceae), *Berchemia discolor* (Rhamnaceae) and *Halleria lucida* (Scrophulariaceae). Of these species only *Searsia lancea* occurs in the area today. The majority of the woods found can tolerate dry conditions and a wide range of temperatures but the presence of *Berchemia discolor and Halleria lucida* indicates that conditions between c. 14,985 and 13,952 years cal. BP may have been slightly wetter than today.

9.1 INTRODUCTION

Wonderwerk Cave in the Northern Cape Province of South Africa contains a record of approximately the last 2 million years of occupation by hominins and humans although to date no skeletal material has been found. The site was first studied by Malan and Cooke (1941) and later extensively excavated by Peter Beaumont and colleagues from the 1970s to early 1990s (Beaumont, 1990, 2004; Camp, 1948; Thackeray, 1983; Avery, 1981, 2007; van Zinderen-Bakker, 1982) producing large amounts of artifacts, bone and botanical remains from seven excavations within the cave. In the mid 2000s a new team, led by Michael Chazan and Liora Kolska Horwitz, has used high resolution mapping and dating techniques in the cave. Beginning with Excavation 1 which is 30 m from the cave entrance, their multinational and multidisciplinary team is in the process of re-analysing old material and collecting new material (Figure 1), (Chazan *et al.*, 2008, 2012; Rüther *et al.*, 2009; Matmon *et al.*, 2011). Some of the results produced by this team are being collected in a special issue of African Archaeological Review (AAR for 2015). A description of the relatively meagre macrobotanical remains from Excavation 1 Strata 12 to 5 includes only the older remains (Bamford, AAR in 2015) where the rare charcoal pieces are unidentifiable. These older remains comprise grass and sedge culms and seeds of a palm and a legume. Charcoal, however, is abundant and well preserved in the younger strata, 5 to 2, from pre-Holocene to Present. This paper is the first in a series of detailed descriptions and palaeoecological interpretations of the charcoal remains from the various Holocene strata. In due course the charcoals from the other strata will be compared and both the anthropogenic selection of woods and palaeoecology will be assessed.

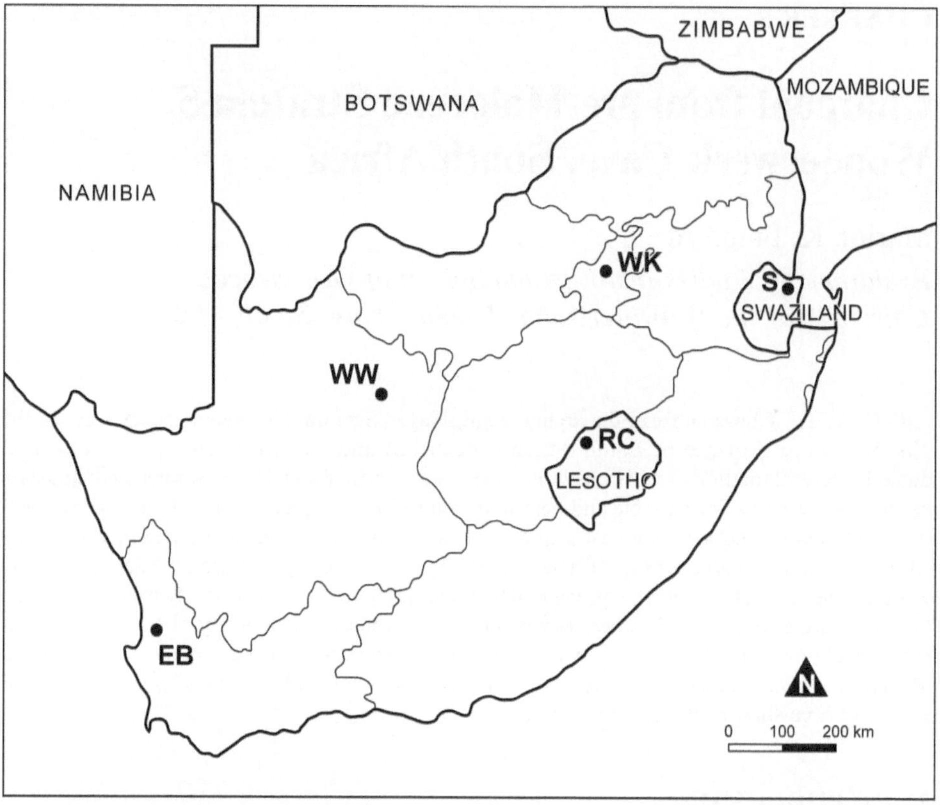

Figure 1. Map of the locality and other sites mentioned in the text. WW = Wonderwerk Cave, WK = Wonderkrater, RC = Rose Cottage, S = Sifiso rock shelter, EB = Elands Bay Cave.

9.1.1 Research history

Wonderwerk Cave is situated in the Northern Cape Province of South Africa at the edge of the Ghaap plateau in the Kuruman Hills (27°50′46″ S, 23°33′19″ E; about 1680 meters above sea level), (Figure 1). With an average annual rainfall of 420 mm (80% occurring during the summer) and relatively flat topography the climate and vegetation are dry to semi-arid (Schulze, 1984). The vegetation biome is savanna but the cave is very close to the grassland biome. According to Mucina and Rutherford (2006) the vegetation type is Kuruman Mountain Bushveld (SVK10) with a wide variety of grasses and some small trees and shrubs on shallow, sandy soils. Trees include *Searsia lancea, Searsia pyroides, Searsia tridactyla, Searsia ciliata, Diospyros austro-africana, Euclea crispa, Olea europaea, Tarchonanthus camphoratus*, and shrubs *Anthospermum rigidum, Helichrysum zeyheri, Wahlenbergia nodosa*. Near the cave entrance are *Grewia flava* and *Boscia albitrunca*, and the common grasses are *Themeda triandra, Cymbopogon plurinoides* and *Aristida* spp.

The cave is long, narrow and almost horizontal with one entrance that faces north-north-west, overlooking the Ghaap Plateau. It is a solution cavity in stratified dolomitic limestones but is overlain by early Proterozoic to late Archaean banded ironstones (2–3 Ga; Kent, 1980). The cave is about 140 m long and 11–17 m wide with

Table 1. Charcoal-rich strata from Wonderwerk Cave with ages and estimated abundances of charcoal pieces.

Archaeo-logical Stratum	Thackeray & Lee Thorp, 1992 (approx. age years BP)	Lee Thorp & Ecker (accepted)	Ecker new charcoal dates	Archaeo-logical technology	Sublayers and estimates of number of pieces of charcoal
1		Last 100 years		Modern	
2a 2b	50 1200	Last 100 years 190–270 (2b–3a)		Late Stone Age	3 sublayers; five boxes of charcoal (pieces not counted)
3a 3b	1890 3990	 2060–4800		Late Stone Age	5 sublayers: >5000 pieces of charcoal
4a 4b 4c 4d	4890 5180 7430 10,200	4300–5890 5500–6500 5960–9800 8600–12,200		Late Stone Age	10 sublayers and spits: >5000 pieces of charcoal
5	12,400 (Vogel et al. 1986)	12,500	14,989–13,952	Late Stone Age	Total of 134 pieces of charcoal

a domed roof. Over time the entrance has receded but light penetrating the cave is partially blocked by a large stalagmite that is about 20 m in from the entrance. Excavation 1 is immediately behind the stalagmite.

One piece of charcoal was recovered from the lowermost layer, Stratum 12, approximately 2 Ma but is too poorly preserved for identification (Bamford, AAR in 2015). The earliest evidence for the controlled use of fire based on four lines of evidence, including burned grasses and sedges, comes from Wonderwerk Cave at 1 Ma (Berna *et al.*, 2012). The subsequent layers have some calcified plant material preserved but none of it has been burned (Bamford, AAR in 2015). The inconsistent macroplant record may be due to gaps in occupation of the site, different uses of the site and/or taphonomic biases. There is a large time gap prior to Stratum 5 (Chazan *et al.*, 2008) and then a rich record of charcoal remains through the Holocene (Table 1).

9.1.2 Dating

The Holocene strata 1–5 in Excavation 1 were isotopically dated by John Vogel using charcoal remains (Thackeray, 1983; Vogel *et al.*, 1986). Later equid teeth were dated from the Holocene layers and produced slightly different ages (Thackeray and Lee Thorp, 1992) and these ages are given in Table 1 together with more recent results based on ostrich egg shell (Lee Thorp and Ecker, AAR in 2015). Ecker and Lee Thorp are in the process of dating more charcoal samples from the Holocene layers but the results have not yet been published.

9.2 METHODS

The Holocene strata (Strata 1–5) of Excavation 1 were excavated by Peter Beaumont in 1978 and Anne and Francis Thackeray in 1979 (Thackeray, 1983). Cave sediments were dry sieved and charcoal fragments were separated from other finds in the McGregor Museum (Thackeray, 1983:40) but there are no further details recorded on the collection methods. Since the charcoal samples have been packaged according to archaeological stratum, square number and depth within the square, and noted in centimetres or as spits, that notation will be used in this work.

Since the sample size was relatively small all charcoal fragments were studied, initially under a binocular microscope (Olympus SZX16, magnification 7–112x) and oriented for hand fracturing to reveal fresh surfaces of the three planes required for study (transverse, radial longitudinal and tangential longitudinal). Under higher magnification (200–500x; Zeiss Axiolab-A1 bright phase/dark phase reflected light microscope) the wood anatomy was studied further and photographed with a digital camera (Olympus DP72 with Stream Essentials® software). Measurements were made of the cell dimensions but precise and averaged measurements of the cells are less useful for identification and comparison with modern taxa because woods shrink between 8–20% depending on the wood type, moisture content and temperature of the fire. Nonetheless when charcoal is compared with charcoal these values are useful. Since diagnostic features such as ray cell type, inter-vessel and vessel-ray pits were visible under the light microscope it was not necessary to use scanning electron microscopy. Charcoal was identified using the modern comparative charcoal collections housed in the Archaeology Department and the Bernard Price Institute (now Evolutionary Studies Institute), both at the University of the Witwatersrand. The computer database for modern woods, InsideWood, was also used. Terminology follows that of the IAWA Committee (1989). The fractured pieces were placed in small individual ziplock bags so they can be retrieved and re-examined. Each fragment has been given a number: locality+(depth)+letter, eg. WW-5 M29 (0–5 cm) I. On completion of the project the material will be returned to the McGregor Museum, Kimberley (Wonderwerk Cave 6508).

9.3 Results

Stratum 5 has over 134 charcoal fragments from many of the grid squares. Most pieces are less than 7 × 7 × 7 mm but a few are up to roughly 20 mm cubes but often the wood is distorted. Approximately one quarter of the pieces are twisted, weathered or too small to fracture and reveal the three planes needed for identification. These specimens have been recorded as "not identifiable." The abundances and identifications of the charcoal are given in Table 2 and the descriptions of the charcoal types are below.

Descriptions of charcoal types from Stratum 5

Anacardiaceae
Ozoroa paniculosa
Described sample: WW-5 M29 (0–5 cm) I
Illustration: Plate 1, Figures 1–4
Total number of pieces: 40

DESCRIPTION: The wood is diffuse-porous and no growth rings are visible (Plate 1, Figures 1, 2). Vessels are arranged in short radial multiples of 1–3, sometimes 4 vessels,

Table 2. Complete list of identified charcoal pieces from Stratum 5, Excavation 1, Wonderwerk Cave and the number of pieces from each grid square. Note that every piece was studied and is listed here.

				Identity of charcoal pieces									
Stratum	Sq	Bl	Depth (cm)	*Ozoroa paniculosa*	*Searsia (Rhus) lancea*	*Commiphora sp.*	*Ehretia sp.*	*Dombeya rotundifolia*	*Olinia ventosa*	*Berchemia discolor*	*Halleria lucida*	Not identifiable	Total number of pieces
5	K	27	0–5	3						3		2	8
	M	29	0–5	12	5				8	1		2	28
	M	29	5–15	5									5
	N	22	0–5									1	1
	N	27	0–5						2			1	3
	O	26	0–5			1	2					2	5
	O	28	5–10	3				4				1	8
	Q	29	0–5	8								4	12
	Q	29	Veg	9						1		11	21
	Q	29	5–10				1				2	3	6
	Q	29	10–15		1							1	2
	Q	31	0–5					3		9		4	16
	Q	33	0–5									1	1
	R	26	0–5							14			14
	R	32	0–5		2					1		1	4
5A	O	24				1						1	2
Totals				40	8	2	3	7	10	29	2	35	134

and in clusters. Vessels have simple, horizontal to oblique perforation plates and the average vessel diameter is about 50 µm, vessel elements are 200–500 µm long, and there are more than 40 vessels per square mm. Paratracheal parenchyma is scanty to vasicentric (Plate 1, Figures 2, 3). Rays are 1–3 seriate, heterocellular with mixed square cells and upright cells. Amorphous dark contents are present in many of the square cells, one per cell and the cells are not enlarged; they do not appear to be crystalline. Fibres are septate. Inter-vessel pits are alternate and minute (2–4 µm); vessel-ray pits are elongated to scalariform (Plate 1, Fig 4). Radial canals were not seen.

IDENTIFICATION AND COMPARISON: Woods of the Anacardiaceae in southern Africa typically have vessels arranged in short radial multiples and not solitary; rays are commonly 1–2 seriate with square and/or upright cells, seldom procumbent; crystals are common in the ray cells; septate fibres; parenchyma is not abundant and can be banded and/or paratracheal. The piece of charcoal described here has these features and so was compared with the modern reference collection and found to be very similar to *Ozoroa paniculosa* (University of the Witwatersrand reference material: Lucy Allott collection sample number A4, and Zimbabwe collection sample number Z97).

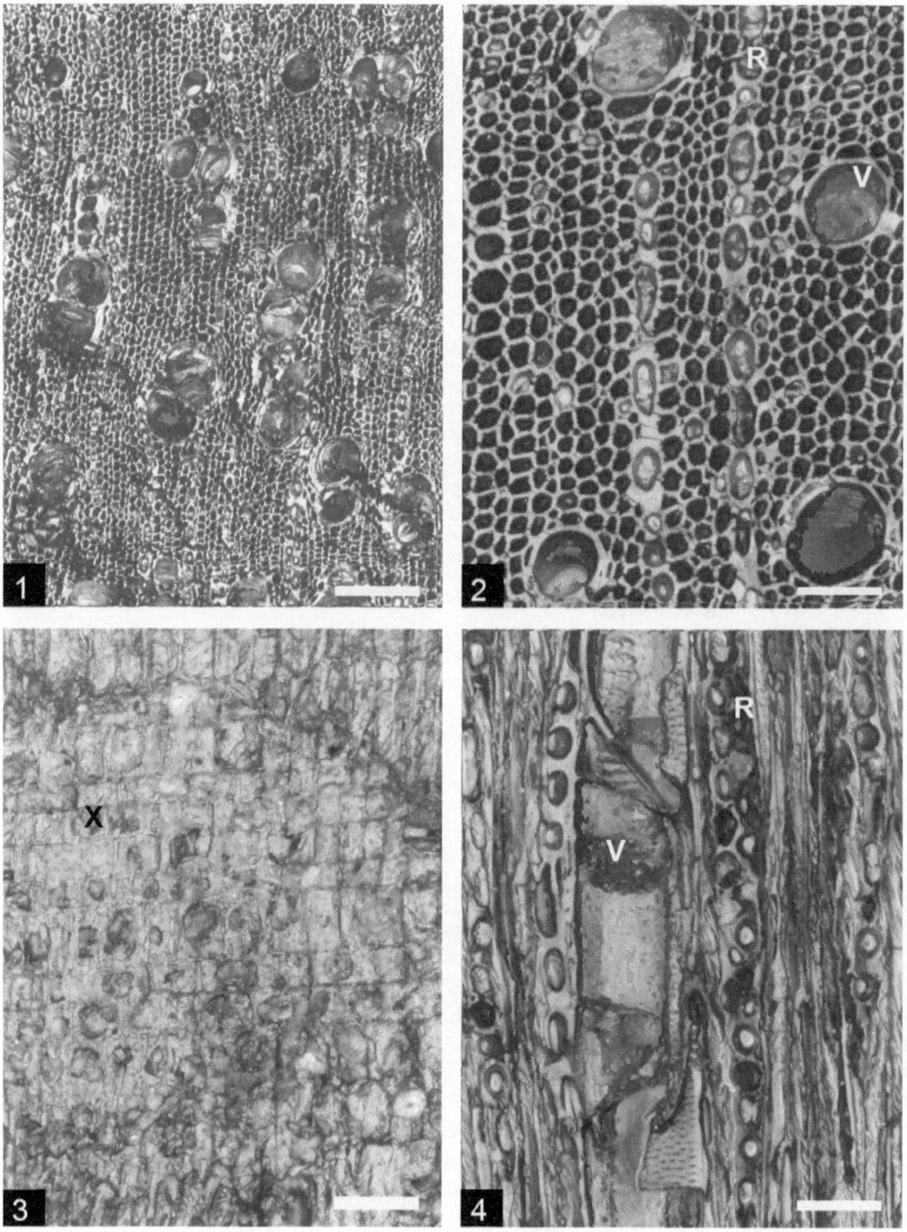

Plate 1. Photomicrographs of charcoal from Stratum 5, Excavation 1, Wonderwerk Cave, taken under bright field at 200x and 500x magnification. Cells are labeled on the photographs of Plates 1–8 for guidance: V = vessel; R = ray, P = parenchyma, X = prismatic crystal in ray cell, F = fibre. TS = transverse section; RLS = radial longitudinal section; TLS = tangential longitudinal section. 1–4 – *Ozoroa paniculosa* (Anacardiaceae). 1, 2 – TS, vessels in short radial multiples of 1–3; longitudinal rays with crystals visible and vasicentric parenchyma (few cells surrounding the vessels; ground tissue comprises fibres). Scale bar 1 = 100 μm; 2 = 40 μm. 3 – RLS, square to upright ray cells containing crystals (angular dark contents). Scale bar = 40 μm. 4 – TLS, 1–2 seriate rays and vessels with simple and oblique perforation plates and small, alternate inter-vessel pits at the bottom of the photograph and more elongated vessel-ray pits above the letter V. Scale bar = 40 μm.

There are fourteen species of *Ozoroa* in southern Africa (Coates Palgrave, 2002) and most of them can be excluded on a geographical basis as they each have a very restricted distribution in either Namibia or the far northwestern part of South Africa or in the eastern and northern parts. *Ozoroa insignis* and *Ozoroa paniculosa* have a much wider distribution and are therefore considered further. The wood or charcoal of these two species can be differentiated by the presence of crystals in the ray cells of *O. insignis* and tyloses in the vessels. The archaeological material has neither of these features so is identified as *O. paniculosa*.

Anacardiaceae
Searsia* (Rhus) *lancea
Described sample: WW-5 R32 A
Illustration: Plate 2, Figures 1–4
Total number of pieces: 8

DESCRIPTION: Growth rings are not visible, wood is diffuse-porous and the simple perforation plates are horizontal or oblique (Plate 2, Figures 1, 2). Vessels are arranged in short radial multiples of 1–3 cells with occasional clusters. The average vessel diameter is about 80 μm, average length is around 100 μm with about 40 vessels per square mm. Parenchyma is scanty paratracheal. Rays are 1–2 seriate, 200–400 μm tall and heterocellular with mixed bands of square, upright and procumbent cells (Plate 2, Figure 3). Radial canals are rare (Plate 2, Figure 4). Prismatic crystals occur in the square ray cells. Fibres have medium-thick walls and are septate. Inter-vessel pits are alternate, minute to very small (4 μm diameter; Plate 2, Figure 4) and vessel-ray pits are elongated to scalariform.

IDENTIFICATION AND COMPARISON: This charcoal also has the typical features of woods of the Anacardiaceae. This wood is very similar to that of *Ozoroa paniculosa* in transverse section but the rays are more heterocellular with procumbent cells included and prismatic crystals in the ray cells. Radial canals are present and there is no diffuse parenchyma. This wood is identical to *Rhus lancea* charcoal reference material (A12).

There are 47 species of *Seasia* (*Rhus*) in southern Africa (Coates Palgrave, 2002; van Wyk and van Wyk, 2013), most of them with restricted distributions that are far distant from Wonderwerk. The 11 species that today either occur in the area or are within a few hundred kilometre radius have been compared with these samples. Only two of these have crystals in the ray cells, *Searsia gueinzii* and *Searsia lancea*. The former species can be excluded as it has diffuse parenchyma whereas *S. lancea* has vasicentric parenchyma, like the archaeological specimens. Additionally, *S. gueinzii* has smaller vessels than *S. lancea.*

Boraginaceae
***Ehretia* sp.**
Described sample: WW-5 Q29 (5–10 cm) B
Illustration: Plate 3, Figures 1–4
Total number of pieces: 3

DESCRIPTION: Wood is diffuse-porous and perforation plates are simple and horizontal. Vessels are arranged in short radial multiples of 1–3 cells and some clusters. Average vessel diameter is about 80 μm, length about 200 μm and there are about 40 vessels per square mm (Plate 3, Figures 1, 2). Parenchyma is aliform to confluent

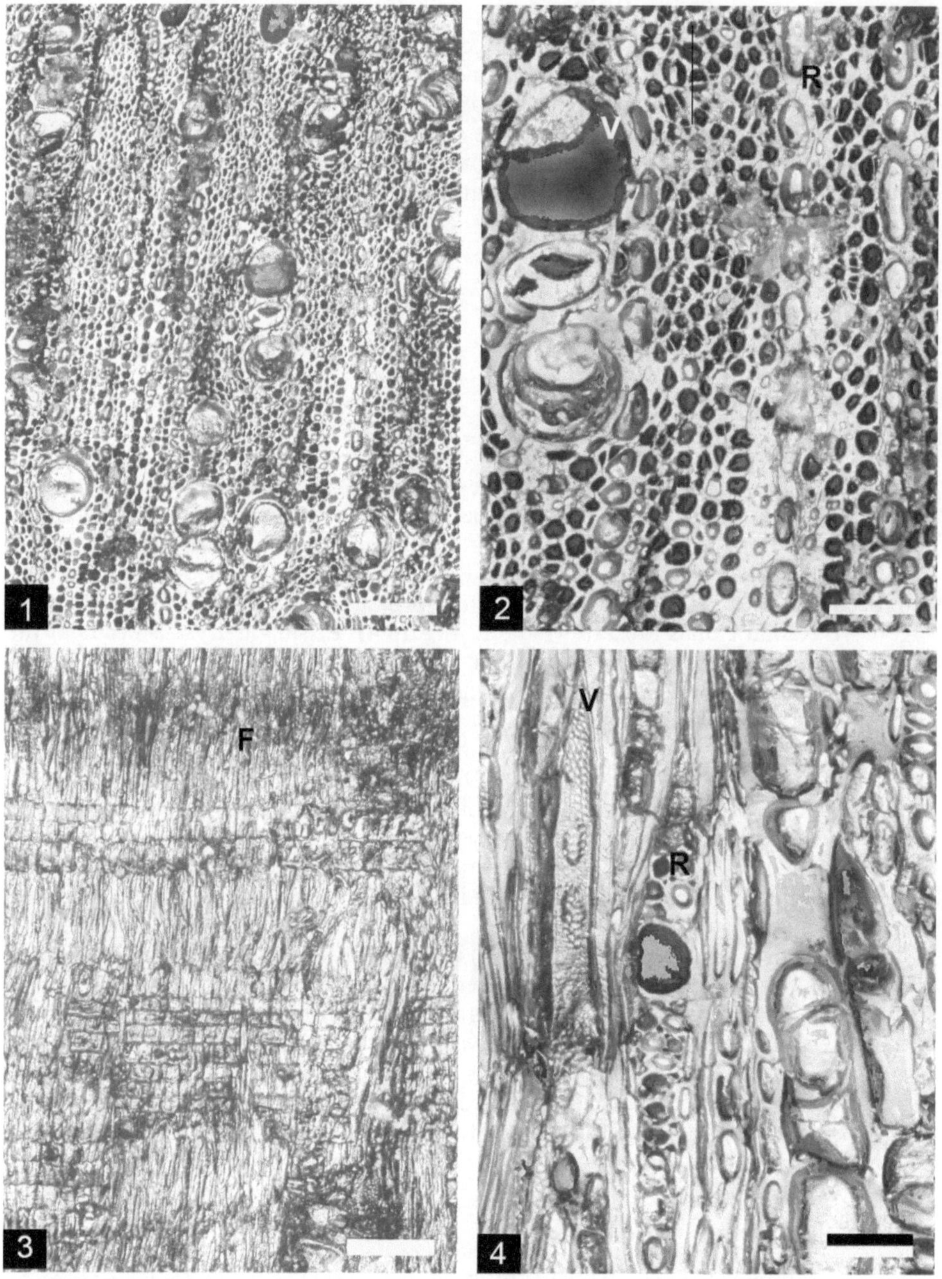

Plate 2. *Searsia* (*Rhus*) *lancea* (Anacardiaceae). 1 – TS, vessels in short radial multiples. Scale bar = 100 μm. 2 – TS, vessels in short radial multiples. Scale bar = 40 μm. 3 – RLS, mostly procumbent ray cells shown with a few crystals present. Scale bar = 40 μm. 4 – TLS, 1–2 seriate rays and central ray has a radial canal (large dark rimmed circle). Inter-vessel pits are small and alternate (below letter V). Scale bar = 40 μm.

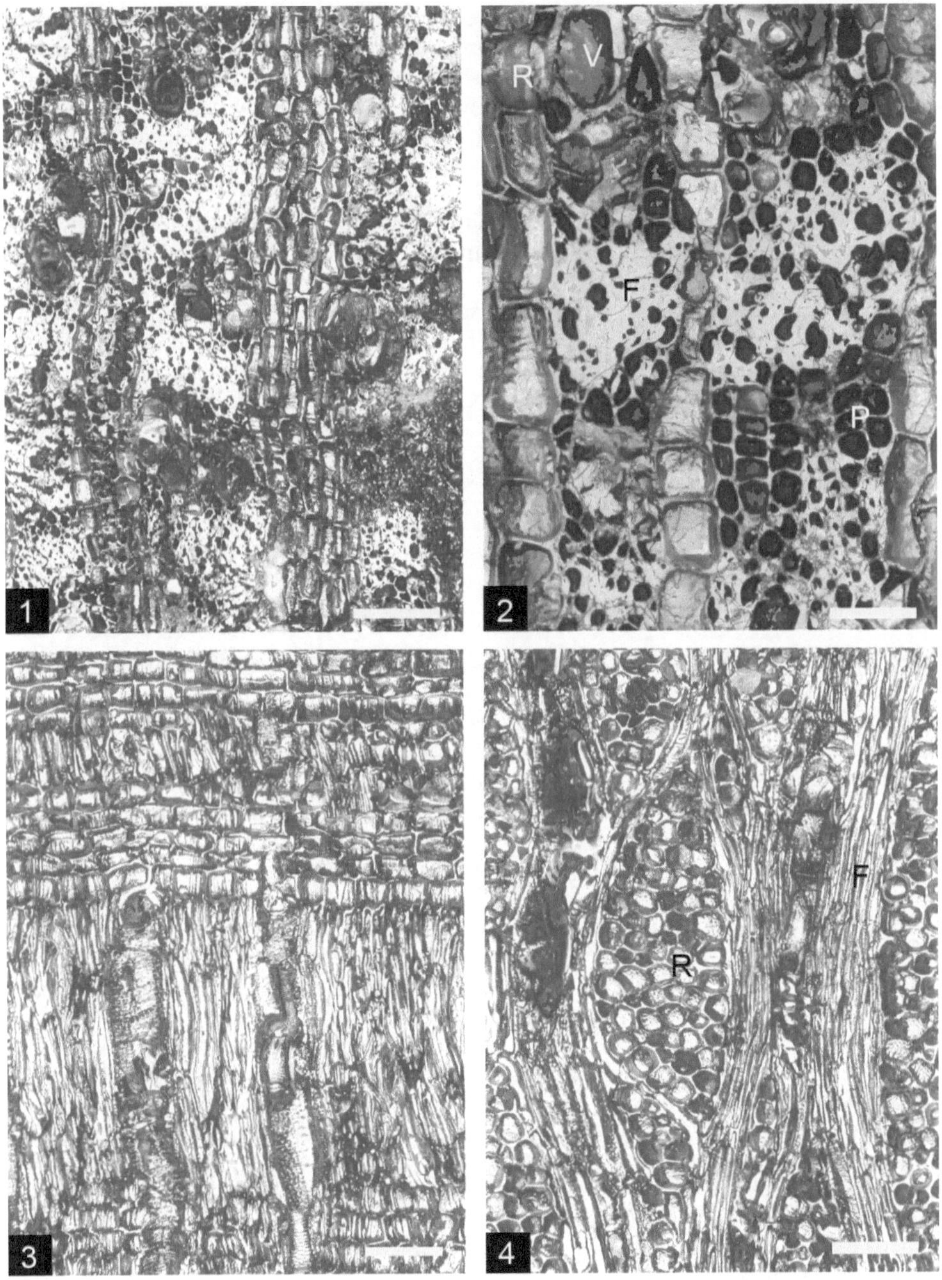

Plate 3. *Ehretia* sp. (Boraginaceae). 1, 2 – TS, vessels in short radial multiples and clusters. Rays are wide and seen as longitudinal lines of brick-shaped cells. Note bands of thick-walled fibres alternating with thin-walled parenchyma cells. Scale bar 1 = 100 µm; 2 = 40 µm. 3 – RLS, mostly procumbent ray cells with 1–2 rows of marginal upright cells (bottom of photograph). Scale bar = 100 µm. 4 – TLS, rays short and 5–6 cells wide and surrounding thick walled fibres. Scale bar = 40 µm.

to almost banded (Plate 3, Figure 2). There are alternating bands of axial parenchyma and thick-walled fibres. Parenchyma strands have 4–8 cells. Rays are clearly visible in transverse section and are 4–8 seriate, short, wide and abundant. They are weakly homocellular comprising procumbent body cells and one, rarely two, rows of upright to square marginal cells (Plate 3, Figures 3, 4). Inter-vessel pits are alternate and small; vessel-ray pits are the same as the inter-vessel pits (Plate 3, Figure 3).

IDENTIFICATION AND COMPARISON: Woods of the Boraginaceae that occur in Africa typically have a few small vessel members, weakly heterocellular rays around 6 cells wide and banded parenchyma. This piece of charcoal is very similar to *Ehretia* spp (InsideWood database) but the four species occurring in South Africa have not been studied. Without more comparative material it is not possible to distinguish the species, *Ehretia alba*, *Ehretia amoena*, *Ehretia rigida* and *Ehretia oppositifolia* so the charcoal is identified only as *Ehretia* sp. Geographically, the charcoal is likely to be *E. rigida*.

Burseraceae
***Commiphora* sp.**
Described sample: WW-5a O24 A
Illustration: Plate 4, Figures 1–4
Total number of pieces: 2

DESCRIPTION: Wood is diffuse-porous and the perforation plates are simple and oblique. Vessels are arranged in pairs or are solitary, about 50 μm in diameter and about 20 per square mm (Plate 4, Figures 1, 2). Parenchyma is rare to absent. Rays are 1–3 cells wide and 10–15 cells high, heterocellular with mixed procumbent, square and upright cells. There are prismatic crystals in the ray cells and sometimes 2–4 crystals per upright cell (Plate 4, Figure 3). Radial canals are rare. Fibres are medium to thick-walled and septa are visible. Inter-vessel pitting is alternate, crowded and small (4–8 μm). Vessel-ray pits are elongated to scalariform (Plate 4, Figure 4).

IDENTIFICATION AND COMPARISON: There are about 35 species of *Commiphora* in southern Africa (van Wyk and van Wyk, 2013) but we do not have charcoal reference material of all of them. The charcoal from Stratum 5 has the typical features of all the species with relatively small and few vessel members, rays 1–4 seriate and heterocellular; and parenchyma rare to absent. The important identifying features are the presence of radial canals and more than one crystal per upright cell. It is tentatively identified as *Commiphora* sp.

Pentapetaceae
Dombeya rotundifolia
Described sample: WW-5 O28 (5–10 cm) E
Illustration: Plate 5, Figures 1–4
Total number of pieces: 7

DESCRIPTION: Growth rings are not visible, wood is diffuse-porous and the perforation plates are simple and horizontal. Vessels are arranged in short radial multiples of 1–3 cells, average diameter 80–100 μm with about 30 vessels per square mm (Plate 5, Figure 1). Vessel elements are comparatively short, about 100–150 μm. Parenchyma is scanty. Rays are 1–3 seriate, rarely 4-seriate, 10–20 cells or 200–500 μm high with mixed upright and square cells (Plate 5, Figures 2–4). Narrow rays seem to be storied.

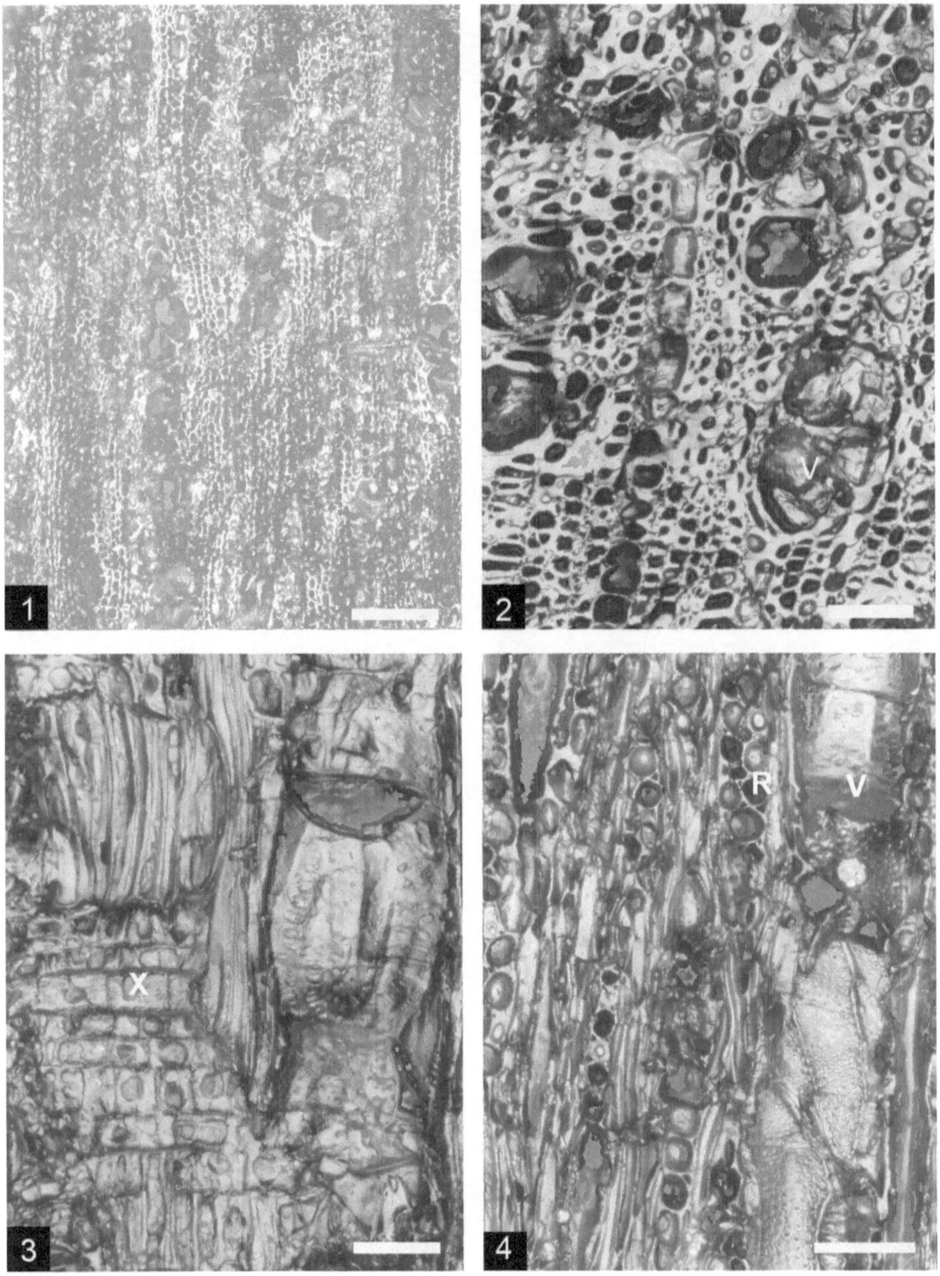

Plate 4. *Commiphora* sp. (Burseraceae). 1, 2 – TS, vessels in 1–2 s and no parenchyma seen. Scale bar 1 = 100 µm; 2 = 40 µm. 3 – RLS, ray cells with crystals. Note 3–4 crystals per upright cell (bottom centre). Ray-vessel pitting (lower right, above the scale bar) elongated to scalariform. Scale bar = 40 µm. 3 – TLS Rays 1–2 seriate. Inte-r-vessel pitting is small and alternate (centre of vessel). Scale bar = 40 µm.

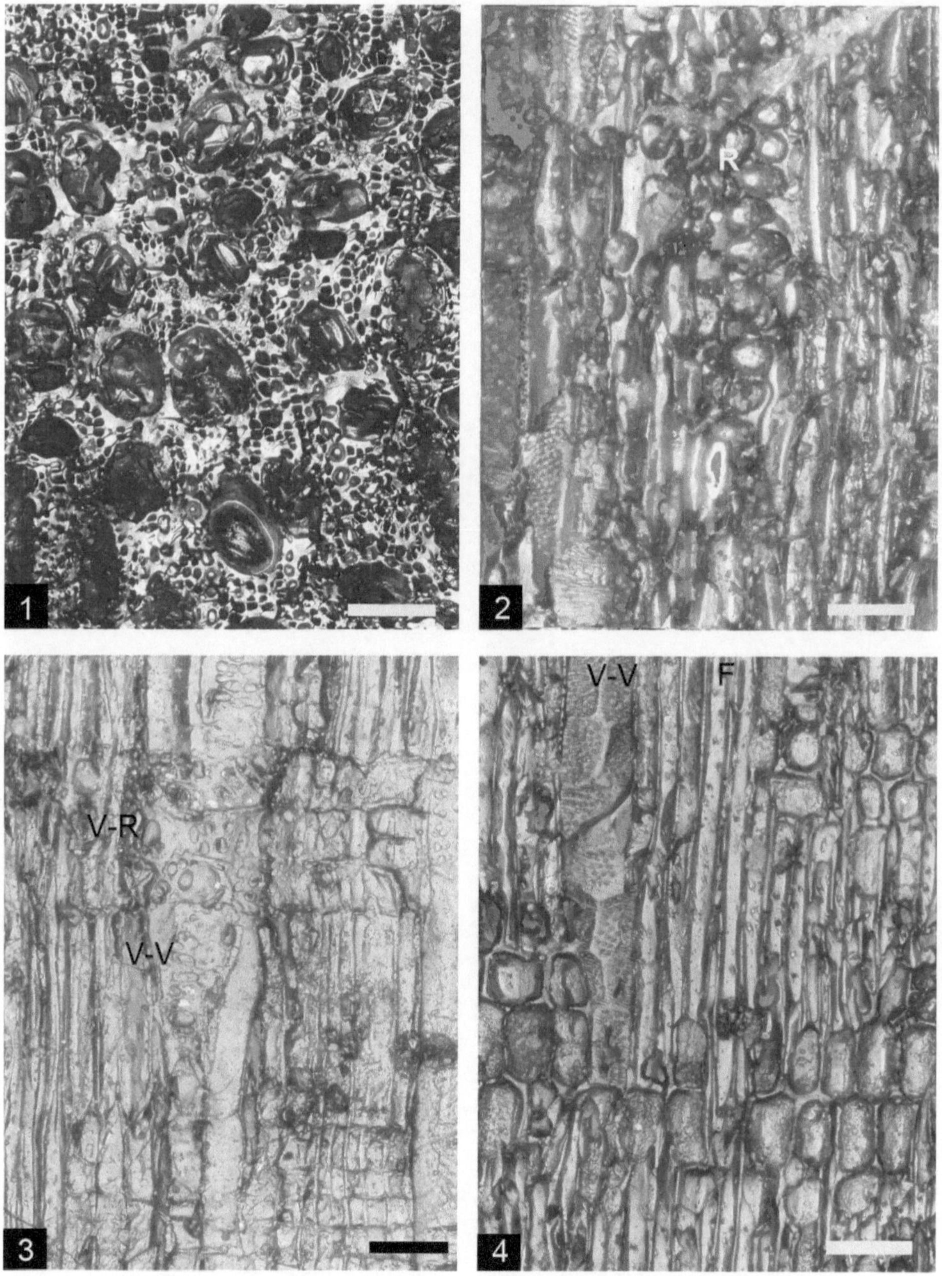

Plate 5. *Dombeya rotundifolia* (Pentapetaceae). 1 – TS, vessels in short radial multiples and fibres are thick-walled. Scale bar = 100 μm. 2 – RLS, upright and square ray cells. Scale bar = 40 μm. 3 – TLS, rays of two sizes, 1 and 3–4 seriate. Fibres are thick-walled. Scale bar = 40 μm. 4 – RLS, upright and square ray cells, inter-vessel pits alternate, and bordered pits in fibre walls. Scale bar = 40 μm.

Fibres are thick-walled and have bordered pits. Inter-vessel pits are round, alternate and 4–6 μm in diameter. Vessel-ray pits are very similar to inter-vessel pits (Plate 5, Figure 3).

IDENTIFICATION AND COMPARISON: This wood, with vessels arranged in short radial multiples, with scanty paratracheal parenchyma and heterocellular rays 1–3 seriate, is typical of many families. Having rays of two sizes and with the unseriate rays being more or less storied is a useful diagnostic feature for the genus. The charcoal is the same as *Dombeya rotundifolia* in the modern reference collection (A162). It is very similar to *Dombeya tiliacea* (A163) but this species has diffuse parenchyma, not scanty parenchyma, and the vessels are fewer in number per square mm. According to the InsideWood database for *Dombeya* spp. parenchyma ranges from diffuse, to diffuse-in-aggregate to scanty paratracheal. The rays are composed of mixed square and upright cells in the reference material, and not with procumbent body cells and 1–4 rows of marginal upright or square cells as described in the more general description in the InsideWood database.

There are nine species of *Dombeya* in southern Africa, mostly with very restricted distributions (Coates Palgrave, 2002; van Wyk and van Wyk, 2013) and very far from Wonderwerk. Only *Dombeya rotundifolia* var. *rotundifolia* is widespread. Based on the limited comparative material and the geographical distribution the charcoal is identified as *D. rotundifolia*.

Oliniaceae
Olinia ventosa
Described sample: WW-5 M29 (0–5 cm) F
Illustration: Plate 6, Figures 1–4
Total number of pieces: 10

DESCRIPTION: Wood is diffuse-porous and vessels have horizontal to oblique, simple perforation plates. Vessels are 40–50 μm wide, about 200 μm long with more than 40 vessels per square mm. They are arranged in short radial multiples up to 4 vessels and clusters are common (Plate 6, Figure 1). Inter-vessel pitting is small and alternate. Parenchyma is scanty paratracheal to vasicentric and with irregular, narrow bands. Rays are numerous, 1–3 seriate, about 10–20 cells high, heterocellular with mostly upright cells but also square and procumbent cells (Plate 6, Figures 2, 3). The multiseriate sections are as wide as the uniseriate sections and the large cells, as seen in tangential longitudinal section, are perforated (Plate 6, Figure 4). There are septate fibres present.

IDENTIFICATION AND COMPARISON: The suite of features, vessels arranged in short radial multiples that have a tendency towards an oblique pattern, heterocellular rays that are 1–2 seriate, with the uniseriate portions as wide as the multiseriate portions, and parenchyma that is scanty paratracheal with narrow bands are typical of the Oliniaceae. This charcoal is the same as *Olinia ventosa* (syn. *O. cymosa*) in the modern comparative collection (A111). Photographs and descriptions in Kromhout (1975) confirm this identification. *Olinia usambarensis* (InsideWood database) is also very similar but in this species the rays have procumbent body cells with 1–4 rows of marginal upright cells.

In southern Africa there are seven species of *Olinia* (Coates Palgrave, 2002; van Wyk and van Wyk, 2013) occurring in low altitude forest or montane forest. It is unexpected to find this charcoal at Wonderwerk so it is treated with caution and will be compared with charcoal from the upper layers.

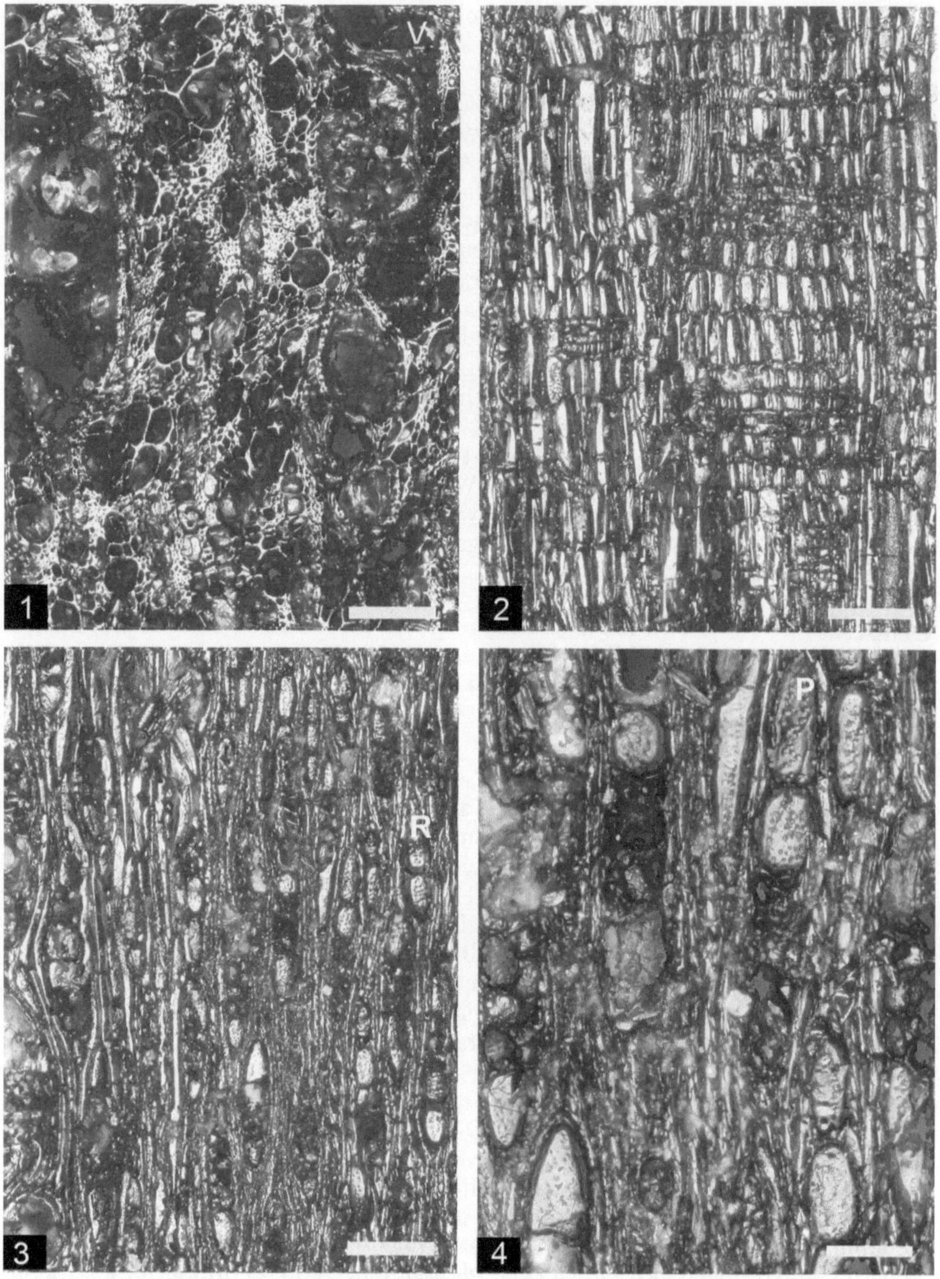

Plate 6. *Olinia ventosa* (Oliniaceae). 1 – TS, vessels in radial lines of up to four cells and in clusters, patches of thick-walled fibres. Scale = 100 μm. 2 – RLS, mostly upright ray cells with bands of square or procumbent cells. Scale bar = 40 μm. 3, 4 – TLS, rays with uniseriate portions as wide as the 2–3 seriate portions. Marginal ray cells are larger and perforated (spotty appearance). Scale bar 3 = 100 μm; 4 = 40 μm.

Rhamnaceae
Berchemia discolor
Described sample: WW-5 M29 (0–5 cm) G
Illustration: Plate 7, Figures 1–4
Total number of pieces: 29

DESCRIPTION: Wood is semi-ring to diffuse-porous with possible ring boundaries made of terminal/initial parenchyma bands. Perforation plates are simple and horizontal. Vessels are arranged in short radial multiples of 1–4 cells, and clusters (Plate 7, Figures 1, 2). Average vessel diameter is 20–40 μm and there are more than 80 vessels per square mm. Vessel clusters comprise small and larger vessels. Parenchyma is vasicentric to aliform and prismatic crystals are visible in transverse section in both the rays and parenchyma cells (Plate 7, Figure 3). Rays are 1–2(–3) seriate and up to 200 μm high. Ray cells are predominantly procumbent but there are 1–2(–4) rows of marginal upright cells that frequently contain crystals. Crystals also occur in the procumbent cells. Inter-vessel and vessel-ray pits are small and alternate (Plate 7, Figure 4).

IDENTIFICATION AND COMPARISON: Typical features of southern African woods of the Rhamnaceae are short to long radial multiples, sometimes with very numerous vessel members, rays 1–3 seriate and paratracheal parenchyma. Local *Ziziphus* species have uniseriate rays so can be excluded. *Rhamnus* species also have narrow rays, 1–2 seriate. The charcoal is identified as *Berchemia discolor* based on comparison with the modern reference material of *B. discolor* (A119) and *Berchemia zeyheri* (Kromhout, 1975) as well as InsideWood descriptions. The charcoal is also very similar to *B. zeyheri* but the latter has only scanty parenchyma whereas *B. discolor* also has narrow bands of parenchyma.

There are only two species of *Berchemia* in southern Africa and they occur in bushveld (Coates Palgrave, 2002; van Wyk and van Wyk, 2013).

Scrophulariaceae
Halleria lucida
Described sample: WW-5 Q29 (5–10 cm) D
Illustration: Plate 8, Figures 1–4
Total number of pieces: 2

DESCRIPTION: Perforation plates are simple and horizontal and the wood is diffuse-porous. Vessel arrangement is short radial multiples of 1–3 pores but predominantly solitary, with an average diameter of about 80 μm and less than 30 per square mm (Plate 8, Figure 1). Parenchyma is rare to absent. Rays are 1–5 seriate, heterocellular with mostly square cells but procumbent and upright cells are present. Rays are 400–800 μm high (Plate 8, Figures 2–3). Inter-vessel pitting is alternate and minute (Plate 8, Figure 4); vessel-ray pitting was not seen. Fibres are thick-walled.

IDENTIFICATION AND COMPARISON: Woods of the Scrophulariaceae are quite variable but tend to have fairly wide and tall rays with vessels arranged in a tangential or oblique pattern. The charcoal is the same as *Halleria lucida* (A158) in the modern reference collection and this was confirmed in the InsideWood key where *Halleria abyssinica* is a synonym of *H. lucida*.

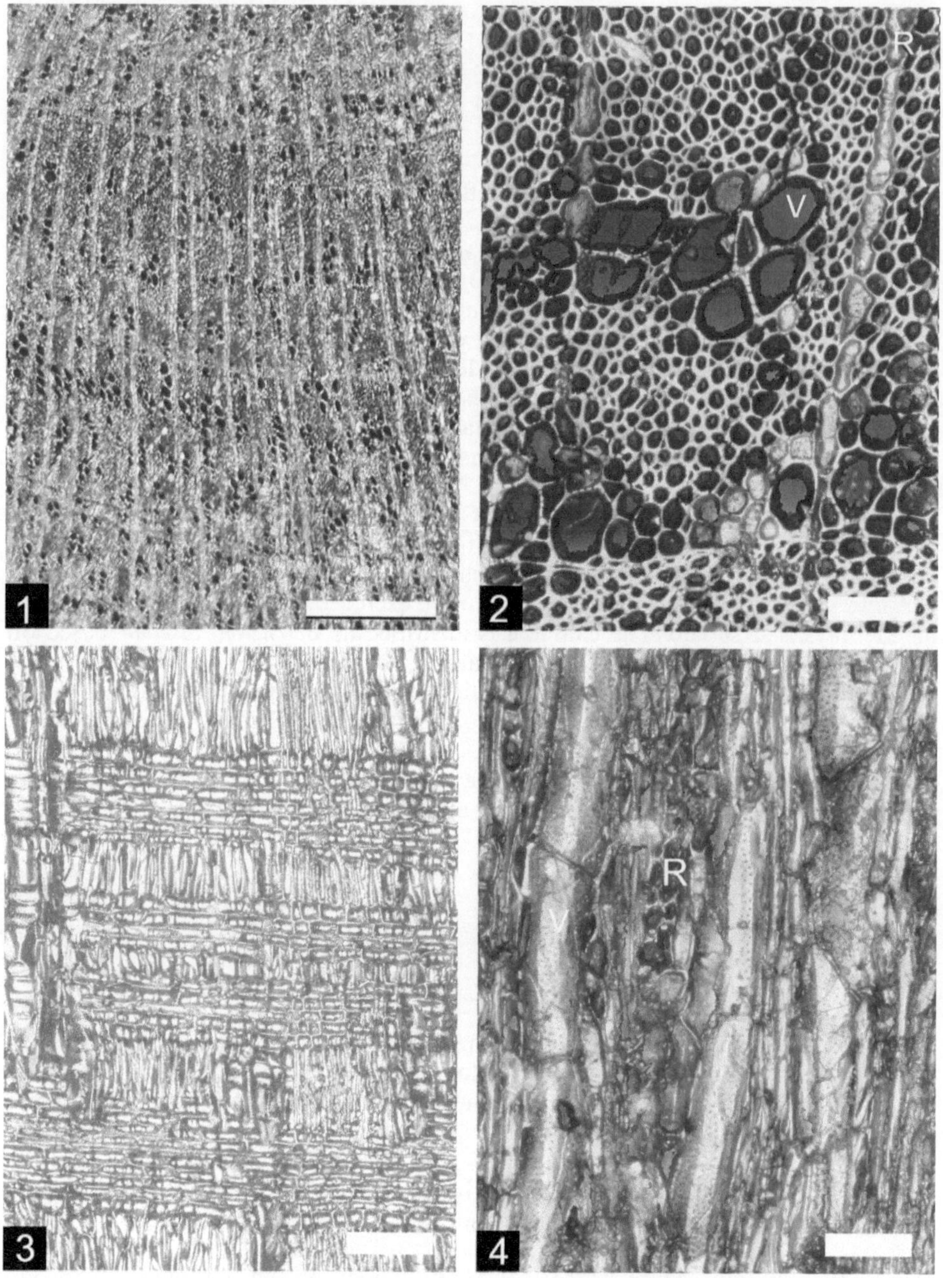

Plate 7. *Berchemia discolor* (Rhamnaceae). 1, 2 – TS, vessels in short radial multiples and clusters, wood semi-ring porous (horizontal bands). Scale bar 1 = 200 μm; 2 = 40 μm. 3 – RLS, 1–2 rows of marginal upright cells and body procumbent ray cells some of which have 3–4 crystals per cell CHECK. Scale bar = 100 μm. 4 – TLS, rays 1–2 seriate. Vessels long with small alternate inter-vessel pits. Scale bar = 40 μm.

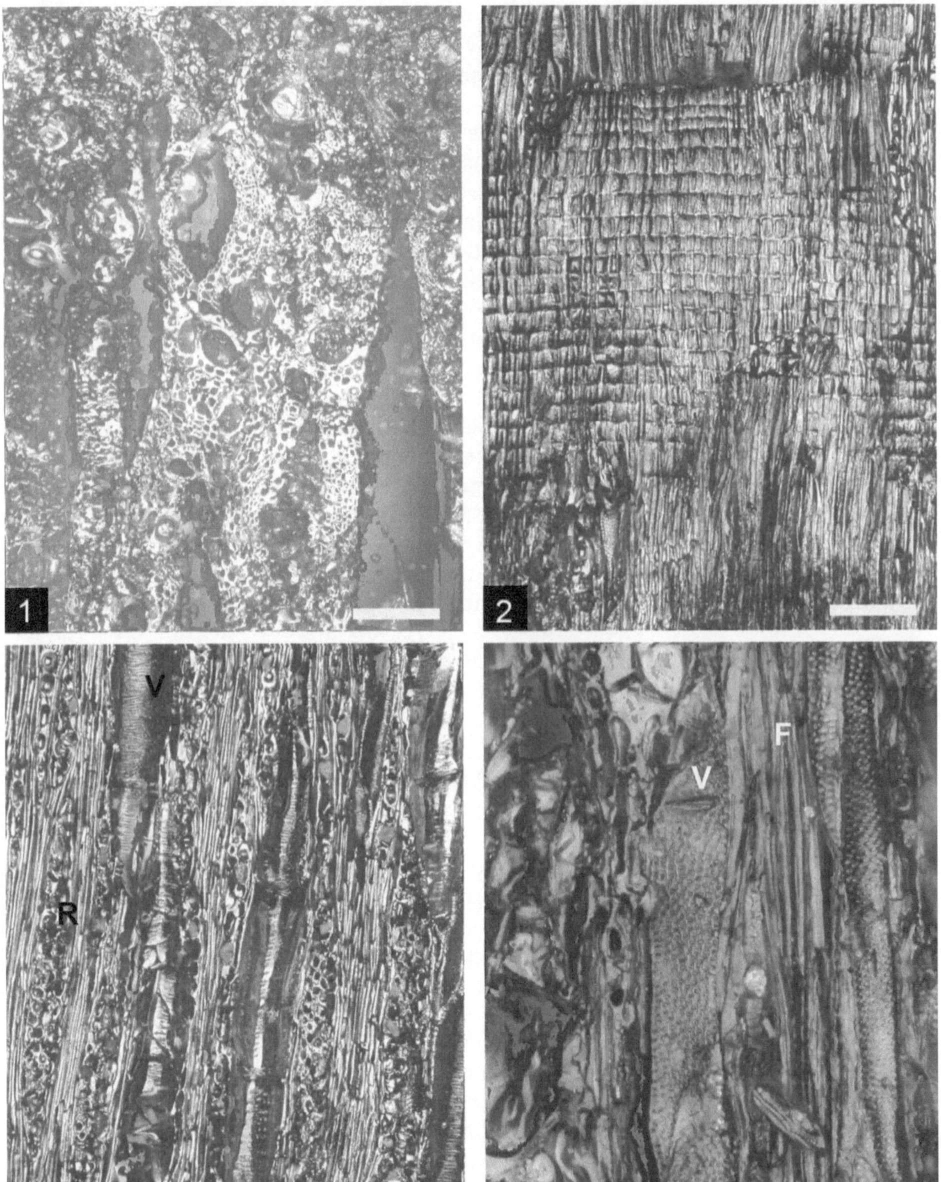

Plate 8. *Halleria lucida* (Scrophulariaceae). 1 – TS, vessels in short radial multiples, fibres thick walled (section cracked). Scale bar = 100 μm. 2 – RLS, square ray cells. Scale bar = 100 μm. 3 – TLS, tall rays 3–5 seriate. Inter-vessel pitting is alternate and minute but appears scalariform at lower magnification. Scale bar = 100 μm. 4 – TLS, small, alternate inter-vessel pits. Scale bar = 40 μm.

9.4 DISCUSSION

9.4.1 Distribution of charcoal within Stratum 5

Numbers of pieces of charcoal cannot be taken to reflect the abundance or preferential use of that wood because charcoal is fragile and can easily be fractured during excavation thus inflating the number of pieces. Furthermore, different woods preserve differently, some forming robust charcoal and others disintegrating or leaving no recognizable trace. Nonetheless presence and absence of taxa, and changing taxa over time can be interpreted to reflect changes in selection or wood use, and/or changes in wood availability or vegetation.

The most common charcoal in Stratum 5 is *Ozoroa paniculosa* with 40 pieces occurring in 6 of the 19 grid squares excavated (Table 2). Most pieces occurred in square M 29 (0–5 cm depth). This square also had the most pieces of charcoal and highest diversity (total 28; diversity 4 species). The second most common charcoal was *Berchemia discolor* with a total of 29 pieces in 6 grid squares. The least abundant or rare taxa are *Commiphora* sp. and *Halleria lucida* with only two pieces each. One piece of *Commiphora* sp. occurs in sub-Stratum 5a but this taxon is abundant in Stratum 4. With more material to work with from this and possibly other strata the species should eventually be identifiable. Since the total number of charcoal pieces in Stratum 5 is low compared with those of the upper strata no spatial analysis will be done at this stage.

9.4.2 Geographical range, ecology, wood quality and uses of charcoal taxa from Stratum 5

There are two genera belonging to the Anacardiaceae, *Ozoroa* and *Searsia* (*Rhus*). *Ozoroa paniculosa* is a small tree or shrub and grows in bushed grassland or on rocky hillsides in dry, summer rainfall areas (Coates Palgrave, 2002; van Wyk and van Wyk, 2013). *Searsia lancea* is also a small tree to shrub, widespread in southern Africa and grows in a variety of habitats. It occurs today in the Wonderwerk environs (ibid.; Mucina and Rutherford, 2006; pers. obs.).

The three species of *Ehretia* (Boraginaceae) are widespread in the drier regions of southern Africa and grow in a wide range of habitats. *Ehretia rigida* or the puzzle bush, for example, occurs in wooded grassland, karroid vegetation and bushed grassland (Coates Palgrave, 2002; van Wyk and van Wyk, 2013). It is a deciduous shrub or small tree and occurs today in the Wonderwerk Cave area.

Commiphora species are common in dry to very dry areas and are often leafless, but some species occur in bushy grassland and bushland. Many species of the Burseraceae produce a scented sap or resin. The wood is light and generally not used for fuelwood but twigs from *Commiphora schimperi* are used as fire sticks (Coates Palgrave, 2002). This species and five others occur today not too far from Wonderwerk Cave and may have had a wider distribution in the past.

Dombeya rotundifolia, the common wild pear, is a small tree or shrub, widely spread in bushland, and on rocky outcrops, from Namibia, northern Botswana, Zimbabwe, Limpopo Province of South Africa and Mozambique. It does not occur in the Wonderwerk Cave area today.

There are seven species of *Olinia* in southern Africa (Coates Palgrave, 2002; van Wyk and van Wyk, 2013) and they each have restricted distributions but occur in montane or evergreen forest. *Olinia ventosa* occurs along the southern Cape coast and is an evergreen tree with strong, hard wood. Its current habitat and distribution place the identification of the charcoal in doubt so this requires further research and sourcing of more comparative material. Two other species also occur in forest or forest margins but much closer to Wonderwerk. More reference material is needed to confirm the identification of the species.

There are about 20 indigenous species of the Rhamnaceae in southern Africa (Coates Palgrave, 2002) including *Berchemia discolor* and *Berchemia zeyheri*. Today neither species occurs in the Wonderwerk Cave area but *B. discolor* occurs in low altitude bushveld across northern Namibia, Botswana and Zimbabwe and south to Mozambique and eastern Limpopo. *B. zeyheri* also occurs in bushveld in the Northern Province, Limpopo and northern KwaZulu-Natal. The woods of both species are hard and used for making furniture.

Halleria lucida is one of the few woody members of the Scrophulariaceae. It is a shrub to small tree occurring in forest margins, grasslands and along streams in rocky places. It ranges from Northern Province, Limpopo, KwaZulu Natal to the Cape Coast but does not occur in the Wonderwerk area today (Coates Palgrave, 2002; van Wyk and van Wyk, 2013). The wood has many uses including fire sticks (Coates Palgrave, 2002).

9.4.3 Comparison with other records

Older charcoal records from South Africa have been discussed in Bamford (AAR in 2015). Sites that contain preserved charcoal that are contemporaneous with Stratum 5 (i.e. slightly older than 12 ka) are few in number. Elands Bay Cave on the southwestern Cape coast has charcoals of proteoid fynbos in the levels dated to 13.6–12.45 ka BP (Cowling *et al.*, 1999) which is different from the Wonderwerk flora. From Rose Cottage near the Caledon River the charcoals from 12–10 ka indicate cool conditions (Esterhuysen *et al.*, 1999). Charcoal from Wonderkrater in the savanna biome of northern South Africa at ~12–13 ka has predominantly *Phragmites* sp and grass charcoal plus some *Diospyros austro-africana* wood probably indicating fluctuating wetter and drier periods (Backwell *et al.*, 2014). Prior and Price-Williams (1985) described charcoal from early Holocene layers in Siphiso rock shelter in northeast Swaziland: from Stratum 6 (12–9 ka) and indicate moist conditions from Leguminosae, *Combretum apiculatum, Androstachys johnstoni, Maytenus* and *Diospyros* species.

The Rose Cottage charcoal taxa from Level DB dated at 12,690 years are: *Protea* spp., *Leucosidea sericea, Maytenus* spp., *Scolopia mundii, Heteromorpha trifoliata* and *Passerina montana* (Wadley *et al.* 1992; Esterhuysen *et al.* 1999). The site is on the northwestern side of the Drakensberg Mountains, eastern Free State, at a higher altitude than Wonderwerk, and represents a region of higher rainfall.

Pollen records for the pre-Holocene have been summarised by Scott *et al.* (2012) for the various biomes. Wonderwerk falls within the savanna/dry woodland biome and pollen from here and Equus Cave indicate lower temperatures, relative to todays, before 17 ka with general warming thereafter. They also show strongly oscillating moisture between about 25 and 11 ka (with peaks at 17 ka, 14,6 ka and 12.6 ka). After 12.6 ka the drying trend reached a peak in dryness between 12–11 ka (the Younger Dryas) with evidence of a sub-humid woodland sequence (Scott *et al.*, 2012). Recent pollen studies at Wonderwerk Cave show that from about 12,500 to 6500 cal years BP (Strata 5–4c), there were some fluctuations in moisture and temperature, based on the changing proportion of grass, fynbos and Asteraceous elements (Scott and Thackeray, AAR in 2015). They add that the vegetation would have resembled the "Renosterveld" that forms the current transition between the Nama Karroo and Fynbos Biomes and that the vegetation seems to have been growing under moderately humid conditions, although not as humid as in areas in the eastern parts of the country during this period.

9.4.4 Pre-Holocene Wonderwerk vegetation

Charcoals from Wonderwerk would have been the product of human selection and activity. The woods represented are, therefore, a subset of the plants growing in the vicinity and represent an unknown proportion of the flora. Nonetheless they can be used to deduce the past climate. There are about 134 fragments and they represent eight identifiable woody species and 34 unidentified pieces. The diversity is relatively low but seven families are represented. The most common woods are *Ozoroa paniculosa* and *Berchemia discolor*. Of the eight identified charcoal taxa, only two are listed in the modern vegetation (*Searsia* (*Rhus*) *lancea* and *Ehretia rigida*) and they tolerate low rainfall in summer and frost during winter. One species occurs in dry to arid regions (*Commiphora* sp.). Two species occur in fairly dry regions (*Ozoroa paniculosa, Dombeya rotundifolia*). Two species occur in more mesic areas (*Berchemia discolor* and *Halleria lucida*) and one species occurs in moister and more forested conditions, *Olinia ventosa*. This may correspond with the peak in moisture reflected by the pollen spectrum at Wonderwerk at 12.6 ka (Scott *et al*., 2012). There is an anomaly in the dates given by Lee Thorpe and Ecker (AAR in 2015) for Stratum 5 which they cannot fully explain but it could also be a problem with mixing of samples.

Considering the range of woods represented by the charcoal it is possible that their past range was wider than it is today as it is unlikely that the cave inhabitants would have walked great distances to collect firewood. The majority of the woods found can tolerate dry conditions and a wide range of temperatures but the presence of *Berchemia discolor* indicates that conditions between c. 14,985 and 13,952 cal. BP may have been slightly wetter than today. Furthermore, there seem to have been important climatic shifts during the pre-Holocene but higher resolution records are needed to determine the degree and rate of climate change.

ACKNOWLEDGEMENTS

I would like to thank Liora Kolska Horwitz, James Brink and Lloyd Rossouw for organizing the Tribute to Louis Scott in Bloemfontein in July 2014 and for inviting me to participate. I also thank Michael Chazan and Liora for including me in the Wonderwerk project, and PAST for funding for the modern plant reference collection at the Evolutionary Studies Institute.

REFERENCES

Avery, D.M., 1981, Holocene micromammalian faunas from the northern Cape Province. South Africa. *South African Journal of Science,* **77**, pp. 265–273.

Avery, D.M., 2007, Pleistocene micromammals from Wonderwerk Cave, South Africa: practical issues. *Journal of Archaeological Science,* **34**, pp. 613–625.

Backwell, L.R., McCarthy, T.S., Wadley, L., Henderson, Z., Steininger, C.M., deKlerk, B., Barré, M., Lamothe, M., Chase, B.M., Woodborne, S., Susino, G.J., Bamford, M.K., Sievers, C., Brink, J.S., Rossouw, L., Pollarolo, L., Trower, G., Scott, L. and d'Errico, F., 2014, Multiproxy record of late Quaternary climate change and Middle Stone Age human occupation at Wonderkrater, South Africa. *Quaternary Science Reviews,* **99**, pp. 42–59.

Bamford, M.K., (2015), Macrobotanical remains from Wonderwerk Cave (Excavation 1), Oldowan to Late Pleistocene (2Ma to 14ka BP), South Africa. *African Archaeological Review.*

Beaumont, P., 2004, Wonderwerk Cave. In *Archaeology in the Northern Cape: Some Key Sites,* edited by Beaumont, P. and Morris, D. (Kimberley: McGregor Museum), pp. 31–36.

Beaumont, P.B., 1990, Wonderwerk Cave. In *Guide to archaeological sites in the northern Cape,* edited by Beaumont, P. and Morris, D. (Kimberley: McGregor Museum), pp. 101–124.

Berna, F., Goldberg, P., Horwitz, L.K., Brink, J., Holt, S, Bamford, M. and Chazan, M., 2012, Microstratigraphic evidence for in situ fire in the Acheulean strata of Wonderwerk Cave, Northern Cape Province, South Africa. *PNAS Plus,* **9(20)**, E1215–E1220 Doi/10/1073/pnas.1117620109

Camp, C.L., 1948, University of California African Expedition – southern section. *Science,* **108**, pp. 550–552.

Chazan, M., Avery, D.M., Bamford, M.K., Berna, F., Brink, J., Holt, S., Fernandez-Jalvo, Y., Goldberg, P., Matmon, A., Porat, N., Ron, H., Rossouw, L., Scott, L. and Horwitz, L.K., 2012, The Oldowan horizon in Wonderwerk Cave (South Africa): Archaeological, geological, paleontological and paleoclimatic evidence. *Journal of Human Evolution,* **63**, pp. 859–866.

Chazan, M., Ron, H., Matmon, A., Porat, N., Goldberg, P., Yates, R., Avery, M., Sumner, A. and Horwitz, L.K., 2008, Radiometric dating of the Earlier Stone Age sequence in excavation 1 at Wonderwerk Cave, South Africa: preliminary results. *Journal of Human Evolution,* **55**, pp. 1–11.

Coates Palgrave, M., 2002, *Keith Coates Palgrave Trees of Southern Africa,* 3rd Edition, (Cape Town: Struik Publishers).

Cowling, R.M., Cartright, C.R., Parkington, J.E. and Allsopp, J.C., 1999, Fossil wood charcoal assemblages from Elands Bay Cave, South Africa: implications for Late Quaternary vegetation and climates in the winter rainfall Fynbos biome. *Journal of Biogeography,* **26**, pp. 367–378.

Esterhuysen, A.B., Mitchell, P.J. and Thackeray, J.F., 1999, Climatic change across the Pleistocene/Holocene boundary in the Caledon River, southern Africa: results of a factor analysis of charcoal assemblages. *Southern African Field Archaeology,* **8**, pp. 28–34.

IAWA Committee, 1989, IAWA list of microscopic features for hardwood identification. *IAWA Bulletin*, n.s. **10**, pp. 219–332.

InsideWood (accessed 2011–May 2014): http://insidewood.lib.ncsu.edu/

Kent, L.E., ed, 1980, *Stratigraphy of South Africa.* 8. Part I. Geological survey of South Africa Handbook (The South African Committee for Stratigraphy).

Kromhout, C.P., 1975, 'n Sleutel vir die mikroskopiese uitkenning van die vernaamste inheemse houtsoorte van Suid-Afrika. *South African Department of Forestry, Pretoria, Bulletin,* **50,** 124 pp.

Lee-Thorp, J. and Ecker, M., (2015), Holocene environmental change at Wonderwerk Cave, South Africa: Insights from stable light isotopes in ostrich egg shell. *African Archaeological Review.*

Malan, B.D. and Cooke, H.B.S., 1941, A preliminary account of the Wonderwerk Cave, Kuruman District, South Africa. *South African Journal of Science,* **37**, pp. 300–312.

Matmon, A., Ron, H., Chazan, M., Porat, N. and Horwitz, L.K., 2011, Reconstructing the history of sediment deposition in caves: A case study from wonderwerk Cave, South Africa. *Geological society of America Bulletin,* **124**, pp. 611–625.

Mucina, L. and Rutherford, M.C., eds, 2006, *The Vegetation of South Africa, Lesotho and Swaziland,* Strlitzia **19** (Pretoria: South African National biodiversity Institute), pp. 1–808.

Prior, J. and Price Williams, D., 1985, An investigation of Climatic change in the Holocene epoch using archaeological charcoal from Swaziland, South Africa. *Journal of Archaeological Science,* **12**, pp. 457–475.

Rüther, H., Chazan, M., Schroeder, R., Neeser, R., Held, C., Walker, S.J., Matmon, A. and Horwitz, L.K., 2009, Laser scanning for conservation and research of African cultural heritage sites: the case study of Wonderwerk Cave, South Africa. *Journal of Archaeological Science,* **36**, pp. 1847–1856.

Schulze, B.R., 1984, *Climate of South Africa. Part 8: General survey.* **WB 28**, fifth ed. (Pretoria: South Africa Weather Bureau).

Scott, L., Neumann, F.H., Brook, G.A., Bousman, C.B., Norström, E. and Metwally, A., 2012, Terrestrial fossil-pollen evidence of climate change during the last 26 thousand years in Southern Africa. *Quaternary Science Reviews,* **32**, pp. 100–118.

Scott, L. and Thackeray, J.F., (2015), Palynology of Holocene deposits in Excavation 1 at Wonderwerk Cave, Northern Cape (South Africa). *African Archaeological Review.*

Thackeray, A.I., 1983, Archaeological sites in the Kuruman Hills area. In *Ghaap and Gariep. Later Stone Age Studies in the Northern Cape,* edited by Humphries, A.J.B. and Thackeray, A.I. (Cape Town: The South African Archaeological Society), pp. 33–47.

Thackeray, J.F. and Lee-Thorp, J.A., 1992, Isotopic analysis of equid teeth from Wonderwerk Cave, northern Cape Province, South Africa. *Palaeogeography, Palaeoclimatology, Palaeoecology,* **99**, pp. 141–150.

Van Wyk, B. and van Wyk, P., 2013, *Field Guide to Trees of Southern Africa.* Second edition, (Cape Town: Struik Publishers).

Van Zinderen Bakker, E.M., 1982, Pollen analytical studies of the Wonderwerk Cave, South Africa. *Pollen et Spores,* **26**, pp. 235–250.

Vogel, J.C., Fuls, A. and Visser, E., 1986, Pretoria radiocarbon dates III. *Radiocarbon,* **28**, 1133–1172.

Wadley, L., Esterhuysen, A. and Jeanneret, C., 1992, Vegetation changes in the eastern Orange Free State: the Holocene and later Pleistocene evidence from charcoal studies. *South African Journal of Science,* **88**, pp. 558–563.

CHAPTER 10

Rediscovering the Intriguing Patrimonies depicted in rock shelters of Iringa, Tanzania

Makarius Peter Itambu
Department of Archaeology and Heritage, University of Dar es Salaam, Dar es Salaam, Tanzania

ABSTRACT: For many years, systematic archaeological studies in the Iringa region specifically on rock paintings were not undertaken. Even the existence of these patrimonies in this region was not known until very recently, when the Iringa Archaeological Project led by Canadian and Tanzanian scientists reported its existence. They reported rock-shelters with paintings, but they didn't undertake detailed and systematic investigations of the paintings. Hence, this research project was conducted in August 2012 to survey, record and document the paintings in detail. The status of preservation status of the paintings was also assessed in order to promote archaeo-tourism in the region. We re-examined two sites that were reported in 2006 and studied two other new rock-shelters with paintings. The study reveals that the rock art of Iringa belongs to two rock art traditions: Hunter-forager and Bantu-speaking art traditions. The former is dominated by naturalistic animal and human figures executed in dark and red pigments while the latter consists of schematic animal and human figures, as well as geometric designs executed in dirty-white pigment. Most of the studied paintings share some artistic traditions such as stylistic motifs, aspects, techniques of execution, subject matter and depicted colour with the rock art of central and north central Tanzania.

10.1 BACKGROUND INFORMATION

The Iringa region is located in the southern highlands of Tanzania lying between 7°46′12″ S and 35°41′24″ E and covering an area of about 58,936 km^2 (Figure 1). The region is surrounded by three major mountainous ranges of the Kipengere and Livingstone Mountains to the south and the Udzungwa Mountain in the north-east. Most of the region is composed of highlands and escarpments ranging from 1200 m to 2700 m, though there are a few lowland areas ranging between 900 m and 1200 m above sea level. Highlands and escarpments are mainly composed of Precambrian migmatites, granite and Konse group outcrops that form most of the rock-shelters and overhangs (Howell *et al.*, 1962 cited in Bushozi, 2012; Harpum, 1970; Harris, 1981). These rock-shelters and overhangs served as prehistoric human settlements as they accommodated humans during the glaciations phases of the Middle and Upper Pleistocene, the period in which most of tropical Africa was characterized by prolonged series of dry and harsh environmental conditions (Willoughby, 2007).

Until very recently, the archaeological potential of Iringa was best known from the palaeoanthropological site of Isimila dated by uranium series to about 270,000 BP (Howell *et al.*, 1962; Hansen and Keller, 1971; Cole and Kleindienst, 1974). New initiatives are undertaken to confirm the age estimate of the Acheulian assemblage at Isimila as other such sites, i.e., Olorgesailie in Kenya, have been re-dated with new

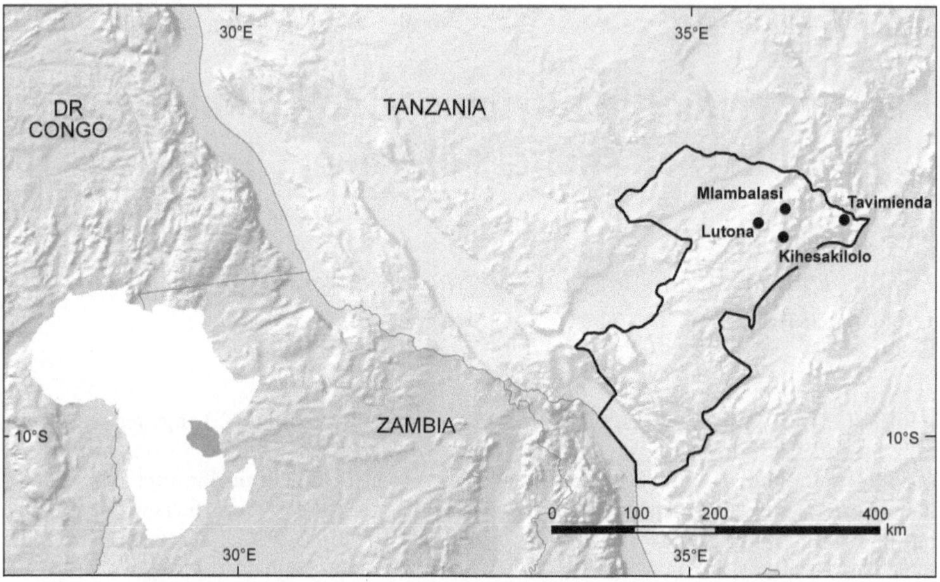

Figure 1. Iringa map showing location of rock art sites and archaeological sites (Source: Adapted and modified from Atlas map of Tanzania, 2013).

dating techniques to 900,000 BP (Evernden and Curtis, 1965; Bushozi, in press). However, ongoing archaeological investigation has revealed substantial information for the post-Acheulian sites, in particular the MSA and LSA occurrences at Magubike and Mlambalasi. These two rock-shelters have shaped our current knowledge about subsistence economy, ecological adaptation, mobility systems and socioeconomic organization of foraging communities who inhabited the Iringa region before the invasion of farming and iron smelter communities around 3,000 BP (Bushozi, 2012; Willoughby, 2012; Itambu, 2013). It is likely that Iringa served as a refuge where people could get potential subsistence requirements during unfriendly environmental conditions such as the one that characterized the world for the most of Pleistocene (Willoughby, 2007).

10.2 METHODOLOGICAL APPROACH

The broad intention of the study basically was to explore four major aspects in detail: identification, documentation and description of the existing and new sites with rock paintings and their cultural authenticity. Based on the time and financial constraints, it was impossible to survey the entire region, instead we sampled a few areas based on the natural landscape whereby all Precambrian rock boulders and outcrops with prospective shelters and overhangs were purposely sampled. Accessible rock-shelters, overhangs and caves were examined in detail so as to discover and document the rock art sites. Rock paintings were traced, described and classified according to their subject matter, stylistic characteristics, composition, action, aspect and techniques of execution. In some cases ethnographic inquiries were involved to attain additional information about the current practice among the local communities and their reciprocal understanding on the rock-paintings.

10.3 RECORDING AND DOCUMENTATIONS

Recording involved the identification of site names and its geographical location, photographing, categorizing the form and type of the rock-shelter, as well as the contemporary use of the site. We also tried to count the number of painted figures, subject matter and types or stylistic representation of the paintings, i.e., naturalistic vs. stylized. All data concerning the paintings and sites were documented based on the documentation scheme developed by Mabulla (2005). The geographical location of a rock art site was documented using a hand held Global Positioning System (GPS) device. A *Sony* digital camera was used for taking photographs. We adopted the Vinnicombe (1976) Rock Paintings Analysis Scheme for our study as it is widely used in northern and north-central Tanzania and it offers a unique opportunity for qualitative and quantitative analysis of rock art. Paintings were also traced using transparent tracing papers (Figure 2). At least four sites with rock paintings were identified and documented in the Iringa and Kilolo districts. These included Mlambalasi, Kihessakilolo, Lutona and Tavimienda. Furthermore, paintings were evaluated and analyzed in detail to determine the state of preservation and the sources of potential threats to their integrity and survival, as well as to ascertain their suitability for public display. This approach was part of a conservation and management strategy aimed at saving these priceless cultural heritage assets (Mabulla, 2005; Deacon, 2007).

Figure 2. Tracing the paintings in Kihessakilolo rock-shelter.

10.3.1 Mlamabalasi Hill

Mlambalasi Hill (7°43′7″ S/35°42′6″ E) is a famous site for archaeological and historical information of Iringa. This is a place where chief Mkwawa, a former leader of Hehe people, killed himself rather than surrendering to German colonialists (Willoughby, 2012). After the drawn out anti-colonial war, he killed himself in 1894, he was beheaded and his head was taken to the Bremen Anthropological Museum in Germany, while the rest of his body was buried at Mlambalasi Village (Bushozi, in press). His head was returned to his family in Tanzania in 1954 and now rests in the Chief Mkwawa Memorial Museum at Kalenga along with other personal belongings and items representing the cultural and economic activities of the Hehe people. In addition, this hill is surrounded by a number of rock-shelters with abundant scatter of LSA and Iron Age archaeological occurrences. One of these rock-shelters was excavated in 2006 and 2010 respectively and was found to be composed of dense and stratified archaeological sites characterized by human burials and other symbolically revealing artefacts. Human remains and the associated LSA assemblage were radiocarbon dated to about 13,490 BC.

During our surveys at least five rock-shelters with paintings referred as Mlambalasi 1 to 5 were documented, but most of them are composed of faded rock-paintings. These shelters are surrounded by thick natural vegetation, thorns and thick bushes, but they are located in the north-eastern Precambrian granite and Konse group outcrops. Only one rock-shelter (Mlamabalsi 2) was found painted with a number of identifiable anthropomorphic and zoomorphic figures, particularly on the eastern wall. Figures are depicted in red-lines and strips that are difficult to interpret. One of the identifiable figures was a very large symmetrical figure with thick-lines, filled-in linear patterns and red symmetrical lines, probably representing a vertebrate animal (Figure 3).

10.3.2 Lutona rock-shelter

Lutona (7°46′7″ S/35°26′2″ E) is a large circular rock-shelter found in the Kitwiru area about 12 km west of Magubike Village. It is surrounded by numerous kopjes, forests, small hills and granite boulders of the Precambrian era. It is located adjacent to the Kitwiru ephemeral river that may have served as a source of water for artisans. The shelter has a lot of archaeological material scattered on the surface. This includes potsherds, slag, bones and lithic artefacts. The site is highly vandalized by iron smelters and treasure hunters. There is a big hole in shelter, probably dug by treasure hunters. The rock-shelter contains a few Late-White geometric paintings, at least five figures were recorded, all of them executed on the roof. The majority of the paintings are faded due to anthropogenic agents rather than biological agents (Figures 4, 11).

10.3.3 Tavimienda Rock-shelter

This site is located in the Kilolo District (7°36′13″ S/36°30′26″ E), about 198.4 km from Iringa on the peak of the Udzungwa Mountains. Due to the mountainous nature of the site, the rock paintings are in a very good state of preservation. We were informed by the local people that there are many rock-shelters with paintings in the Udzungwa Mountains, but we failed to access them because of uneven terrain as the survey was carried out during the rainy season. A single painted rock-shelter was recorded, mapped and documented at Udzungwa. The documented shelter has a high number of paintings on the roof and we documented over 350 figures (Figure 4). The majority of them are in a

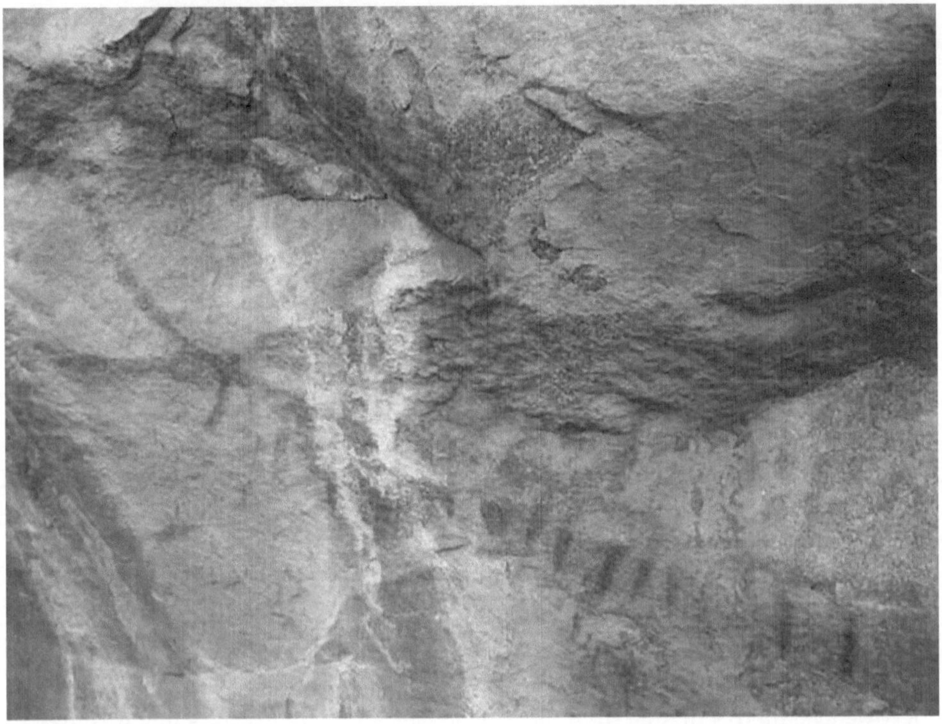

Figure 3. Red-geometric lines painted in red color at Mlambalasi 2 (photo by P.G.M. Bushozi 2008).

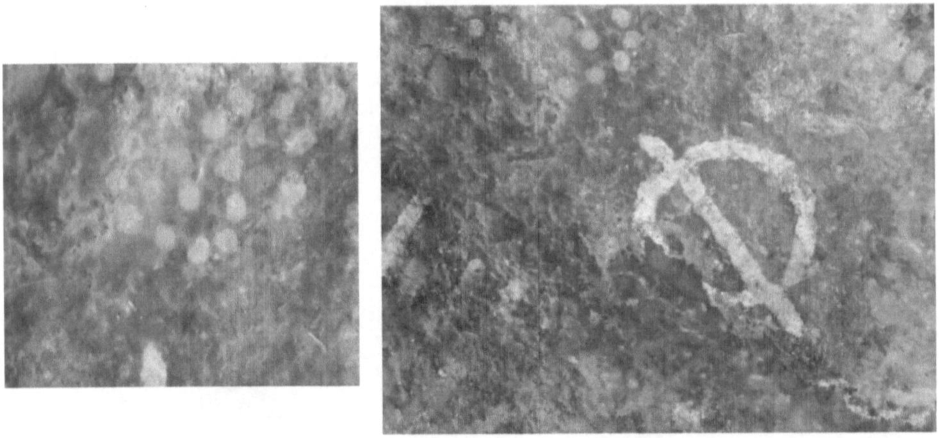

Figure 4. Late White paintings at Lutona rock-shelter.

good state of preservation, but some of them were superimposed over others. It is likely that the age estimates of paintings at Tavimienda are not similar with some painted earlier than others. The majority of the paintings represent animals or organisms that are locally found, but some symbolic expression signs representing cultural expectations of artists are also present. All the paintings in the Tavimienda rock-shelter are depicted in white pigment and cover an area of about 108.10 m^2.

The paintings subject matter includes animal figures, human figures, letter-like signs and amorphous or abstract figures (Figure 5). Circles and concentric circles with all-round, externally radiating lines dominate. There are also depictions of reptilians such as alligator, lizard and crocodile figures. Stylized humans painted vertically, horizontally and sometimes kneeling down with stretched hands and legs are common. The site is very suitable for public display and archaeo-tourism as a good number of the paintings observed are well preserved and visible. Nonetheless, most of the concentric circles and schematic human figure designs show faded colours; suggesting that they are older than the others, representing the first phase of painting. Circles, bold white-dots, concentric circles, geometric squares and triangles depicted in dusky white may represent a second phase of paintings (Figure 5).

10.3.4 Kihessakilolo rock-shelter

This rock rock-shelter is located on the peak of Igeleke Hill (7°43′32″ S/35°30′22″ E) at an elevation of 1556 m, about 100 m west of Igeleke Primary School. The site is surrounded by granite-kopjes and *Euphorbia candelabrum* (candelabra) trees. The surrounding landscape is composed of LSA stone artefacts, slag and pottery. The painted shelter's wall covers an area of about 7.31 m^2. We recorded over 50 figures, most of which represent naturalistic humans and animals. Animals represented include giraffes, antelope and elephants. Animals are depicted in a naturalistic style and some of them are placed over others (Figure 6). Other depicted figures include candelabra trees, parallel lines, as well as whites and black dots, concentric circles and two head-dresses. White dots extended up to 4 m high from the ground, suggesting that scaffolding was used in the process of execution. The majority of the paintings are done in red, but few of them are in black (Figure 6). Although the rock-shelter of Kihessakilolo is located on the outskirts of Iringa town, most of the painted figures are in a good state of preservation.

The southern Tanzanian highlands consist of a massive inland plateau with isolated mountains and numerous granite exposures from the Precambrian era (Itambu, 2013). In the Iringa region shelters with rock art occur mainly in the highlands, and they are commonly scattered along the Precambrian granite rock outcrops. These

Figure 5. Right is dusky white geometric paintings, left are traced figures at Tavimienda rock-shelter.

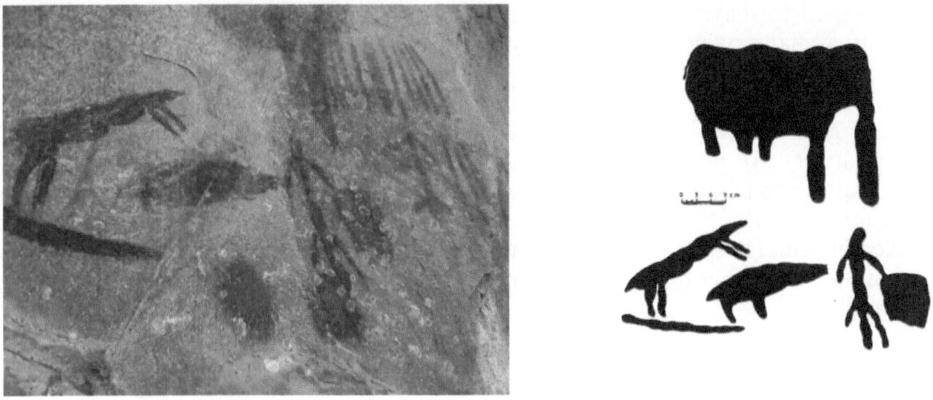

Figure 6. Left is various red friezes painted at Kihessakilolo rock-shelter and right are traced animals and naturalistic humans.

rocky highlands are surrounded by dry forests, bush savannah and occasionally grassy plains. Most of the depicted animals such elephants, giraffes, antelope, crocodiles and ostriches are locally abundant in the National Parks, especially at Ruaha and Udzungwa national parks. The artists probably selected mountainous landscapes for arty expression as the painted arts can probably survive longer as the geomorphology of the landscape discourages and limits accesses to potentially destructive people such as iron smelters, hunters and stone quarrying activities.

10.3.5 Description of the painting subject matters and styles

The painting subject matters were both representational and non-representational. The former is represented by naturalistic and semi-naturalistic animals, semi-naturalistic humans and semi-realistic inanimate figures. The latter depicts schematic, geometric and amorphous (SGA) signs. In this study, only monochrome paintings in red, white, black and yellow were encountered. Neither bichrome nor polychrome styles were used for paintings in Iringa. Red pigment dominated the paintings e.g. the monochrome dusky-red (stale and fresh) is the apparent dominant colour. Over 50 individual painted images at Kihessakilolo were done with dusky red pigments, followed by white (16 images) and about three images were painted using black pigment. At Lutona rock-shelter, at least four faded figures were in white while two faded figures were depicted in a yellowish pigment.

Variations in colours i.e. red (stale and fresh), dusky white and white, white and yellow might may have been the result of fading and weathering of pigment through time or the ingredients used to process the pigments (Namono, 2010; Rasolandrainy, 2011). Weathering processes and deterioration might also have altered the pigments at Lutona rock-shelter from white to a yellowish colour (Figure 4). Ingredients used to process the pigments are one of the leading factors for the variation in colour among the identified rock paintings. For instance, red pigments could have been made from soft rocks containing oxidized iron such as hematite or ochre since these are in abundance in the vicinity, while the black colour was more likely made from charcoal, and white was probably made from gypsum or lime (Masao, 2007; Itambu, 2013). The liquids used to make the binder were not clearly understood, but water was likely used since many paints were powdery (Vinnicombe, 1976: 130). Furthermore, since all the rock art sites in Iringa are located at the streamside or along the water sources, water could have been used for binding by mixing it with other materials such as latex or glue. Interestingly, we found a figure of a candelabra tree at Kihessakilolo, tree paintings are not common in Tanzania. However, candelabra trees have symbolic expression in some of the local communities, for instance among the Wanyaturu (Warimi people) of Singida, where this tree species is used to make latex for binding broken utensils (Itambu, 2013). It is likely that in the past, latex from candelabra trees was mixed with other ingredients such as ochre to produce colouring and/or binding materials.

10.3.6 Order of superimposition

The order of superposition was mainly represented at Kihessakilolo, Lutona and Tavimienda rock-shelters. At Kihessakilo at least three phases of superposition were identified: in the first phase, red are overlaid by black paintings. The black are also superimposed by white paintings. This order of superimposition shows the genealogy of three cultural sequences. At Lutona rock-shelter, white pigment was found overlaid by the yellow pigment meaning that the white pigment was painted much earlier than others. A similar trend of superimposition also appears at Tavimienda (Figure 7).

10.3.7 Traditions and styles

Two cultural traditions are apparent in the rock paintings of Iringa. These include hunter-foragers and Bantu traditions. The Hunter-Forager (HF) paintings show clarity, simplicity and their edges are well refined, suggesting that they may have been painted with fine brushes. Mabulla (2005) and Mabulla and Gidna (2013) subdivided this art tradition into two broad categories: the Hunter Forager Figurative Fine Line (HFFFL) and the Hunter Forager Red Geometric (HFRG). In most cases the HFFFL and HFRG arts occur simultaneously, but HFRG are always placed on top

Figure 7. Order of superimposition (a) red overlaid by black and white pigments at Kihessakilolo; (b) white overlaid by yellow pigments at Tavimienda.

of HFFFL paintings, suggesting that the latter are relatively older. This HFFFL art represents mainly naturalistic and semi-naturalistic wild animals or humans in stylized forms (Mabulla and Gidna, 2013). Other subject matter included executions of birds, vegetation, handprints, and anthropomorphic, therianthropic and shamanistic figures expressing the dynamics of local environment and the essential needs of foraging communities. Humans are depicted engaged in activities including hunting, singing and dancing (Mabulla and Gidna, 2013).

Naturalistic animals are characterized by wild animals such as giraffes, elephants, antelope and lions. At Kihessakilolo, like other places of central and north central Tanzania, giraffes are the most dominantly painted animal and they are executed in such a way that it is easy to identify. The dominance of giraffes on the recorded rock art sites may suggest that they played a certain cultural role or they were the most dominant animal species. In modern context, giraffes are wild animals that are normally associated with either religious or ritual activities. For instance, in some societies like the Wanyaturu people of central Tanzania, giraffes symbolize beauty and/or calmness (Itambu, 2013). Among the San community in the Kalahari Desert in Botswana, giraffes are considered to be potent animals and play an important role in the rituals and folklore (Bisele, 1993; Bwasiri, 2011). At Kihessakilolo other animals widely depicted are antelope, in particular eland and wildebeest, followed by elephants. They are painted with well elaborated body parts making them easy to identify. Most elephants miss ears and tusks while the tails are well elaborated.

Stylized human figures were documented among the hunter-gatherers art at Kihessakilolo as well as among of the farming communities rock art at Tavimienda (Figure 5). Stylized red human figures depicted with longer arms and legs than trunks. Also, the schematic human figures painted with square/box-like trunks in open lines sometimes with stylized legs while others have thin-lined trunks or schematized tails. Human figures represent males and females, but males predominate. Males can easily be identified by their slender and upright bodies contrary to females, who are represented by reverse articulation of the legs, exaggerated lordosis of the spine and frequently by breasts attached to the thorax. In this study, female bodies were represented in reverse articulation of legs and exaggerated lordosis of the spine, but had no breasts attached to the thorax. However, some female figures recorded at Kihessakilolo are executed in various shades of red including the head-dresses and plumes.

The last hunter-gatherer category is the HFRG arts characterized by geometric designs such as concentric circles, parallel lines, horizontal lines, ladder-like lines, scaffolding-like, millipede-like and snake-like lines thought to represent the later phase of hunter-gatherers traditions and they are constantly laid over the HFFFL paintings (Mabulla and Gidna, 2013). In most cases, figures of this group appeared in association with other figures like human, animal or both. They are painted in various shades of red, especially, light, reddish brown and dark red colours. In Iringa, such paintings appear at Kihessakilolo and Mlamabalasi rock-shelters. Elsewhere in Tanzania, the HFRG art tradition is widely represented in Dodoma, Singida, Manyara, and in the Lake Victoria Basin (Leakey, 1983; Mturi and Bushozi, 2002; Mabulla, 2005; Masao, 2007; Mahudi, 2008).

The Bantu rock art tradition or Late White art is attributed to iron working or farming communities. The paintings appear crude and the tradition is dominated by geometric designs including dots, lines, circles, squares and smears (Mabulla and Gidna, 2013). Also present are schematic depictions of anthropomorphic figures such as spread-eagled, stylized humans and stylized animals such as reptiles and insects (Mabulla, 2005). Depicted animals include crocodiles, alligators, lizards and insect-like figures. In some cases reptiles, birds and insects are painted in light and dark red, while

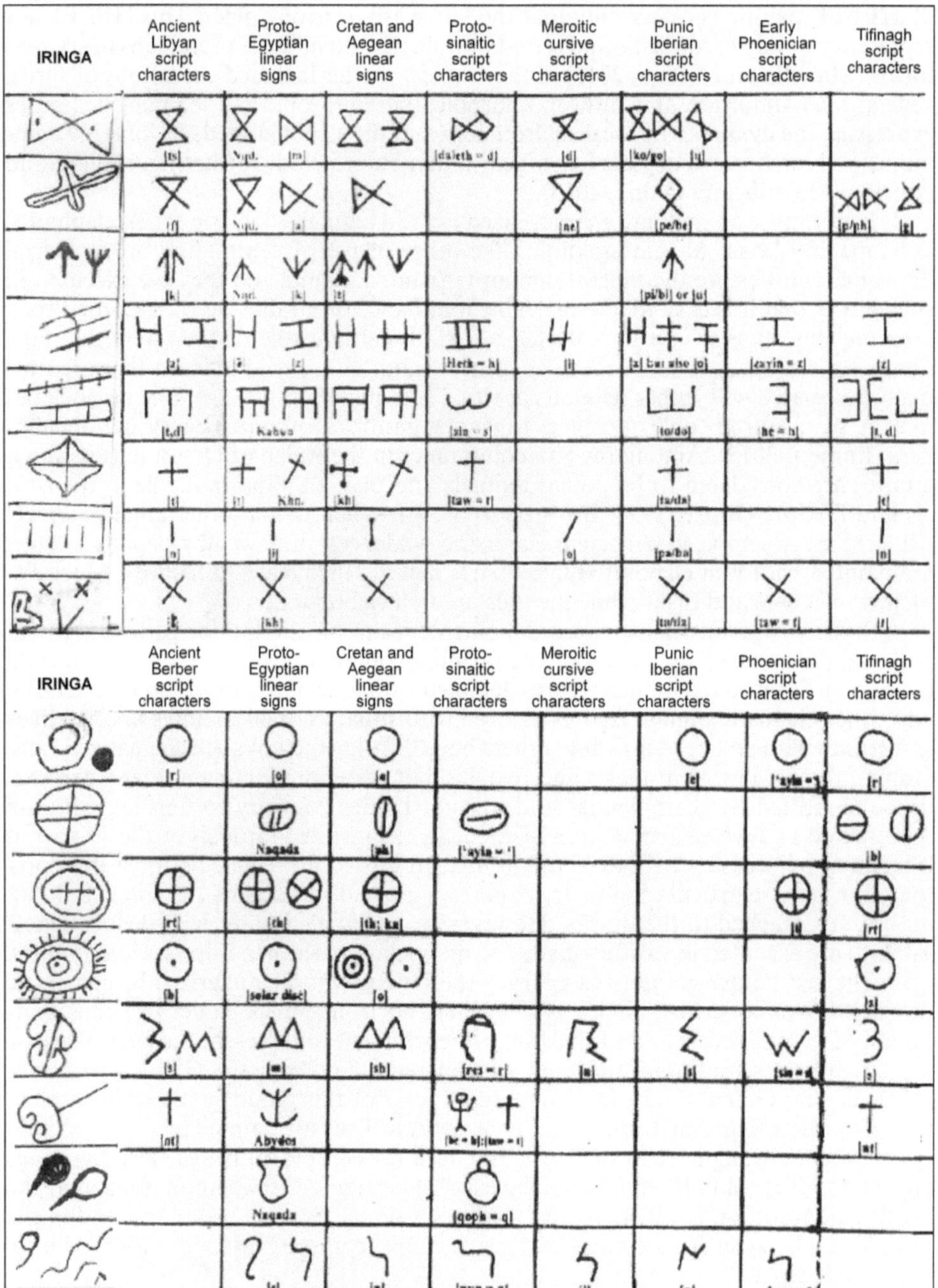

Figure 8. Alphabet-like signs from Iringa compared to some Ancient Script Alphabets from other parts of the world (Evans, 1897: 384, 386; quoted and re-drawn from Rasolondrainy, 2011 and www.ancientscripts.com). Phonetic value of the letter is not included.

all reptiles were painted in dirty white (Figures 6, 7). However, there is a sub-tradition under this category known as art Script/letter-like designs characterized by geometric, script-like designs (letters, numbers, signs or symbols) and non-representational paintings or abstract arts (Chami, 2008).

Non representational paintings or script-like designs appear as alphabetic and numerical signs. They represent alphabetic, syllabic or ideographic characters, in particular, those from ancient writing systems of early African civilizations. Others resembled numerical patterns that are similar to numbering systems of today (see Chami, 2008 and Rasolandrainy, 2011). This category includes vertical lines, parallel vertical lines, parallel horizontal lines, crossed X signs, X barred on top, V signs, Chinese hat signs or upside down V's, arrow signs, squares open at the base, inverted U, Up-side down E, H signs, B signs, line and dots signs, *Y*-like signs, numerical signs such as 9, cobra-like signs, cow's head-like signs and pottery/gourd vessel-like signs (Figure 8).

These script/letter-like signs/patterns, signs or symbols were also compared with ethnographic signs painted on ceremonial gourds of the Wanyaturu people of central Tanzania (Figure 9). Among Wanyaturu signs engraved or painted on ceremonial gourds have symbolic meaning. For instance, cattle figures symbolize wealth; parallel and curved lines symbolize peace, fruitful marriage and as a symbol of fertility as well (Itambu, 2013). Such signs and patterns are widely depicted at Kihessakilolo and Tavimienda, and are comparable to the ethnographic objects recorded among the Wanyaturu where it can be used as a sign of a link between the past and the present communities.

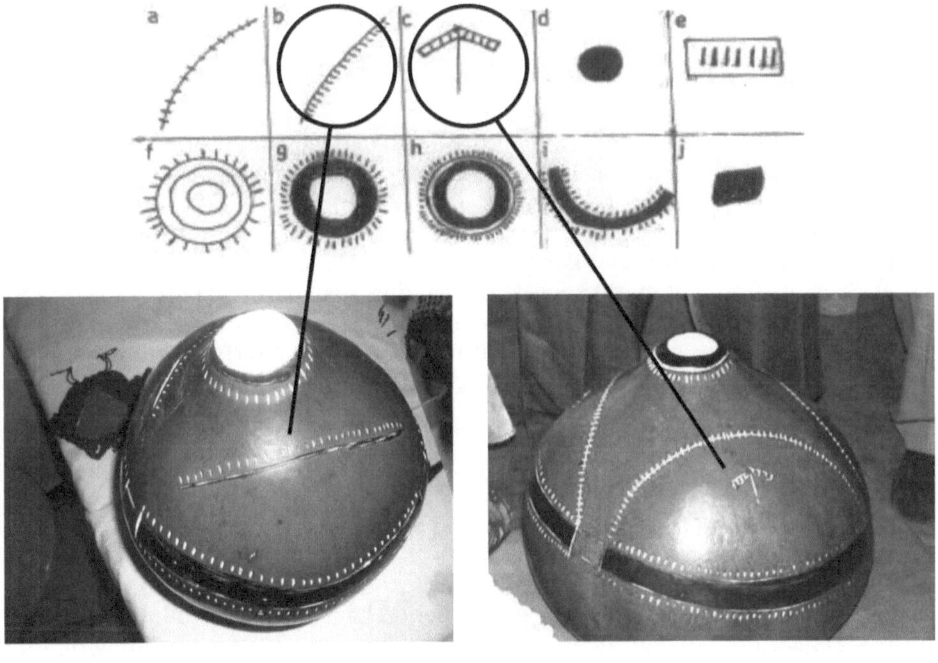

Figure 9. Paintings from Tavimienda rock-shelter in comparison with signs engraved on Wanyaturu ceremonial gourds.

10.4 RESULTS FROM ARCHAEOLOGICAL EXCAVATIONS

During the process of surveying and making assessment on rock art sites, we observed that there was an abundant scatters of artefacts on surface. Kihessakilolo rock shelter site contains a large scatter of artefacts on surface, hence was purposely selected for archaeological excavations. Excavation was done following arbitrary levels of 10 cm apart and it went down to the sterile layer about 100 m below the surface. A wire mesh of 5 mm was used to sieve small archaeological materials uncovered from the trench. We excavated three excavation units whereby we recovered LSA lithic artefacts, faunal remains, red ochre as well as land snail shells.

10.4.1 Lithic artefacts

In terms of lithic raw materials, quartz is the most dominant lithic raw material accounting for about 84% of the total stone tools. Basalt accounts for about 8%, quartzite 6% and chert 2% of the total materials. These raw materials were constantly distributed in all three excavated units. Today, local vein quartz and quartzite outcrops are locally found in the Precambrian granite outcrops that characterize most of the landscape in the Iringa Region (Bushozi, 2011). Chert is not locally found. This was obtained from early archaeological sites such as Isimila (Bittner, 2011; Bushozi, 2011). The abundance of quartz and a few quartzite artefacts suggest a pattern of localized raw material procurement in the Iringa Region during the LSA (Itambu, 2013).

10.4.2 Faunal remains

Wild animal remains mostly belonging to small and medium sized bovids were recovered. These include two astragalus and two distal metapodials of a bovid, one long shaft of a large mammal, one rib of a large mammal and one tooth of a canid. Other identifiable bones include two pieces of ribs, a femur and humerus of rodents. Rodents inhabit rocky hills and kopjes (Itambu, 2013). All the identified bones show the evidence of stone cut marks suggesting that they were processed using stone tools, but some of the fractures resulted from natural processes. The presence of highly fragmented bone remains, plus those with cut marks could be interpreted as a reflection of food residues.

10.4.3 Red ochre

This study recovered a total of 16 pieces of red ochre, most of them ochre pencils (Figure 9). Red ochre was retrieved from about 20–40 cm below the surface. The use of red ochre has been linked to the executions of rock paintings (Leakey, 1983). Intensification of the use of red ochre during the LSA period has been accepted by the majority of rock art scientists to indicate that these were commonly used for painting on rock walls (Mabulla, 2005). Human burials, rock art, red ochre and ostrich eggshell beads are among the symbolically revealing objects that date back to the MSA and LSA periods (Leakey, 1983; Bushozi, 2011; Itambu, 2013). The ochre pencils from Kihessakilolo are characterized by pointed, facetted and striated tips suggesting they were used for painting (Figure 10).

Figure 10. Ochre pencils with striated surfaces from Kihessakilolo rock shelter.

10.5 CHALLENGES FACING THE ROCK ART SITES AND THEIR PAINTINGS

This study indicates that most of the rock art sites in the southern highlands are in danger and are highly affected by physical weathering, as well as biological and anthropogenic actions. The major physical agents threatening paintings in this place are rock weathering, exfoliation and oxidization. The major biological threats include vegetation growth on rocks, as well as birds and hyrax dropping. Threats related to anthropogenic actions include *graffiti* (Figure 11) and smoke that cause the formation of soot on paintings. The situation is worse at Mlambalasi rock-shelter where most of the paintings are faded due to rock weathering and biological actions. Reoccupation of the rock shelters was evident, for instance in this study repeated utilization of rock-shelters was recorded at Lutona, Mlambalasi and Tavimienda where anthropogenic actions were among the leading threats facing rock paintings. However, at Kihessakilolo and Tavimienda, a moderate number of paintings were found in a fair state of preservation as the majority of paintings were still visible, decipherable and identifiable. The major problem observed at Kihessakilolo rock-shelter is the attempt made by the local government to fence the shelter. It is likely that the government mission was to keep the local people outside the site catchment. This is contrary to UNESCO's regulations that local people are the immediate custodians and stewards of the sites. In Tanzania, most of government institutions responsible

Figure 11. Graffiti on rock paintings and iron smelting in the rock shelter.

for cultural heritage management believe that local people have no role to play in the sustainable heritage management. They have forgotten their role of educating the local community about the scientific aesthetic and economic values of heritage resources. For instance, the lack of sensitization has made most local people in the Iringa Region believe that rock paintings are signs left by German colonialists to locate or indicate places of buried treasures. Therefore, treasure hunters are looking for German possessions in rock shelters with paintings. Such threats need serious mitigation measures; otherwise most of the paintings will disappear due to vandalism by treasure hunters and other activities of present day modernization such as encroachments.

10.6 CONCLUDING REMARKS

It was revealed that painted rock-shelters are emotionally attached to the cultural belief system of indigenous people. Sometimes indigenous people use rock-shelters with paintings for sacrifices, rituals and offerings. For instance, we were informed that the local community of Ikula Village occasionally uses rock-shelters with paintings for religious practices, in particular, rain-making rituals, blessings and forgiveness or appeasing gods and/or ancestors. A similar trend in the use of rock-shelters with paintings for rituals and religious aspects was recently recorded among of the Warangi people in Kondoa (Bwasiri, 2011). In addition, most of the rock-shelters surveyed are occasionally used for overnight stays during the rainy season, sometimes they are used as workshops or industry by iron smelters or blacksmithing. Finally, this study has contributed to our current understanding of the spatial-temporal distribution of rock art sites in Tanzania. This artistic expression is not limited to the central, north central and interlacustrine regions as it was previously thought, it extends to other places including southern Tanzania. The existence of rock art sites in the southern highlands that links the Eastern Rift Valley Region and the Zambezian Region could suggest that artists from similar or closely related communities inhabited the eastern and southern Africa landscape for most of the Pleistocene and Holocene. More studies on rock arts in southern Tanzania and the Zambezian regions are highly encouraged in order to come up with meaningful assertions about human adaptation, subsistence and relatedness between this priceless heritages depicted on rock shelters of Tanzania and that of Southern part of Africa.

REFERENCES

Bisele, M., 1993, *Women like meat: the folklore and Foraging Ideology of the Kalahari*, (Johannesburg: Witwatersrand University Press).

Bittner, K., 2011, Characterization and utilization of Stone Age Lithic raw materials from Iringa region, Tanzania. PhD Dissertation, University of Alberta, Edmonton.

Bushozi, P.G.M., 2011, Lithic technology of and hunting behaviour of during the MSA in northern and southern Tanzania. PhD Dissertation, University of Alberta, Edmonton.

Bushozi, P.G.M., 2012, Middle Stone Age (MSA) points form and function: Evidence from Magubike rock shelter, Southern Tanzania. *In Studies in the African Past, The Journal of African Archaeology Network,* Volume **10** (E and D Publishing Limited).

Bushozi, P.G.M., (in press), Towards sustainable management of cultural heritage resources in Tanzania: a case study from Kalenga and Mlambalasi sites in Iringa southern Tanzania.

Bwasiri, E.J., 2011, The challenge of managing intangible heritage: problems in Tanzania legislation and administration. *The South African Archaeological Bulletin,* **66(194)**, pp. 129–135.

Chami, F.A., 2008, "The Great Lakes: a complexity of cultural wellsprings", In *Art in Eastern Africa*, edited by Arnold, M. (Dar es Salaam: Mkuki na Nyota), pp. 47–64.

Cole, G.H. and Kleindienst, M.R., 1974, Further reflections on the Isimila Acheulian. *Quaternary Research,* **4**, pp. 346–355.

Deacon, J., 2007, African Rock-Art: The Future of Africa's Past. In *TARA 2004,* (Nairobi: Proceedings of the International Rock Art Conference).

Evernden, J.F. and Curtis, G.H., 1965, The potassium-argon dating of the late Cenozoic rocks in East Africa and Italy. *Current Anthropology,* **6**, pp. 343–385.

Hansen, C.L. and Keller, C.M., 1971, Environment and Activity Patterning at Isimila Korongo, Iringa District, Tanzania: A Preliminary Report. *American Anthropologist,* **73(5)**, pp. 1201–1211.

Harpum, J.R., 1970, *Summary of the Geology of Tanzania,* (Dodoma: Tanzania Mineral Resource Division).

Harris, J.F., 1981, *Summary of the Geology of Tanganyika,* (Dar es Salaam: Tanzania Government Printers).

Howell, F.C., Cole, G.H., Kleindienst, M.R. and Haldemann, E.G., 1962, Isimila, an Acheulian occupation site in the Iringa Highlands. In *Actes du IV Congres Panafricaine de de Préhistoire et de l'Etude du Quaternaire,* edited by Mortelmans, J. and Nenquin, J. (Tervuren: Musée Royale de l' Afrique Centrale), pp. 43–80.

Itambu, M.P., 2013, The Rock Art of Iringa Region, Southern Tanzania. A Descriptive and Comparative Study. Unpublished MA Dissertation, University of Dar Es Salaam

Leakey, M.D., 1983, *Africa's Vanishing Art: the Rock Paintings of Tanzania* (New York: Doubleday and Company).

Mabulla, A.Z.P. and Gidna, A., 2013, The Dawn of Human Imagination: The Rock Art of North-Central Tanzania.

Mabulla, A.Z.P., 2005, *The Rock Art of Mara Region, Tanzania*, (Azania XL), pp. 19–42.

Mahudi, H., 2008, The Use of Rock Art in Understanding of Socio-Economic Activities and Cultural values; The Case of Matongo-Isanzu in Iramba Disrict, Tanzania. Unpubished MA Dissertation. University of Dar es Salaam.

Masao, F.T., 2007, *The Rock Art of Singida and Lake Eyasi Basin in Tanzania*, (London: Duggan Foundation).
Mturi, A.A. and Bushozi, P.G.M., 2002, *The documentation of Lake Victoria rock art in Bukoba area: A preriminary report* (unpublished), (Dar es Salaam: Department of Antiquities).
Namono, C., 2010, Surrogative Surfaces: A Contextial Interpretative to the Rock Art of Uganda. PhD Thesis, University of the Witwatersrand.
Rasolondrainy, T.V.R., 2011, Archaeological Study of the Prehistoric Rock Paintings of Ampasimaiky Rock-shelter in the Upper Onilahy, Isalo Region, Southwestern Madagascar, Unpublished MA Dissertation, University of Dar es Salaam.
Vinnicombe, P., 1976, Rock Paintings Analysis. *South African Archaeological Bulletin*, **22**, pp. 129–141.
Willoughby, P.R., 2007, *The evolution of modern human in Africa: A comprehensive guide* (Lanham: Altamira Press).
Willoughby, P.R., 2012, The Middle and Later Stone Age in the Iringa Region of southern Tanzania, *Quaternary International,* **270**, pp. 103–118.

CHAPTER 11

The effects of global warming on the rock art in the uKhahlamba-Drakensberg Park World Heritage Site, South Africa

Alvord Nhundu
Anthropology and Archaeology Department,
University of Pretoria, South Africa

ABSTRACT: In the uKhahlamba-Drakensberg Park (UDP) World Heritage Site, South Africa, global warming poses a major threat to the preservation of rock art sites. Rock shelters in particular are at risk because they are highly dependent on the integrity of bedrock and linked geomorphological landforms. The principal agents in rock surface weathering are precipitation and temperature; both are subject to significant changes at present. Results of this study indicate that the warm climate is accelerating weathering processes leading to the fast deterioration of the art. The extent of recent and future damage of art in rock shelters is currently unknown. However, there are at present a number of techniques (e.g. drip lines) that reduce the impact of global warming rock art.

11.1 INTRODUCTION AND METHODS

The United Nations Framework Convention on Climate Change (1994) views global warming as a threat to environment and to societies in the future, and therefore it states that the causes must be dealt with now (Kemp, 1998). Clark (1998) stated that global warming is one of the most serious environmental issues of our time. Climate change is a threat to humans' heritage. In 2005 The United Nations Educational, Scientific and Cultural Organisation (UNESCO's) World Heritage Committee placed it on record that climate change was having an adverse impact on cultural values of a considerable number of heritage sites. Cognisance was also taken of the continuance of these negative effects for generations to come (Heath, 2008).

In this context this paper has two main objectives; the first is to determine how global warming is impacting upon the climate in the uKhahlamba-Drakensberg Park (UDP) World Heritage Site, and the second is to explore the general extent to which the predicted effects will impact rock art in the area. Whilst a large range of geomorphological, geological and biological factors lie behind rock art deterioration of any specific site, the goal is to consider the general impact upon Drakensberg rock art by the effects of global warming. To achieve this, a number of methods were applied. First a literature review on climate change was done to identify a series of predicted models for the impacts of global warming upon in the Drakensberg. This was followed by a visit to the South African Weather Service to collect temperature and rainfall data for the study area. Secondly investigations on the likely impacts of the different scenarios on Clarens Sandstone, the substrate on which the rock art is found, have been evaluated. Finally, general mitigation measures were proposed to limit the effects of global warming on Drakensberg rock art. Comparisons with other practical conservation trials and experiences from Australia, Canada, France and South Africa were drawn.

The rock art of the uKhahlamba-Drakensberg Park (UDP) World Heritage Site in South Africa is at the mercy of global warming, but the extent and nature of this threat is yet to be determined. Heath (2008) pointed out that this could give rise to permanent alteration to geological and geomorphological processes, which can never be reversed. He went on to say that unique cultural heritage sites such as rock art and cave art sites are at special risk because their continual existence is reliant on the preservation of the landforms upon which the art is found. The main changes brought by global warming are expected to be temperature and precipitation, and since these are the principal agents in rock art weathering this is of concern for future rock art management. These processes can damage the substrate on which is the rock art is on and this will lead automatically to the destruction of the art.

11.2 THE UKHAHLAMBA-DRAKENSBERG: LANDSCAPE AND THE ART

11.2.1 Geographical location

The UDP lies on the west of KwaZulu-Natal province on the Lesotho border (Figure 1) (Porter, 1999).The 2400 km^2 park stretches from Royal Natal Park which is located from the north of Bushmen's Nek in the southern direction. Generally it is divided into three sections; the southern part which comprises places such as Mzimukulu Wilderness area and Kamberg, the central region which includes places such as the Giants Castle and Cathedral Park, and lastly the northern section that consists of the Royal Natal Park (Porter, 1999). In terms of geographical coordinates the western side of the park lies at 29°45′ E and extends to 28°52′ E, the most northern area is located at 28° 38′ S and extends to 28°46′ S, and the southern edge lies between 28°55′ S and 29°55′ S (Derwent, 2006). The park was formally recognised as a world heritage site on 29 November 2000 (UNESCO, 2000).

11.2.2 Geology and geomorphology

South Africa's geological history is uniquely complex and predates to 3.6 billion years (Johnson *et al.*, 2006). These old geological 'landscapes' are characterised by an interior highland at its eastern direction, and by an escarpment at its south western margin (Truswell, 1970).The south eastern escarpment is constituted by the Drakensberg which forms the highest mountain range in South Africa (Eriksson, 1983). The mountain range extends for 1000 km from south to north east (Truswell, 1970).

The geomorphic shape of the Drakensberg was primarily caused by tectonic uplift (Truswell, 1970). Initially the mountain was a constituent of Gondwanaland which separated from the original supercontinent of Pangaea (Caircross, 2004). Gondwanaland separated 300 to 100 million years ago to constitute the continents of Africa, Asia and South America (Lapidus, 1990). This continental drift that culminated in the formation of today's major continents has simultaneously taken place with geological processes like arching, faulting and extensive vulcanism (Truswell, 1970).

The Drakensberg area is drained by two major river catchments. The Orange and the Vaal River system on the western slopes of the mountains: the eastern

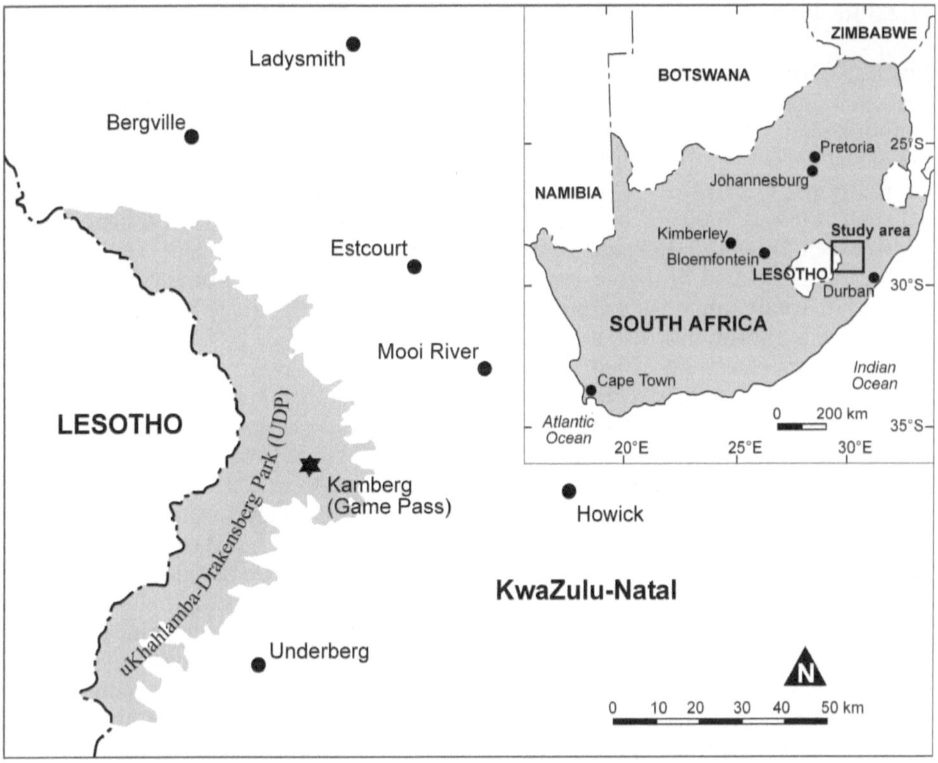

Figure 1. Geographical location map of UDP (modified from Hoerlé, 2005).

and southern slopes are drained by the Thukela River (Eriksson, 1983). The rivers played an important role in shaping up the geomorphology of the Drakensberg. Lateral, vertical and headward erosion exposed varying resistant sediments of the Karoo Supergroup such as desert dune deposits of the Upper Karoo Sandstones (Caircross, 2004). As the Drakensberg was comprised of different phases, the layer that consisted of fluvially deposited sandstones, mudstones and siltstones in the lower parts of the Clarens Formation weathered and eroded at a faster rate than aeolian deposits above resulting in the formation of the unique rock overhangs (Mazel and Watchman, 2003). These are the caves which have been inhabited by the San people (Lewis-Williams, 1981). The range divides the eastern part of South Africa with the state of Lesotho (Porter, 1999).

The Clarens sandstone consists of relatively fine grained quartz-rich feldspatic wackes where most of the feldspars are plagioclases, and the clay minerals of the matrix are mainly illite and dark mica (Eriksson, 1983). The Clarens sandstone is a product of an initial alluvial fan deposition followed by an aeolian dune deposition and is located between the Elliot Formation sandstone and the Stormberg basaltic lava of the high Drakensberg (Eriksson, 1983). Within the Clarens sandstone, other layers include clay minerals, notably kaolinite and illite which can shrink and swell due to changes in temperature and moisture (Meiklejohn, 1994). The latter fact makes the Drakensberg substrate susceptible to the environmental modifications.

11.2.3 Rock art

The Drakensberg is well known for its rock art attributed to the San hunter-gatherers (Lewis-Williams, 1981). Evidence from archaeological excavations and other dating methods shows that the existence of humans on the UDP predates a millennium, the oldest dates from the southern part of the Drakensberg is ±8000 BP whereas one for the north is ±5000 BP (Mazel and Watchman, 2003). Some of the art is dated between 100 and 4000 BP (Blundell and Lewis-Williams, 2001), but a general dating between 2000 and 3000 BP is given (Mazel and Watchman, 2003:71). This fine line art is associated with the Later Stone Age hunter-gatherers (Lewis-Williams, 2000). The total number of paintings in the UDP World Heritage Site is not known, since there is continuous discovery of sites (Blundell and Lewis-Williams, 2001), but recent studies (Derwent, 2006) have recorded 600 sites with paintings containing more than 40,000 images. The UDP has the second highest number of rock art panels in the world; first is Australia (Smith, 2008, personal communication). Subjects in the San paintings range from animals dominated by elands to human therianthropes, to ox wagons, men armed with rifles, stretching a quite long historical record (Mazel and Watchman, 2003). The paintings are dominated by shamanistic trance dances showing the religious and social life of the San people. The UDP rock paintings are unique distinctive, in particular this art was designed through the use of shaded polychrome technique, pursuant to which figures of human beings, elands, snakes amongst other animals are depicted using more than two colour shades, mostly through the grading of red and white shades. There are many of these at sites such as Game Pass, Main Cave, Battle Cave and the Giants Castle (Lewis-Williams and Dowson, 1992). The style of painting has impressed many researches and viewers, for instance an eland is depicted with a height of 35 cm, with clearly designed eyes, legs, hooves, tail, mouth and ears. Interestingly, even hair no more than 1,5 mm can be easily identified (Blundell and Lewis-Williams, 2001). The depiction of the animals is also amazing; not only that they are drawn side on, but in different styles; walking, running, lying down, leaping, limping, even looking back over their shoulders. Animals are depicted with human features, for instance you can find an eland with a human head or human legs, these animals can be viewed from different angles, from the front and rear (Lewis-Williams, 1981, 2000). Human paintings are also shown in a grotesque fashion (Blundell and Lewis-Williams, 2001); for example therianthropes paintings which are constituted of both animal and human features that are interpreted by Lewis-Williams (1981, 2000) as representing people blend with animal potency that are experiencing trance. There are also figures creating humorous impressions of animals and humans exiting on the rock surface (Lewis-Williams and Dowson, 1992).

Some rock art sites in the UDP are living heritage sites and this means they are utilised by people for various purposes (Derwent, 2006). There is the manifestation of the current indigenous knowledge systems in the park. The people see rock art as a connection to their ancestors (Lewis-Williams, 2000). This is playing a significant role as far as the preservation of the sites and rock art heritage is concerned. The Zulu-speaking healers and San descendants still use some painting sites in their rituals, for example Pernwan Cave is used as an initiation site, Twalelani rock outcrop is used for women fertility rituals, women who struggle to conceive are taken there. Inkanyamba Cave is used for rainmaking and the Waterfall Shelter is visited by Zion Christian Church who believe that their prayers will be answered if they are made at the Waterfall (Derwent, 2006).

11.3 THE UKHAHLAMBA-DRAKENSBERG AND GLOBAL WARMING

11.3.1 Current climate

The Drakensberg experiences a tropical continental or a savannah type of climate (Nel, 2008). Summers are hot and winters are cool and dry (Nel and Sumner, 2006). The mean temperature is about 17°C in summer and 12.5°C in winter. The overall mean temperature is about 16°C (Nel, 2008). The highest temperature is in the summer season and reaches 35°C. At the slope on the northern direction, situated at lower altitudes, the lowest temperature of about –20°C are experienced at winter nights on the summit plateau. Annual rainfall distribution in the UDP is not uniform (Nel, 2008). The south-western region receives 450 mm; and it ranges between 1000 and 1100 mm in the northeast and the high Drakensberg yearly receives 1800–1900 mm (Wright and Mazel, 2007). Most rainfall occurs in January and February. Snow and hail also contribute to the overall moisture content (Jury, 1997). Snowfalls take place at an average of eight times annually, predominantly at the apex of the mountain range (Nel and Sumner, 2006). If due regard is paid to the localised snow-falls, the frequency could actually be higher. Frost is common in winter, where it occurs in the 180 day period from the middle of April up to the month of October (Sumner, 2008).

Modifications/Trends in temperature

The world average temperature has been changing; by the 1980s it was estimated that a doubling of carbon dioxide would cause an average warming of 1.3 to 4°C. In 1990, the Intergovernmental Panel on Climate Change (IPCC, 2007) assessment produced values of 1.5 to 4.5°C with a most likely figure of an increase of 2.5°C. By the time of their second assessment in 1995, the estimate had been reduced to an increase of 2°C with maximum of 3.5°C to be reached by 2100.

The Fourth Assessment Report (IPCC, 2007) concluded that there is a possibility exceeding 90% that human activities such as burning of fossil fuels and deforestation are having a negative impact on the earth's climate system. Observations made confirm the changes that have been captured in the IPCC projections. There is a high probability that the noted trend will continue for a considerable period of time unabated, notwithstanding various measures being put to curb climate change due to lags and inertia contained in the global biosphere response (Jury, 1997). Current climate models predict a global increase in temperature of between 1.2 and 5.8°C rise by 2100 (IPCC, 2007).

For Africa, the IPCC (2007) argued that warming will pose a risk, greater than the global annual mean. IPCC (2007) predicted that warming will occur over the continent and in all seasons. However, subtropical regions will be exposed more to these phenomena than the moister tropics. In the four regions of Africa the median temperature increase is expected to range between 3 and 4°C up to 2100, which is greater compared to the global mean figure by approximately 1.5 times. Southern Africa will experience the largest temperature response in the months of September, October and November (IPCC, 2007).

In common with the global picture, the climate of the UDP has changed since 1980 (IPCC, 2007). Regional climate predictions are used to give climate predictions for countries and provinces. The IPCC (2007) noticed changes in climate in the region of the UDP. There are various climate signals which aid in climate prediction for the UDP. Hewitson (1992) considers winds that blow over the tropical east Atlantic and North Indian Ocean as a significant tool in the prediction. Projections of the KwaZulu-Natal climate are based on El Nino signals from Pacific temperatures and the southern oscillation index (Jury, 1997). The manner in which the various factors alluded to, gives rise to a deduction that predictions for KwaZulu-Natal should be dictated by how the local circulation responds to large tropical polar temperature gradients. Considering the likely seasonal cycle, there is a high probability that a unimodal distribution of rainfall and temperature will occur, reaching its highest level between December and March (Nel and Sumner, 2006).

In South Africa temperature has been analysed using records from 1053 stations which passed quality control, but of these only 23 have records extending for more than 50 years of daily temperature data and 4 stations exceed 75 years of daily data (Sumner, 2008). The network of stations is densest in KwaZulu-Natal, Gauteng, and Western Cape and sparsest in the Northern Cape. Information gathered from these stations has shown that air temperature for the past three decades has drastically increased (Jury, 1997). Statistics reveal an important sequence of increased maximum temperature, being recorded in the western, eastern and southern half of the country (IPCC, 2007). According to the IPCC (2007) there has been a steady increase in temperature in the Drakensberg since 1910. A plot of meteorological values shows a slight upward trend by 0.18°C underlining the fact that the Drakensberg is slowly warming. According to the IPCC (2007) during the period 1920–1995, South Africa experienced a significant warming of 0.12°C. Temperature levels are as a matter of course argued to retain an increased rate of between 1 and 3°C throughout the country (IPCC, 2007). The IPCC (2007) report states that 11 of the past 12 years, recorded temperatures were the warmest since 1850, the year in which worldwide measurement of temperature ranges was introduced. Incidence of hot days, hot nights and heat waves will also escalate.

Modifications/Trends in precipitation

Precipitation just like temperature is affected and will continue to be affected by global warming. The IPCC (2007) stated that the influence of global warming on precipitation will not be uniform. In some regions, global warming will result in dry conditions and prolonged periods of dry weather (drought) whereas in other regions there will be an increase in the amount of rainfall due to high evapotranspiration rates. There is high probability that there will be a decrease in rainfall received annually in most areas of the Mediterranean Africa and the northern Sahara. The trend will continue as the Mediterranean coast is approached. Southern Africa is also at risk of witnessing a decrease of rainfall in winter, particularly in the western regions. In east Africa, however, predictions state that annual rainfall will increase (IPCC, 2007). Hewitson (1992) stated that there is a high probability of an increase in precipitation in areas between east Africa and southern Africa in the months of June, July and August. In South Africa, IPCC (2007) says the eastern part of the country where the UDP is found which experiences high rainfall will experience more rainfall at the beginning of the next century whereas the drier western region will continue to receive less rainfall.

Moving to precipitation in the Drakensberg, changes in South African precipitation has been analysed using a data set of approximately 3000 rainfall stations. In KwaZulu-Natal—due to the remoteness and inaccessibility of the place—there are not many weather stations operating in the province, let alone reliable ones (Nel, 2008). However, there are 11 well established weather stations of the South African Weather Service that have provided reliable rainfall pattern statistics and data for the last 100 years (Nel and Sumner, 2006).

Although it has been difficult to detect any fluctuation of mean annual rainfall in the last half of the 20th century (Nel, 2008), historical records show that precipitation has been steadily increasing in the Drakensberg. This is augmented by the fact that the east part of South Africa is experiencing rainfall increase whereas the western part is experiencing rainfall decrease (Hewitson, 1992; IPCC, 2007). The gradual increase in PCI, an index which is used to calculate precipitation changes at four weather stations situated in the UDP region has been of significant statistical value (Nel, 2008). The 95% confidence level captured portrays a seasonal increase in rainfall being received monthly in the Drakensberg region while there had been a steady increase in rainfall in the Drakensberg (IPCC, 2007), there had been a decrease in autumn rainfall by 30 mm from 1955 to 2000 (Nel, 2008).

Other meteorological conditions have also changed. The phenomenon of extremely high clouds like cumulonimbus and cumulus, now characterises the tropics. This has also been observed by scientists at NASA's Propulsion Laboratory, who have observed a direct link between increase in coverage ocean surface temperatures, and the frequency of storms per decade (Jury, 1997). In particular, they have concluded that a degree centigrade increase in coverage ocean temperatures will lead to 45% increase in the frequency of storms (IPCC, 2007). The said frequency is predicted to increase by 6% for every ten years to come (IPCC, 2007). Linked to the high frequency of clouds is lightning. Hoerlé (2005) describes lightning ground-flashes as regular in the UDP and are the common cause of veld fires.

According to the IPCC (2007), there has been a global increase of the frequency at which heavy rainfall storms have occurred between 1900 and 2005. In South Africa, there has been clear increase in the intensity of heavy rainfall from the period 1931–1960 to the period 1960–1990. The weather scenario when high rainfall is received for ten years (10 year high rainfall events) are predicted to increase by 10% in many regions of South Africa, save for those parts in the north–west, and those regions which receive winter rainfall (Jury, 1997).

The IPCC (2007) considers climate change, and recurred flooding as the major risks to settlement structures and posterity. The intensity of floods and frequency are predicted to escalate in many regions as a direct result of an increase in the frequency of heavy rainfall events. Correlatively these phenomena will result in the escalation run-off excess over and above groundwater recharge through floodplains. Extreme weather events are already a feature of the regional UDP climate and spatio-temporal dynamics are high. Heavy rainfall on the 19th of June 2008 (The Star Newspaper) prompted extreme flooding and triggered mudslides that caused the downslope movement of rock boulders in the Drakensberg. According to the South African Weather Service (2008), KwaZulu-Natal received a total of 128 mm rainfall in one day, the highest rainfall amount in 24 hours on record. In 2006, there were also floods in the Drakensberg, three people were killed and seven were left homeless after severe flooding. This happened after it had rained for the sixth consecutive weekend (The Star newspaper, 2008).

11.4 RESULTS

11.4.1 The effects of global warming on sandstone

Rock art is prone to the ravages of global warming because it is dependent on the nature and stability of the substrate (Heath, 2008). Hoerlé (2004, 2005), Hoerlé *et al.* (2007, 2008, 2009), Meiklejohn (1994) and Meiklejohn *et al.* (2009) state that weathering of the Clarens sandstone is causing the loss of indigenous rock art heritage in the UDP and will continue to do so in the future. Although weathering processes have been occurring for centuries, global warming is likely to accelerate weathering processes causing the faster deterioration of the art. Trends in temperature and rainfall point to a situation whereby increasing rainfall and the warming of the UDP are likely to exacerbate the deterioration of the rock art. The influence of global warming on the rock art will happen through increased salt weathering, flaking, block collapse, granulation and bio-deterioration

The increase in temperature and precipitation is likely to accelerate salt weathering in the study area. Salt weathering is defined by Ollier (1984) as the growth of salt crystals from solution which result in the disintegration of rocks. There are several ways in which the increase in salt weathering will aggravate the deterioration of the paintings: salt weathering causes the disintegration of rocks by causing three types of expansion forces; thermal expansion, hydration and crystallisation. For rock breakdown to occur, the thermal expansion coefficient of salt must exceed that of the rock (Bland and Rolls, 1998). At Game Pass Shelter (Figure 2), one of the most popular sites in the UDP the roof of the shelter oozes salt rich water. The salt is causing the fading of the images. The problem of water increase is likely to worsen the situation because salt weathering is more effective when the weathering agent salt enters the rock pores in solution.

Although it can be argued that global warming will result in the decrease of salt weathering basing the argument on the fact that more heat will mean more evaporation giving salt weathering no chance to take place, the presence of wash zones negates this point. Salt-laden water will just cause the images to fade even before weathering takes place underlining the fact that an increase in precipitation will cause further deterioration of the paintings. Wash zones are already a problem on many panels in the UDP. Figure 3 shows an example of the damage caused to rock art by wash zones containing salts.

Rock flaking which is rife in the UDP is likely to be accelerated by global warming. Flaking works in association with salt weathering (Benito *et al.*, 1993). Flaking

Figure 2. Landscape view in UDP at Kamberg showing sandstone escarpments of the Clarens Formation.

Figure 3. A buck in the Cederberg, Western Cape (photo by K. Crause).

is the expansion and separation of the rock due to external forces (Ollier, 1984). This process creates the detachment or falling off of rock fragments due to the influence of salt, water and heat. This is slowly destroying the art and is one of the most common forms of rock art decay. Salt crystallisation which is brought by increasing the concentration of a solution at a constant rate usually by evaporation (Bland and Rolls, 1988) cause the weakly bonded part of the rock to fall off from the main rock causing the art to deteriorate. At some sites in the UDP salt is seen on paintings or where flaking has taken place confirming the fact that the two processes are inseparable. Hoerlé (2005) detected a large flake on an eland at Game Pass. This problem is likely to get worse with global warming since more water will mean more flaking. Figure 4 depicts how flaking has destroyed sections of an eland at a site in the UDP.

The rock art in the UDP is also at the mercy of block collapse. A warmer climate is likely to accelerate block collapse since an increase in precipitation will make the ground wet most of the time and this will hasten block collapse. Meiklejohn *et al.* (2009) noted that at Main Cave and Battle Cave rock shelters have a lot of pores and bedding planes which are actually a threat to the paintings. The ground penetrating radar which was used by Hoerlé *et al.* (2007), a technique used to detect the extent of rock cracks showed that there were a lot of cracks at Game Pass. More precipitation

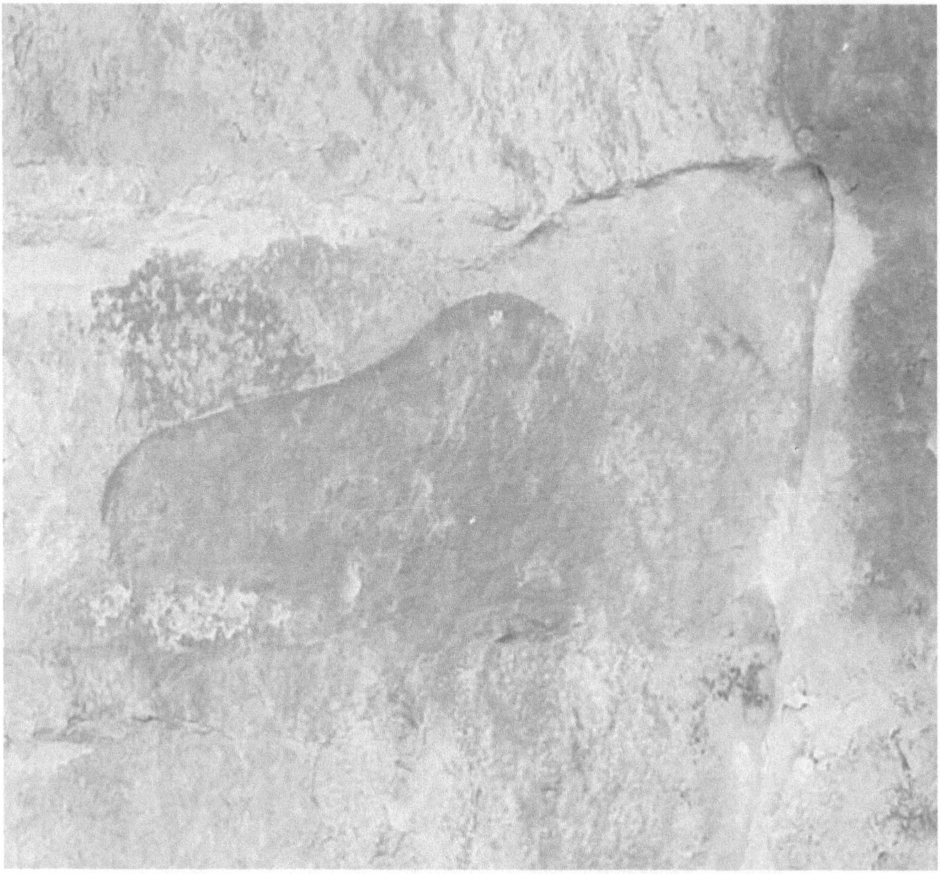

Figure 4. An eland at Game Pass which has been partially destroyed by flaking.

and an increase in temperature will produce more blocks collapsing resulting in further damages to the paintings (Figure 5).

Granular disintegration is another problem facing rock art in the UDP. Granulation refers to the deterioration of the rock matrix and cement which holds the rock together (Lapidus, 1990). It is the detachment or falling off of granules or small particles of rock. The main cause of this problem is salt and water. The increase in precipitation and temperature associated with global warming is likely to accelerate granulation. This increase in temperature is causing the minerals to expand at different rates resulting in them falling off the rock. More precipitation means the minerals are going to be wet thus weakening the bond and causing the rock to crumble. This weathering process also works in association with salt weathering. In the Giants Castle at Injasuthu (Figure 6), granulation has resulted in the deterioration of the images.

Global warming will also create thicker vegetation in the study area. The increase in precipitation (IPCC, 2007) is likely to create vegetation thickets. Vegetation will thrive under high rainfall and high temperature. Lichens, algae and mosses are contributing to the deterioration of the tangible heritage in the UDP. Tree roots grow in

Figure 5. Block collapse at Game Pass.

Figure 6. Destruction of rock art (eland painting) by granulation at Giants Castle, UDP (courtesy T. Forssman).

rock cracks and as they grow they expand widening and deepening cracks which result in the breakdown of the substrate (Chesnut, 1972). In addition to that tree roots produce humic acids which react with rock minerals and this causes the deterioration of the paintings. Mosses grow in the darker and wetter parts of the rock shelter and have physical and corrosive effects on the rock surface (Fry, 1924). Lichens grow right on the paintings interfering with the aesthetic beauty of the site. There are a lot of lichens and mosses already causing a lot of bio-deterioration (Figure 7).

11.4.2 Mitigation measures

There are a number of measures to save rock art that have been successfully used in Australia, Canada, France, USA and also in some parts of South Africa. These measures could be used to allay the effects of global warming on the UDP. What is important to note is that when considering the long term cultural value of a site, it is essential to choose the least intrusive, most easily reversible strategy available (David, 2003). The strategies must be consistent with the identified cultural significance of the site and its statement of goals and objectives. In implementing the strategies, it is always advisable to guard against negative impacts. Conservators always say

Figure 7. Bio-deterioration at Game Pass in the UDP.

"Bad intervention is often worse than no intervention at all" (Loubser, 2001:105). Hands-on conservation intervention is effective when it conforms to all the necessary steps such as consultation with all stakeholders, recording assessment and reviewing these potential mitigation measures. Although Silver (1989) (cf. Loubser, 2001) argues that the treatment of each site should be site specific, it has been proved that the techniques that have been applied at different sites with similar conditions can work.

It is always important to apply the methods that do not interfere with the ambience of the site (Bednarik, 1989). In some cases conservators have realised that in trying to solve a problem they had actually created another that has worsened the situation. It is not easy to establish an appropriate method to solve a problem at a site and it is very difficult to isolate the rock shelter from the natural environment (Meiklejohn, 1994). However, the measures proposed for the UDP will not stop the impacts of global warming but will lessen the effects.

The principal agent in rock art weathering is water. This moisture is mainly in the form of rainwater, capillary moisture in porous rocks and condensation in caves and shelters (Meiklejohn, 1994). This water culminates in what are known as wash zones in many rock shelters in the UDP. There are a number of measures that can be used to control moisture in and around rock shelters. The most commonly used is drip line (Gillespie, 1983).

There are a number of universal principles that govern the installation of drip lines. If a drip line is put it must not spoil the appearance of the site. It must not be put near to paintings. It must also be assessed whether the drip line is the most appropriate or the best potential solution to the problem identified at the site that is to keep water away from the rock panels (Gillespie, 1983).One must always guard against drip lines spoiling the ambience of the site. There are some cases where a drip line has been installed and the deterioration of the paintings has become worse (Deacon, 2009, personal communication).

In Australia drip lines have been used to lessen the effects of water and salt. There they use silicone drip lines whereas in South Africa the putty drip line is preferred. In Australia, the silicone drip line is applied with a pressure gun (Bednarick, 1989). Before the drip line is installed, the substrate is thoroughly cleaned so that the drip line is put on a clean surface. The drip line is normally put along the junction of the vertical and horizontal places at the upper edge of the rock shelter (Gillespie, 1983). They make sure that the silicone has good bonding ability, but at the same time must be able to be removed without damaging the paintings when it is not needed. The silicone is also expected to have a high thermal ability and to be damp proof and resistant to ultra-violet radiation. The issue of colour is also important in Australia; the clear silicone is preferred (Bednarick, 1989).

In South Africa a drip line has been used on many sites. One of the sites is Bushman's Kloof Shelter in the Cedeberg, Western Cape. The drip line was installed in the 1970s and had been effective to a certain extent (Deacon, 2009, personal communication). The epoxy putty drip line is used to divert water away from reaching the paintings (Figure 8).

The method of water control is based on the fact that a thin laminar flow can be halted, broken or diverted from flowing directly on the images. The preferred position of the drip line is on the outer most horizontal surface just under the roof where the water still has high velocity (Gillespie, 1983).

The drip line could assist in providing a solution to the problems of many sites in the UDP. At sites that are open to the public such as Game Pass, Battle Caves, Main Caves and the Giants Castle it is advisable to use coloured silicone to match the colour of the substrate. Bednarick (1989) reported that in Australia tourists sometimes pull

Figure 8. Drip line at Bushman's Kloof in the Cederberg.

the silicone threads off the rock out of anxiety. In some cases where the drip line has not been used properly it has worsened the problem so there is always need to follow all the procedures properly. The use of drip lines could be useful in the UDP. To assess the effectiveness of the drip line it should be monitored frequently (Rosenfield, 1988).

Salt crystallisation in most cases leaves salts on the paintings and this can obscure the images. Global warming will cause more evapotranspiration that will leave more salts on the paintings. This salt is normally brought by rain water, capillary or gravitational water. According to (Schwartzbaum, 1985) the salt could be removed by compresses or poulstices. In this case sheets of long fibre, wet strength tissue are applied with water and a flat hard brush is used from the bottom of the substrate. This could be effective in the UDP where there is a lot of salt on wash zones.

11.5 DISCUSSION

Flaking could also be slowed down in the study area. Some form of consolidation is required to maintain the integrity of a flaking surface. Methyl methacrylate and polyvinyl can be used to avert the problem of flaking. Care must be taken when using

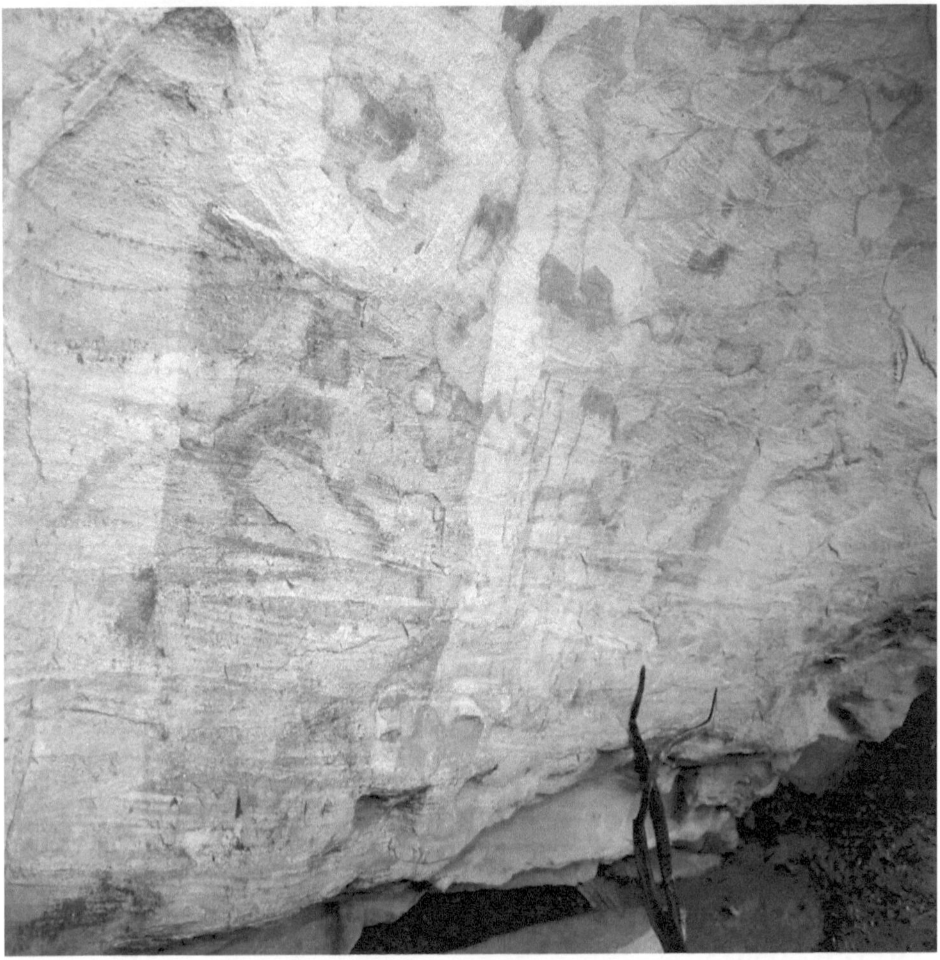

Figure 9. Negative effects of a Wacker H at Bleeding Nose, Cederberg, Western Cape: note the darkening of the paintings in the left section.

these products in that sometimes they can block pore spaces in the rock resulting in the crumbling of the substrate worsening the condition. Bednarick (1989) reports that a product called "paraloid" has been used successfully at many sites in France. Ethyl silicate could also be used as a cementing agent for quartz grains. The disintegration of sandstone from the substrate results in the loss of the art. Trials should be put in place, on rock surface away from the art to test the viability and effectiveness of all these potential consolidants. Only after they have been thoroughly tested on the Clarens sandstones should they be used at any rock art site (no regret measures).

Block collapse can also be slowed down. The disintegration of sandstone in blocks that was detected by Hoerlé *et al.* (2007)'s ground penetrating radar at Game Pass is a threat to heritage since it exposed huge cracks in the rocks. This problem can be slowed down by grouting and other physical means. Robert and Wallace (1970 cf. Bednarick, 1989) attempted to arrest exfoliation of a massive layer of sandstone by using sealants. To solve the problem of rock shelters threatened by block collapse, large pins of stainless steel could be installed in the gaps between the blocks. In this case small exfoliating rock fragments of a painted panel could be fixed into position by epoxy resin. This was successfully done at Trotman's Cave in the Great Sicily desert (Bednarick, 1989). In this situation Clarke (1978) reattached exfoliated fragments of a pictogram panel with epoxy resins. The same measure has been used in Siberia at a site called Shishkino. It has also been used in India and the USA. Fry (1924) suggested that for the method to be effective there is need to put the resin into the layers to make them impervious.

The tangible heritage is also under threat from biological weathering. Bio-deterioration agents include mosses, lichens and algae. Due to global warming lichens, mosses and other plants thrive under such conditions because there is high temperature and high rainfall. Fry (1924) stated that the easiest and cheapest way of removing lichens is just depriving them of light. This can be achieved by covering them with a black plastic or to suffocate them with sand. Another way to deal with the lichens which is not encouraged by conservators is the use of chemicals. The chemicals that are used to remove lichens and mosses include orthophephenyl phenol, dehydrated ethanol, ammonium hydroxide, fluorosilicates of zinc and magnesium. Other chemicals that have been used in other parts of the world are bleach, compressed water, jets, manual abrasive removal (Fry, 1924). For the UDP it is recommend to apply the cleaning method or suffocating it with sand since that does not involve the use of chemicals.

The problem of cleavage and cracking could be solved by the use of a sealant called Wacker H. This sealant is put into cracks to stabilise the substrate. Its purpose is to fill in the gaps in the rocks so that there is minimal further cracking. There is need to use the sealant properly so that it does not worsen the problem. The use of Wacker H in the Cederberg, Western Cape at a site called Bleeding Nose brought more problems than solution. If not properly used the sealant does not bond well with the rock and can darken the substrate which will eventually leads to further deterioration of the paintings (Figure 9).

11.6 CONCLUSION

The IPCC (2007) is presently predicting increasing temperatures in the UDP over the next century. There is less unanimity on rainfall (Jury, 1997; Nel and Sumner, 2006; Nel, 2008), but some authors predict a slight increase and with more high rainfall events (IPCC, 2007). The increase in temperature and precipitation is accelerating the deterioration of the substrate and in the process is destroying the rock art. This is likely to get worse with global warming. This is going to put a lot of pressure on the government and heritage managers. It has been seen that whenever conservation-intervention has been done in a hurry in response to emergency situations, results have been disastrous. The increase in temperature and precipitation will cause the stakeholders to implement some measures in hurry and this can worsen the situation. The problem often comes with the use of sealants. Most of them are still experimental and they must be used as a last resort. These need thorough testing before being used on rock art sites in the UDP. The problems brought by global warming are difficult to combat since it is a natural process and takes time, however, there is need to test, understand the science and experiment with potential mitigation measures that may prove useful in helping us to preserve and extend the life of our rock art. Confidence can only be achieved through monitoring and further research.

ACKNOWLEDGEMENTS

I would like to thank my supervisor Professor Ben Smith for his guidance throughout the project. I am also grateful to the editors for their useful comments. Financial support for this research came from the Rock Art Research Institute of University of the Witwatersrand and the African World Heritage Fund.

REFERENCES

Bednarick, R.G., 1989, Rock art conservation in Australia. In *Preserving our rock art heritage* edited by Crotty, H.K. (San Miguel: American Rock Art Research Association), pp. 23–37.

Bland, W. and Rolls, D., 1988, *Weathering: an introduction to scientific principles,* (London: Arnold).

Benito, G., Machando, M. and Sancho, C., 1993, Sandstone weathering processes damaging prehistoric rock paintings at Albarracan Cultural Park, N.E. Spain. *Environmental Geology,* **22,** pp. 71–97.

Blundell, G and Lewis-Williams, J.D., 2001, Storm Shelter: a rock art discovery in South Africa. *South African Journal of Science*, **97**, pp. 1–4.

Caircross, B., 2004, *Field guide to rocks and minerals of southern Africa*. (Johannesburg: Desmond and Sacco).

Chesnut, W.S., 1972, Geological investigations of deterioration problems affecting Aboriginal art works at Mootwingee (Broken Hill) and Mount Grenfell (Cobar), (Sydney: Geological Survey of N.S.W Department of Mines).

Clark, A.N., 1998, *Penguin Dictionary of Geography* (London: Longman).

David, L., 2003, Pioneering Stewardship: New challenges for Cultural Resource Management. *The Journal of Stewardship,* **1**, pp. 7–13.

Derwent, S., 2006, *KwaZulu-Natal heritage sites: A guide to some great places* (Claremont: David Phillip publishers).

Eriksson, P.G., 1983, A palaeoenvironmental study of the Molteno, Elliot and Clarens Formation in the Natal Drakensberg and north-eastern Orange Free State. Unpublished Ph.D thesis (Pietermaritzburg: University of Natal).

Fry, E., 1924, A suggested explanation of the mechanical action of lithophytic lichens on rocks (shale). *Annals of Botany,* **38**, pp. 175–196.

Gillespie, D.A., 1983, The practice of rock art conservation and site management in Kalandu National Park. In *The rock art sites of Kalandu National Park*, edited by Gillespie, D.A. (Canberra: Australian National Parks and Wildlife Service).

Heath, L., 2008, Impacts of climate change on Australia's world heritage properties and their values. (Sydney: Institute for Environment).

Hewitson, B., 1992, Regional climate in the GISS General Circulation model: surface air temperature. *The supplementary report to the IPCC scientific assessment* (Cambridge: Cambridge University Press), pp. 1–17.

Hoerlé, S. and Solomon, A., 2004, Microclimatic data and rock art conservation at Game Pass shelter in the Kamberg nature reserve. KwaZulu-Natal. *South African Journal of Science,* **100,** pp. 340–341.

Hoerlé, S., 2005, A preliminary study of weathering activity at the rock art site of Game Pass shelter, KwaZulu-Natal, South Africa in relation to conservation. *South African Journal of Geology,* **108,** pp. 297–308.

Hoerlé, S., 2006, Rock temperature as an indicator of weathering processes affecting rock art. *Journal of Earth Processes and Landforms,* **31,** pp. 383–389.

Hoerlé, S., Huneau, F., Solomon, A. and Dennis, A., 2007, Using ground penetrating radar to assess the conservation condition of rock art sites. *C.R. Geosciences,* **339,** pp. 536–544.

IPCC, 2007, *The supplementary report to the IPCC scientific assessment* (Working Group 2), (Cambridge: Cambridge University Press).

Johnson, M.R., Anhaeuser, C.R. and Thomas, R.J., eds, 2006, *The Geology of South Africa*, (Johannesburg: Council of Geosciences).

Jury, M.R., 1997, Statistical analysis and prediction of KwaZulu-Natal climate. *Climatology,* **60,** pp. 1–10.

Kemp, D., 1998, *The environment dictionary*, (New York: Routledge).

Lapidus, D., 1990, *Collins dictionary of Geology*, (London: Collins).

Lewis-Williams, D., 1981, Believing and seeing. Symbolic meanings in southern African rock art, (London: Academic Press).

Lewis-Williams, D., 2000, *Discovering southern African rock art,* (Cape Town: David Phillip Publishers).

Lewis-Williams, D. and Dowson, T.A., 1992, *Rock paintings of the Natal Drakensberg uKhahlamba series*, (Pietermaritzburg: University of Natal press).

Loubser, J., 2001, Management planning for conservation. In *Handbook of rock art research*, edited by Whitley, S.D. (New York: Rowman and Littlefield publishers), pp. 80–115.

Mazel, A.D. and Watchman, A.L., 2003, Dating rock paintings in the uKhahlamba-Drakensberg and the Biggersberg, KwaZulu-Natal, South Africa. *Journal of Southern African Humanities,* **15,** pp. 59–73.

Meiklejohn, K.I., 1994, Some aspects of the weathering of the Clarens Formation in the KwaZulu-Natal Drakensberg: Implications for preservation of indigenous rock art. Unpublished Ph.D. thesis. University of KwaZulu-Natal.

Meiklejohn, K.I., Hall, K. and Davis, J.K., 2009, Weathering of rock art sites in the KwaZulu-Natal Drakensberg, South Africa. *Journal of Archaeological Science,* **36,** pp. 973–979.

Nel, W. and Sumner, P.D., 2006, Trends in rainfall total and variability (1970–2000) along the KwaZulu-Natal Drakensberg foothills. *South African Geographical Journal,* **88** (2), pp. 130–137.

Nel, W., 2008, Rainfall trends in the KwaZulu-Natal Drakensberg region of South Africa during the twentieth century. *International Journal of Climatology,* **10,** pp. 15–22.

Ollier, C., 1984, *Weathering* (London: Longman).

Porter, R., 1999, *Nomination proposal for the Drakensberg Park alternatively known as the uKhahlamba-Drakensberg Park to be listed as a world heritage site*, (KwaZulu-Natal: KwaZulu-Natal Conservation Service).

Rosenfield, A., 1988, *Rock art conservation in Australia*. (Sydney: Government Publishing).

Schwartzbaum, P.M., 1985, The role of conservation techniques in rock art conservation. *Rock Art Research,* **2 (1)**, pp. 65–70.

South African Weather Service, 2009, Temperature and rainfall data for the KwaZulu-Natal Province, (Pretoria).

Truswell, J.F., 1970, *An introduction to the historical geology of South Africa* (Cape Town: Purnell).

UNESCO, 2000, *Report of the 24th session of the World Heritage Committee*, 27th November 2000 to 2nd December 2000. (Cairns: UNESCO).

Whitfield, G. and Norman, N., 2006, *Geological journeys. A traveller's guide to South Africa's rocks and landforms*, (Cape Town: Struik publishers).

Wright, J. and Mazel, A., 2007, *Tracks in a mountain range: Exploring the history of the uKhahlamba-Drakensberg*, (Johannesburg: Wits University Press).

CHAPTER 12

Climate Change Adaptation and Mitigation through Agroforestry systems in Wolaita zone, Southern Highland of Ethiopia

Wondimu Tadiwos Hailesilassie
National Meteorological Agency, Addis Ababa, Ethiopia

ABSTRACT: Climate change is one of the most serious threats the world faces. It will have disproportionate impacts on millions of poor rural people in Ethiopia. For development work to be effective, it should not only help rural people emerge from poverty, it should also enable them to cope with and mitigate the impact of climate change. Agriculture is the human enterprise that is most vulnerable to climate change, because of the subsistence nature of the farming practices. Farmer's adaptive capacity is constrained by a lack of economic and technical resources, and they are vulnerable due to a strong dependence on rain-fed crops. While agroforestry may play a significant role in mitigating the atmospheric accumulation of Greenhouse Gases (GHG), it also has a role to play in helping smallholder farmers adapt to climate change. This paper examines the mitigation and adaptation potential of home garden and tree-based agroforestry systems. The planting of fast growing and nitrogen fixing trees and shrubs releases nitrogen for crop growth and save fertilizers. They also buffer climate extremes and temperature increases. Within the Sodo reforestation project mainly *Grevillea robusta* is used due to its improvements by fixing nitrogen without chemical effects on the soil. The research focuses on the role of agroforestry in both mitigation and adaptation to climate change. It is recommended that in low-income and food-deficit regions livelihoods including food security and climate change cannot be tackled in isolation.

12.1 INTRODUCTION

12.1.1 Background

Climate change is probably the most complex and challenging environmental problem facing the world today. Currently, the intriguing questions include weather uncertainties, persistent climatic abnormalities, rampant environmental degradation and imminent food insecurity. Some of the complexities are exacerbated by increasing human population and demand for more agricultural land for food production, resulting in the destruction of the vegetation cover and subsequently rampant environmental degradation (Ojwang *et al.*, 2010). The Physical Science basis of IPCC (2007b) noted that climate models predict that climate change will lead, among other things, to an increase in unpredictability of rainfall, higher temperatures, and an increase in the severity and frequency of extreme weather events. The report of IPCC (2007a) states that climate change will have a major impact on food and water resources and suggests that adaptive measures must be developed.

Agroforestry provides an example of a set of innovative practices designed to enhance overall productivity, increase carbon sequestration, and strengthen the ability of a system to cope with adverse impacts of changing climate conditions. Agroforestry management systems offer important opportunities creating synergies between actions undertaken for mitigation and for adaptation (Verchot *et al.*, 2007). There is increasing interest to combine adaptation and mitigation measures that provide win–win solutions to climate change (Lasco *et al.*, 2014).

Carbon is particularly useful in agricultural systems, making agroforestry a quantitatively important carbon sink. Worldwide it is estimated that 630×10^6 ha are suitable for agroforestry (IPCC, 2000). Agroforestry provides opportunities for smallholder farmers to adapt to climate change. Sustainable agricultural development is widely acknowledged as an important component in a strategy to respond to the challenges of poverty and environmental degradation and adaptation to climate change (Antle and Diagana, 2003). Temperature, radiation, rainfall, soil moisture and carbon dioxide (CO_2) concentration are important variables to determine agricultural productivity, and their relationships are not simply linear. For example, the modelling studies discussed in the IPCC report (2007a) indicate that moderate to medium increases in mean temperature (1–3° C), along with associated CO_2 increases and rainfall changes, are expected to benefit crop yields in temperate regions. However, in low-latitude regions, moderate temperature increases (1–2° C) are likely to have negative yield impacts for major cereals. Warming of more than 3° C would have negative impacts in all regions.

IPCC (2007a) uses "global warming" in a precise way, to mean "a tendency for the globe to warm over a given period". As indicated in Figure 1 average global temperature is increasing. IPCC (2007a) noted that the real increment of average global temperature began around the time of the Industrial Revolution i.e., after 1750. When fossil fuels are burnt or combusted, CO_2, CH_4, and N_2O are given off as gases. Today the use of fossil fuel for power and electricity is thousands of times more than what it was in the 1800s. In Figure 2 it can also be observed that the average minimum temperature increment in Wolaita zone is increasing.

In developing countries, annual agricultural sector adaptation costs are estimated by the World Bank to be USD 2.5 to 2.6 billion a year between 2010 and 2050. A higher estimate, from the United Nations Framework Convention on Climate

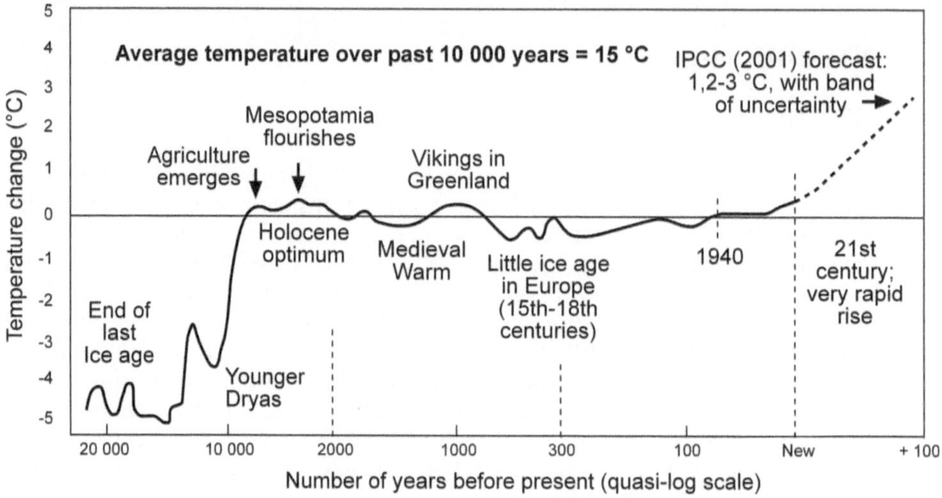

Figure 1. Variations in earth's average surface temperature over the past 20,000 years (adapted from McMichael *et al.*, 2003).

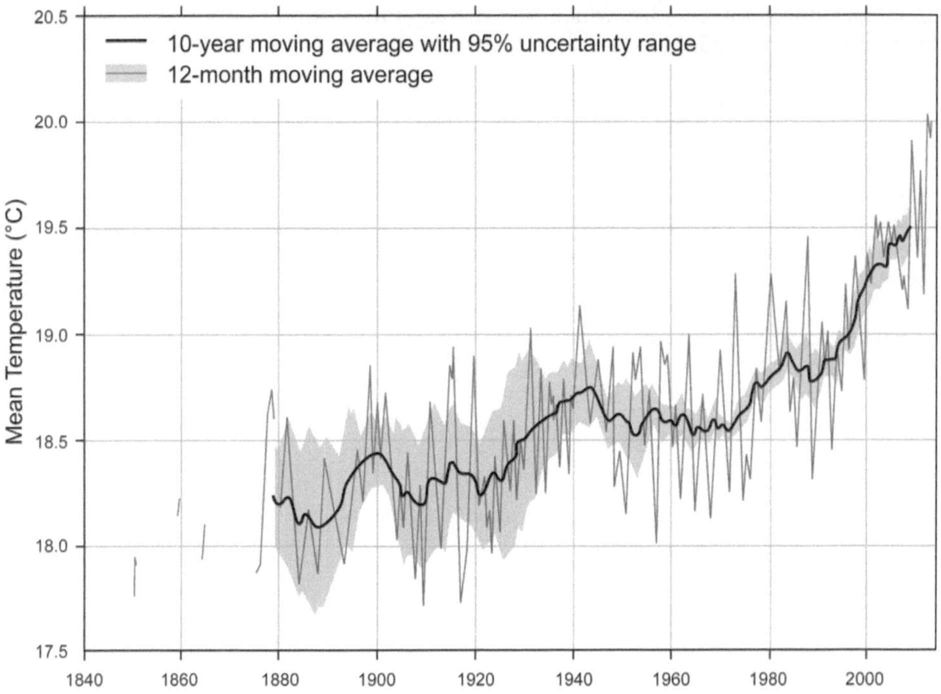

Figure 2. The mean annual minimum temperatures change in Wolaita zone in southern Ethiopia highlands (adapted from BEST, 2015).

Change (UNFCCC), is that incremental investments and financial flows required for agricultural adaptation will total USD 7 billion a year in 2030 (Ojwang *et al.*, 2010). Agroforestry is an increasingly important adaptation strategy for enhancing resilience to adverse impacts of rainfall variability, shifting weather patterns, reduced water availability and soil erosion (Taonda *et al.*, 2001). Understanding the impacts and vulnerabilities of local communities and ecosystems to climate variability and change as well as generating indigenous and science-based information for mitigation and adaptation options will enhance the adaptive capacity of local communities and help build a climate-resilient green economy in Ethiopia (Bishaw *et al.*, 2013). Antle and Diagana (2003) noted that Payments for Ecosystem Services (PES) are the reward made to farmers for sustainable land use practice.

Like much of Africa, Ethiopia has become warmer over the past century and human induced climate change will bring further warming over the next century at unprecedented rates (Figure 2). Climate models suggest that Ethiopia will see further warming in all seasons of between 0.7° C and 2.3° C by the 2020's and of between 1.4° C and 2.9° C by the 2050s. It is likely that this warming will be associated with heat waves and higher evapotranspiration. More regular heavy rainfall events are expected which result in increased flooding. Agriculture is particularly sensitive to climate change. Greater total or more intense rainfall across Ethiopia may increase soil erosion and the incidences of crop damage. The Ethiopian government launched in 2011 the Strategy Document for Ethiopia's Climate Resilient Green Economy (CRGE). The CRGE is a green growth economic plan that is being developed with the support of the Global Green Growth Institute and its partners, the Ethiopian government's ambitious development plan which sets the aspiration for Ethiopia to reach middle income levels by 2025. It is based on three complementary objectives: fostering economic development and growth, ensuring mitigation of Greenhouse Gases (GHGs), and supporting adaptation to climate change (CRGE, 2011).

Bishaw *et al.* (2013) concluded that promoting and scaling up evergreen agriculture, which is based on agroforestry and conservation farming, is a promising approach toward promoting sustainable land-use systems in Ethiopia and Kenya in order to address food security, environmental degradation, and climate change.

The main objective of this study is to identify options and scale up approaches to facilitate smallholder farmers' adoption of sustainable practices that provide benefits to individual households (e.g. food production) while also helping them to adapt to climate change and generating ecosystem services (e.g. reduction of net Green House Gas emissions for climate change mitigation).

12.1.2 Study area

12.1.2.1 Human population and geographical location

Wolaita is one of the administrative zones in South Nations and Nationalities Peoples' Regional State (SNNPRS). It is located between 6° 40′ and 7° 58′ N and 37° 14′ and 37° 56′ E (Figure 3). With a total area of 438,370 ha, Wolaita is inhabited by

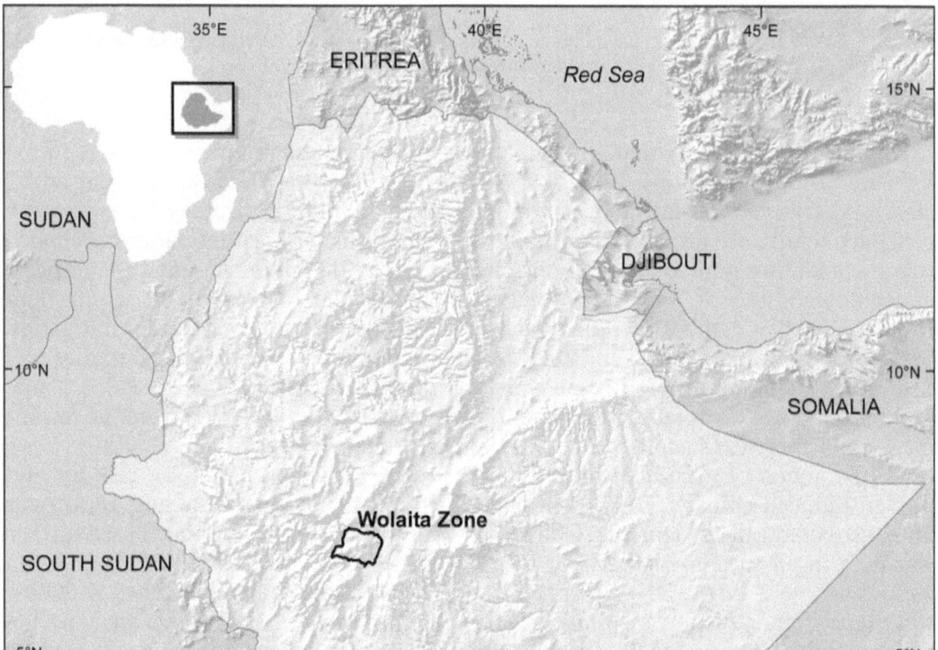

Figure 3. Geographical location of the study area (adapted from Olango *et al.*, 2014).

over 1.7 million people. Wolaita zone represents one of the most densely populated parts of the country. In 2006, rural density varied from 167 to 746 inhabitants/km^2 from Humbo District in the lowlands to Damot Gale District in the highlands (CSA, 2006). The capital town of the zone is Sodo/Wolaita Sodo located 380 km away from Ethiopia's capital Addis Ababa. It has a total population of 76,050 (CSA, 2007).

12.1.2.2 Climate

The altitude of Wolaita zone ranges between 1200 m and 2500 m asl. Mean annual rainfall in the area varies between 800 mm and 1400 mm. Average temperature varies between 15°C and 20°C. Rainfall occurs in two distinct rainy seasons, 'kremt' rains (also called the 'big rains') occurring in summer (roughly June, July and August) and 'belg' rains (also called the 'small rains') occurring in spring (roughly the mid-February to mid-May period). Kremt is the main production season, but the occurrence of rain during the belg season is equally important as it has implications on the food security of the households. Ayalew (2011) noted that the highest rainfall is experienced during the months of July and August and causes high erosion rates. Soil fertility gradient decreases from homestead to the outfield due to management effects.

12.2 DATA AND METHODS

This study used the data of different agroforestry systems or activities in the study area gathered from Wolaita bureau of Agriculture and World vision Ethiopia (WVE) in Wolaita zone. These are mixed cropping activities in home garden, crop productivities in outside field (far from the home garden), and tree-planting agroforestry systems based on PES (Payment for Ecosystem Services). The data is consisting of household surveys which were used to examine adaptation and mitigation strategies. In this case, it can be noted that agroforestry systems improve resilience of smallholder farmers through improved microclimate, and improved farm productivity, and diversified and increased farm income while at the same time sequestering carbon. The home garden agroforestry systems generate a microclimate within the agro-ecosystems that preserve the function and resilience of the larger ecosystem. Increasing productivity relates directly to the ability of a system to accumulate and retain carbon, enhancing the capacity to such systems to cope with adverse climatic changes, and to tree species improvement to increase biomass productivity and carbon sequestration.

12.3 RESULTS AND DISCUSSION

12.3.1 Home garden agroforestry systems

The home garden is a small-scale traditional ecosystem and is locally known by the name Darkkuwa in Wolayta language and has played an important role in conservation of plant biodiversity as well as in adaptation to the change in climatic conditions. The most important characteristics of home gardens are their location adjacent to homes, close association with family activities and a wide diversity of crop and livestock species to meet family needs. Figure 4 depicts exemplary drawing of the home garden system. According to Talemos *et al.* (2013) the size of home gardens in Wolaita zone in average is 600 m^2. The home garden has played a central role in household food security, fuel, fiber, materials and even land ownership, as people changed from an

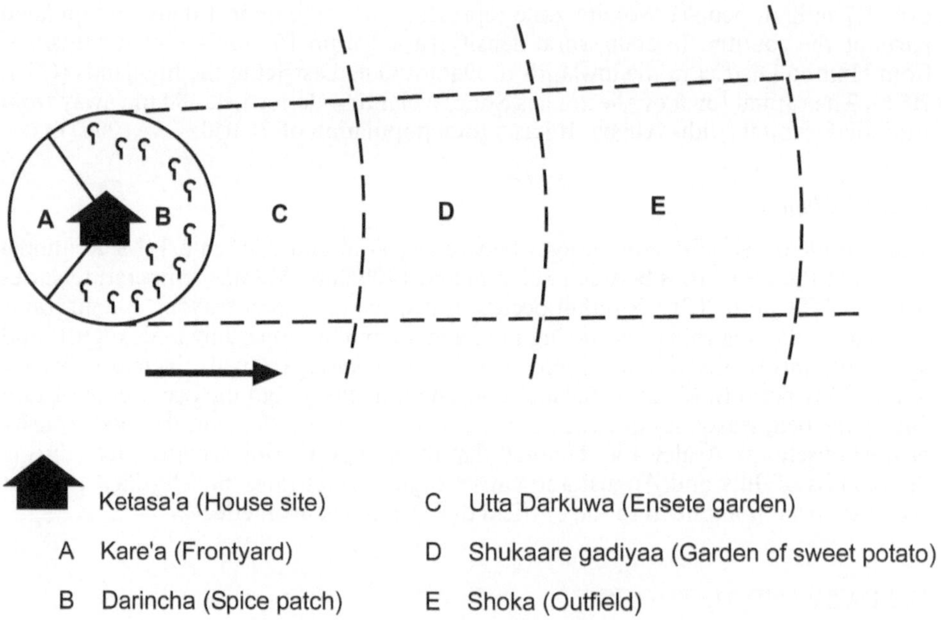

Figure 4. The spatial horizontal structure of the home garden and outfield in Wolaita zone (adapted from Talemos *et al.*, 2013).

exclusively hunting and gathering lifestyle and settled in small communities (Landon-Lane, 2004).

The farming systems in the study area broadly represent a mixed crop-livestock production, more specifically, it is enset-coffee-cereal-livestock system that combines annual and perennial crops with livestock production. Enset cultivation occupies a central position in the agricultural systems of the Wolaita, and every farming household cultivates enset in its home garden (see Table 1).

As it can be seen in Table 1 and 2 the mixed farming involving the production of cereals, root crops, enset (*Ensete ventricosum*), coffee, shade trees, and others are practiced in farmers' home garden. Shade trees buffer climate extremes (e.g. heavy winds, rainfall) and temperature increases while sequester carbon at the same time. According to farmers in the study area the reasons for the importance of enset cultivation in Wolaita is as follows: Flexibility in farming systems as an intercrop with annual and perennial crops, drought tolerance (extreme weather event), storability of enset products for long periods, possibility of harvesting at any time of the year, use for integration of crop-livestock system, and generating income (Table 2).

12.3.2 Productivity of major crops in agroforestry systems

Improvement in agricultural productivity here defined as crop yield (i.e., metric tons of grain production per hectare of land), is a particular emphasis of the plant science community, as researchers and farmers seek to sustain the impressive historical gains associated with improved genetics and agronomic management of major food crops (Lobell and Gourdji, 2012). Table 3 depicts the comparison of productivity of

Table 1. Plant species which were mentioned by local people as the most important food crops in the Homegarden, their percentage preference, use values and, preference ranks (adapted from Talemos *et al.*, 2013).

Plant species	% preference	Use value	Ranked
Ensete ventricosa	100	5	1st
Ipomoea batatas	97.5	4.6	2nd
Zea mays	91.6	4.3	3rd
Coffea arabica	86.7	4	4th
Persea americana	83.4	3.4	5th
Colocsin esculenta	66.6	2.9	6th
Brassica carinata	64.3	3.2	7th
Manihot esculenta	62.5	2.2	8th
Solanum tuberosum	62.4	3.6	9th
Phaseolas vugaris	58.3	3	10th

Table 2. Types of perennial crops and their productivities (data obtained from Wolaita Zone Bureau of Agriculture).

Types of crop	Area (ha)	Annual yield (Qt)	Productivity (Qt/ha)
Enset	2380	182070	77
Coffee	11905.1	88100	7
Avocado	2030	406000	200
Mango	751	119580	159
Pineapple	15	2700	180

Table 3. Comparison of the productivity (yield, t/ha) of the major crops in 2004 and 2011 (the productivity in quintals per hectare) (data obtained from Wolaita Zone Bureau of Agriculture).

Crop name	Yield in 2004	Yield in 2011
Maize	10.5	36
Wheat	9.7	47
Teff	4.7	16
Barley	6.0	26
Potato	54.0	187

the major crops per hectare in 2004 and 2011. According to Wolaita Zone Bureau of Agriculture (2011) the reasons for the significant difference in productivity between 2004 and 2011 are as follows: farmers used the improved agricultural practice; planting crop varieties that are resilient to adverse climatic conditions; planting strains of crops that have reasonable using minimum tillage and mulching the soil with crop residues to conserve soil moisture; mixed cropping to maximize soil nutrient use; useful minimum tillage and mulching the soil with crop residues to conserve soil moisture.

Table 4. The farm inputs (DAP and UREA) in quintals (Wolaita Zone Bureau of Agriculture, n.d.).

Year	DAP	UREA
2005	30829	1680
2006	35630	3555
2007	28907	3222
2008	63326	9736
2009	77786	24984

Table 4 shows the inputs of chemical fertilizers (which are urea, Urea Ammonium Nitrate (UAN) and Diammonium Phosphate (DAP)) for increasing the productivity of crops mainly funded by credits to the farmers of the Ethiopian government. Matsumoto and Yamano (2010) evaluated the impact of fertilizer credit on crop choice, crop yield, and income using two-year panel data of 420 households in Ethiopia. They indicated that the fertilizer credit is found to increase input application for crop production. Lobell and Gourdji (2012) have also confirmed that globally yields of most major crops have increased markedly over the past half century, largely due to greater use of irrigation, chemical inputs and modern crop varieties.

Fertilizers are essential when efficiently and effectively used to produce maximum increase in crop yields so that farmers receive the best outputs from their expenses. The chemical fertilizers should be used judiciously and along with manures for improving the crop yield and soil productivity in a sustainable way.

Farmers practiced leguminous trees/shrubs agroforestry systems. This involves planting fast growing and nitrogen fixing trees/shrubs. They release nitrogen for crop growth and save expensive fertilizers. Climate change may translate into reduced total rainfall or increased occurrence of dry spells during rainy seasons. In addition, according to Wolaita Zone Bureau of Agriculture, 810,750 quintals of major crops were produced using irrigation and rain water harvesting in 2010.

12.3.3 Tree-based agroforestry systems

Tree-based agroforestry systems have some obvious advantages for maintaining production during wetter and drier years. Beside the biophysical resilience, which allows the various components of the agroforestry systems to withstand the shocks related to climate variability, the presence of trees in agricultural croplands can provide farmers with alternative or additional sources of income thus strengthening the socio-economic resilience of rural population (Verchot *et al.*, 2007).

According to Wolaita Zone Bureau of Agriculture and World Vision Ethiopia report (World Vision Ethiopia, 2013) there is a project named "Sodo reforestation project". This long term project (35 years) started in 2006 and is aimed at restoring and protecting some 503 hectares of native forests and selected non-indigenous species on Mount Damota (Figure 5) through a combination of tree planting and natural regeneration. Mount Damota is part of the southern Ethiopian highlands, and was previously a source of water and vital agricultural and community land for the Sodo community and region. The initiative was developed between the Sodo community and World Vision Ethiopia, with support from World Vision Australia. In the project zone communities are now able to sell carbon credits from agroforestry, reforestation and afforestation over the life time of the project, which should ensure the

Figure 5. The Mount Damota site of reforestation project (nursery project) (adapted from World Vision Ethiopia, Sodo reforestation project, 2013).

long-term regeneration of the Mount Damota project area (Figure 5) through offsets of an estimated sequestration of 189,026 metric tons of carbon dioxide. IFAD (2008) noted that one way of effectively engaging smallholders in the mitigation process is to expand the concept of carbon trading to include compensating rural communities for soil conservation and reforestation.

Based on the report document of World Vision Ethiopia (2013), Sodo-forest regeneration project, communities are planting *Grevillea robusta* shown in Figure 6 which is accounted for 39% of the entire plantation. This tree is known for its soil nutrient improvement by fixing nitrogen. It has no chemical effect and thus, grows well and is friendly with other plants.

The community forest cooperatives and carbon credits in Sodo reforestation project

The population living within the project zone is 18,592 out of which 7,300 are already enrolled by the forest cooperatives– nearly 40% of the total population. The project zone spreads over seven kebeles (villages) administrations namely, Kokate, Gurumu Woide, Delbo Wogene, Damot Waja and Kunassa Fullassa, Woraza Lassho and G/Koisha. These target areas have been degraded for years due to population pressure for fire wood, timber, poles and additional farm land, increasing pressure on natural forests existing in the area. The project zone comprises of seven kebeles that constitute 4593.34 ha, with the closed project area located within five of them with an area of 503 ha. Table 5 depicts the population number and size of the project zone where the five respective forest cooperatives reside (World Vision Ethiopia, 2013).

Figure 6. Nursery establishment of the *Grevillea robusta* tree (photo by Wolaita Zone Bureau of Agriculture, 2015).

As indicated there are two kebeles in the project zone which do not have any land as part of the closed project area. The five forest development and protection cooperatives have been created with the aim of establishing responsible institutions to manage the project in an ongoing and sustainable manner. In addition, agroforestry practices are implemented on individual farm lands to boost productivity, decrease pressure on the natural closed area, and create alternative income sources from the various types of agroforestry practices. Consequently, this will increase the fertility of farm land soil and crop yield automatically. The community livelihood at the same time will be improved. A number of community members are engaged on rearing of bees, cultivating apple fruit on nurseries and individual homestead areas, from which many have started to gain income. Some of the impacts of this project are: To sustainably improve the environmental and economic conditions of farmers within the project site, through the conservation of soil resources, the protection of water resources and the enhancement of biodiversity and indigenous species; protection of agricultural lands, and improving the economical situation of the surrounding villages; and to facilitate the socio-economic development of the local communities through direct environmental benefits from reforestation and the co-benefits from carbon revenue (World Vision Ethiopia, 2015).

According to the World Vision Ethiopia (2015) report the five legally organized and certified forest cooperatives received 41,559 US$ payment from carbon market

Table 5. The population number and size of the project zone (adapted from World Vision Ethiopia, Sodo reforestation project, 2013).

Cooperatives/ Kebele names	Project area (in ha)	Project zone exclusive of project area (in ha)	Project zone (in ha)
Kokate	181.10	648.12	829.22
G/Woide	126.55	742.96	869.51
Dalbo Wogane	104.51	399.91	504.51
Damot Waja	34.00	601.101	635.01
K/Fullasa	57.12	925.55	982.67
Waraza Lasho	0.00	435.09	435.09
G/Koisha	0.00	337.42	337.42
Total	503.28	4096.06	4593.34

selling through World Vision Ethiopia (WVE) Sodo Forestery and Agroforesrty project, in Sodo District, Wolaita zone. This has been working to improve resilience capacity of household, children and community to the negative effects of climate change through improved livelihoods, resilience to shocks and recovery from disaster as a result of improved ecosystem and household food security.

12.4 CONCLUSIONS

Agriculture plays an important role as a carbon sink through its capacity to sequester and store greenhouse gases, especially in the form of carbon in soils, plants and trees. Agroforestry promises to create synergies between efforts to mitigate climate change and efforts to help vulnerable populations adapt to the negative consequences of climate change. Home garden agricultural activity, productivity of major crops and tree-based agroforestry systems have played a great role in adaptation and mitigation strategies for climate change. Responses to the challenges of climate change should be made within the context of sustainable development taking into consideration livelihood and food security needs of subsistence farmers. Adaptation activities leading to increased agricultural and forestry system resilience, as well as improved natural resources management and productive practices, may be attractive to carbon markets because of their associated mitigation value. Increasing productivity relates directly to the ability of a system to accumulate and retain carbon, enhancing the capacity to such systems to cope with adverse climatic changes. A key question is how the wider society- community, national and international- can motivate farmers to reduce negative side-effects and adapt to climate change while continuing to meet the increasing demand for agricultural products.

ACKNOWLEDGEMENTS

I sincerely acknowledge the Wolaita Zone Bureau of Agriculture and World Vision Ethiopia for providing the necessary data for this study. I am very much thankful to two anonymous reviewers for providing useful suggestions, which helped to improve the quality of this contribution.

REFERENCES

Antle, J.M. and Diagana, B., 2003, Creating Incentives for the Adoption of Sustainable Agricultural Practices in Developing Countries: The Role of Soil Carbon Sequestration. *American Journal of Agricultural Economics,* **85(5)**, pp. 1178–1184.
Ayalew, A., 2011, Construction of Soil Conservation Structures for improvement of crops and soil productivity in the Southern Ethiopia. *Journal of Environment and Earth Science,* **1(1)**, pp. 21–30.
BEST, 2015, Berkeley Earth Surface Temperature. Compiled by Berkeley Earth team, available at http://berkeleyearth.lbl.gov/regions/ethiopia (11.05.2015).
Bishaw, B., Neufeldt, H., Mowo, J., Abdelkadir, A., Muriuki, J., Dalle, G., Assefa, T., Guillozet, K., Kassa, H., Dawson, I.K., Luedeling, E. and Mbow, C., 2013, *Farmers' Strategies for Adapting to and Mitigating Climate Variability and Change through Agroforestry in Ethiopia and Kenya,* (Corvallis: Oregon State University).
CRGE, 2011, Ethiopia's Climate-Resilient Green Economy. Green economy strategy. Federal Democratic Republic of Ethiopia. Addis Ababa, p. 188.
CSA, 2006, National statistics. Central Statistical Agency of Ethiopia.
CSA, 2007, National statistics. Central Statistical Agency of Ethiopia.
Ewnetu, Z. and Bliss, J.C. 2010. Tree growing by smallholder farmers in the Ethiopian highlands. Proceedings of the Conference Small Scale Forestry in a Changing World, IUFRO Conference, 06–12 June 2010, Bled, Slovenia, pp. 166–187.
IFAD, 2008, Climate change and the future of smallholder agriculture. Discussion paper prepared for the Round Table on Climate Change at the Thirty-first session of IFAD's Governing Council, 14 February 2008. International Fund for Agricultural Development, available at http://www.ifad.org/climate/roundtable/index.htm (24.07.2014).
IPCC, 2000, *Land-use, land-use change and forestry. Special report of the intergovernmental panel on climate change,* (Cambridge: Cambridge University Press).
IPCC, 2007a, Climate Change 2007: Impacts, Adaptation and Vulnerability. Contribution of Working Group II to the Fourth Assessment Report of the Intergovernmental Panel on Climate Change, edited by Parry, M.L., Canziani, O.F., Palutikof, J.P., van der Linden, P.J. and Hanson, C.E. (Cambridge: Cambridge University Press).
IPCC, 2007b, Climate Change 2007: The Physical Science Basis. Contribution of Working Group I to the Fourth Assessment Report of the Intergovernmental Panel on Climate Change, edited by Solomon, S., Qin, D., Manning, M., Chen, Z., Marquis, M., Averyt, K.B., Tignor, M. and Miller, H.L. (Cambridge: Cambridge University Press).
Landon-Lane, C., 2004, *Livelihoods grow in gardens. Diversifying the rural incomes through home gardens.* FAO Diversification booklet, **2**, (Rome: Food and Agriculture Organization of the United Nations).
Lasco R.D., Delfino, R.J.P. and Espaldon, M.L.O., 2014, Agroforestry systems: helping smallholders adapt to climate risks while mitigating climate change. *WIREs Clim Change,* **5**, pp. 825–833, doi: 10.1002/wcc.301.
Lobell, D.B. and Gourdji, S.M., 2012, The Influence of Climate Change on Global Crop Productivity. *Plant Physiology*, **160**, pp. 1686–1697.
Matsumoto, T. and Yamano, T., 2010, *The Impacts of Fertilizer Credit on Crop Production and Income in Ethiopia.* GRIPS Discussion Paper 10–23 (Tokyo: GRIPS Policy Research Center).
McMichael, A.J., Campbell-Lendrum, D.H., Corvalán, C.F., Ebi, K.L., Githeko, A.K., Scheraga, J.D. and Woodward, A., eds, 2003, *Climate change and human health. Risks and responses,* (Geneva: World Health Organization).

Ojwang, G.O., Agatsiva, J. and Situma, C., 2010, Analysis of Climate Change and Variability Risks in the Smallholder Sector. Case studies of the Laikipia and Narok Districts representing major agro-ecological zones in Kenya. *Environment and natural Resources management working paper*, **41** (Rome: Food and Agriculture Organization of the United Nations).

Olango, T., Tesfaye, B., Catellani, M., Pè, M.E., 2014, Indigenous knowledge, use and on-farm management of enset (*Ensete ventricosum* (Welw.) Cheesman) diversity in Wolaita, Southern Ethiopia. *Journal of Ethnobiology and Ethnomedicine,* **10(41)**, p. 18.

Talemos, S., Sebsebe, D., Zemede, A., 2013, Home gardens of Wolayta, Southern Ethiopia: An ethnobotanical profile. *Academia Journal of Medicine Plants,* **1(1),** pp. 14–30.

Taonda J.-B., Hien, F. and Zango, C., 2001, Namwaya Sawadogo: The ecologist of Touroum, Burkina Faso. In *Farmer innovation in Africa: a source of inspiration for agricultural development,* edited by Reij, C. and Waters-Bayer, A. (London: Earthscan), pp. 137–143.

Verchot, L., Van Noordwijk, M., Kandji, S., Tomich, T., Ong, C.K., Albrecht, A., Mackensen, J., Bantilan, C., Anupama, K.V. and Palm, C.A., 2007, Climate change: linking adaptation and mitigation through agroforestry. *Mitigation and Adaptation Strategies for Global Change* 12: 901–918.

World Vision Ethiopia, 2013, The Sodo Community Managed Reforestation (Forest Regeneration) Project. Wolayita Zone, Sodo Zuria and Damot Gale Woredas. Available at: https://s3.amazonaws.com/CCBA/Projects/Sodo_Community_ Managed_ Reforestation_%28Forest_Regeneration%29_Project/Validation+-+Sept+2013/ Sod+Ethiopia+CCB+PDD+13.pdf (23.03.2015).

World Vision Ethiopia, 2015, Carbon Revenue transferred to Five Cooperatives through World Vision Ethiopia. Available at: http://www.wvi.org/ethiopia/pressrelease/ carbon-revenue-transferred-five-cooperatives-through-world-vision-ethiopia (29.02.2015).

Regional Index

Subject Index

Palaeoecology of Africa

International Yearbook of Landscape Evolution and Palaeoenvironments

ISSN: 2372-5907

Volume 1-12 Out of Print

13. Palaeoecology of Africa and the Surrounding Islands
    Editors: J.A. Coetzee & E.M. van Zinderen Bakker
    1981, ISBN: 978-90-6191-203-3

14. Palaeoecology of Africa
    Editors: J.A. Coetzee & E.M. van Zinderen Bakker
    1982, ISBN: 978-90-6191-204-0

15. Palaeoecology of Africa
    Editors: J.A. Coetzee, E.M. van Zinderen Bakker, J.C. Vogel,
    E.A. Voigt & T.C. Partridge
    1982, ISBN: 978-90-6191-257-6

16. Palaeoecology of Africa
    Editors: J.A. Coetzee & E.M. van Zinderen Bakker
    1984, ISBN: 978-90-6191-510-2

17. Palaeoecology of Africa
    Editors: J.A. Coetzee & E.M. van Zinderen Bakker
    1986, ISBN: 978-90-6191-625-3

18. Palaeoecology of Africa
    Editor: K. Heine
    1987, ISBN: 978-90-6191-689-5

19. Palaeoecology of Africa – *Out of Print*
    Editors: K. Heine & J.A. Coetzee
    1988, ISBN: 978-90-6191-834-9

20. Palaeoecology of Africa
    Editor: K. Heine
    1989, ISBN: 978-90-6191-880-6

21. Palaeoecology of Africa – *Out of Print*
    Editors: K. Heine & R.R. Maud
    1990, ISBN: 978-90-6191-997-1

22. Palaeoecology of Africa
    Editors: K. Heine, A. Ballouche & J. Maley
    1991, ISBN: 978-90-5410-110-9

23. Palaeoecology of Africa and the Surrounding Islands
    Editor: K. Heine
    1993, ISBN: 978-90-5410-154-3

24. Palaeoecology of Africa
    Editor: K. Heine
    1996, ISBN: 978-90-5410-662-3

25. Palaeoecology of Africa – *Out of Print*
    Editors: K. Heine, H. Faure & A. Singhvi
    1999, ISBN: 978-90-5410-451-3

26. Palaeoecology of Africa and the Surrounding Islands
    Editors: K. Heine, L. Scott, A. Cadman & R. Verhoeven
    1999, ISBN: 978-90-5410-476-6

27. Palaeoecology of Africa and the Surrounding Islands: Proceedings of the 25th Inqua Conference, Durban, South Africa, 3-11 August 1999
    Editors: K. Heine & J. Runge
    2001, ISBN: 978-90-5809-350-9

28. Dynamics of Forest Ecosystems in Central Africa During the Holocene: Past – Present – Future
    Editor: J. Runge
    2007, ISBN: 978-0-415-42617-6

29. Holocene Palaeoenvironmental History of the Central Sahara
    Editors: R. Baumhauer & J. Runge
    2009, ISBN: 978-0-415-48256-1

30. African Palaeoenvironments and Geomorphic Landscape Evolution
    Editor: J. Runge
    2010, ISBN: 978-0-415-58789-1

31. Landscape Evolution, Neotectonics and Quaternary Environmental Change in Southern Cameroon
    Editor: J. Runge
    2012, ISBN: 978-0-415-67735-6

32. New Studies on Former and Recent Landscape Changes in Africa
    Editor: J. Runge
    2014, ISBN: 978-1-138-00116-9